Technical and Professional Writing

Solving Problems at Work

George E. Kennedy
Washington State University

Tracy T. Montgomery
Idaho State University

Upper Saddle River, New Jersey 07458

Library of Congress Cataloging-in-Publication Data

Kennedy, George E.
Technical and professional writing : solving problems at work / George E. Kennedy, Tracy T. Montgomery.
p. cm.
Includes index.
ISBN 0–13-055072–8
1. Technical writing—Problems, exercises, etc. 2. English language—Technical English—Problems, exercises, etc 3. English language—Business English—Problems, exercises, etc. 4. Business writing—Problems, exercises, etc. I. Montgomery, Tracy T. II. Title.

PE1475.K47 2002
800'.0666—dc21

2001036654

VP, Editor-in-chief: Leah Jewell
Acquisitions Editor: Craig Campanella
Senior Managing Editor: Mary Rottino
Production Liaison: Fran Russello
Editorial/Production Supervision: Bruce Hobart (Pine Tree Composition)
Prepress and Manufacturing Buyer: Mary Ann Gloriande
Marketing Manager: Rachel Falk
Art Director: Jayne Conte
Cover Design: Bruce Kenselaar
Cover Art: IEEE Computer Machine. Jud Guitteau

Acknowledgments for copyrighted material may be found on pp. xvii–xviii, which constitutes an extension of this copyright page.

This book was set in 10/12 New Baskerville by Pine Tree Composition, Inc. and printed and bound by R. R. Donnelley & Sons Company.
The cover was printed by Jaguar Advanced Graphics.

Printed in the United States of America
10 9 8 7 6 5 4 3 2 1

ISBN 0-13-055072-8

Pearson Education Ltd., London
Pearson Education Australia Pte. Limited, Sydney
Pearson Education Singapore, Pte. Ltd.
Pearson Education North Asia Ltd., Hong Kong
Pearson Education Canada, Ltd., Toronto
Pearson Educación de Mexico, S.A. de C.V.
Pearson Education—Japan, Tokyo
Pearson Education Malaysia, Pte. Ltd.
Pearson Education, Upper Saddle River, New Jersey

To Nancy and Hannah
G.E.K.

To my students
T.T.M

CONTENTS

APPENDIX Selected Problems of Usage and Style 479

PREFACE

We based this book on these two premises: solving problems is the primary activity of all professionals, and the process of solving problems is always, in some way, reflected through good communication. From the very beginning and throughout our discussions, we try to support these premises. Readers are constantly reminded of the process of solving problems in all kinds of communication: from the planning of the briefest memo to the proofing of the longest completion report. And we try to show that the process is organic and multidimensional: the demands of definition, research, analysis, resolution and synthesis, and implementation must be met at various stages or levels of investigation. The process is not simply flat and linear. Instead, it is recursive and thoroughly analytical, often flipping back on itself, requiring redefinition and reanalysis of the old in light of the new. It is a demanding process but, if fully engaged in, is also rewarding. And as we point out in Chapter 1, it is one that all professionals, if they are acting professionally, take part in.

This latter notion, that of acting professionally, has also informed our sense of purpose in writing this book. We think it is important that students who are educating and training themselves to join the professional working world know what is expected of them. They need to understand, if they don't already, that the responsibilities they have to their families, friends, and themselves expand not only in number but also in kind when they become members of the professional world. Problems that they will address, no matter how seemingly trivial at times, are part of a pantheon of problems that if left unaddressed or unsolved, will negatively affect the public and private lives of many people. It is not too grand to claim, on the other hand, that professionals who meet their responsibilities conscientiously, energetically, and ethically do much to improve the quality of life for everyone.

In presenting the various problems that professionals have to address, we have chosen situations (we have called them *case studies*) that are typical of the working world, and, in every case, we have either implicitly or explicitly pointed to the responsibilities the participants have to others and to themselves to address and solve those problems professionally. We would also like to note that the case studies we have chosen reflect different kinds of problems in differing venues that are handled by women and men of differing backgrounds and positions of responsibility. It is a diverse professional world through which we hope to reveal the inherent redundancy in the idea of *good* problem solving, since the result of bad problem solving is no solution at all.

One other important element we should mention has to do with the kinds of writing done in the professional world at large. Our notion, which we discuss specifically in Chapter 3, but which is implicit throughout the book, is that distinct rhetorical differences (dependent on purpose, audience, context, and ethics) exist among kinds of writing in the professions. And those differences, which help dictate content and format or presentation, are determined by the degree of professional objectivity that professionals feel they have to attain and demonstrate. On a continuum of objective to subjective, science writing is thought of as the most objective and most subject to conventions; that is, it is dictated largely by fairly rigid expectations, styles, and forms. Technical and professional writing falls in the middle, with varying expectations and ideas of the need for objectivity, styles, and forms. Administrative writing comes at the other end, the most fluid in many senses, displaying a good deal of variety of style, much less objectivity, but instead more subjectivity. We offer this continuum because it helps lay out the realities of solving problems and writing in the professional world. Not all situations are alike, and they shouldn't be seen as such; each situation often requires different approaches, different analyses, and different results. Knowing and understanding this reality make the use and appreciation of writing in the professions much more pertinent and effective.

So much for the philosophical basis of the book. We turn now to a brief review of how we have structured the contents. The first five chapters on professionalism and the nature of solving problems, the basics of technical and rhetorical problem solving, solving problems both through research and through collaboration and the uses of technology, we think of as more theoretical. They deal with the generic processes of solving various kinds of problems, a process which is essential to the purpose of the book, but we have nevertheless tied the discussions to real situations. Thus readers see in the cases of Susan James, Mary Anne Cox, and Patricia Wynjenek, among others, how problems arise, how they are defined, researched, analyzed, and ultimately resolved—and then written about. The ways these professionals go about their work and put theory into practice is clearly an important element of a book like this.

Chapters 6 through 10 discuss more specific kinds of problems that result in more specific kinds of communication or writing: proposals; progress and completion reports; trip, feasibility, and scientific reports; and policy statements, manuals, and procedures. Again, real working-world situations form the basis of the discussion and create practical examples, all the way from formal proposals to letters and memoranda. We believe we have covered, in more or less detail, practically every professional situation that would require some writing response (though we don't claim that every professional will have to write in every situation).

Chapters 11 through 13 discuss different kinds of communication situations, some not immediately associated with specific writing, though they all require very similar analyses to those required of writing. Chapter 11, which discusses the professional job search, presents strategies for applying for jobs: when and how to revise a résumé; the rhetorical needs for writing letters of application, both for first jobs right out of college and for subsequent jobs; and some tips on preparing for interviews. Chapter 12 offers suggestions on the physical design of documents and the use of graphics to meet rhetorical needs of writing. It is vitally important that the physical arrangement of words, paragraphs, headings, and all manner of graphics be produced in a way that promotes readability and comprehension, and this chapter attempts to acknowledge specifically that importance. Chapter 13, on oral presentations, discusses the process of preparing oral presentations and shows how it is similar to the writing process. An Appendix offering a review of selected problems in usage and style completes the substance of the book.

Just a word on our intended readership. We have in mind students of technical and professional writing who will have to approach or solve problems in their future professional lives and will have to write about them to enable decisions to be made. We also want to help students who would benefit from approaching or solving a research problem in a sustained way over the course of a term or semester. The discussions of proposals and of progress, trip, and completion reports are among several discussions of research and collaboration that would be particularly useful for completing the work of sustained projects. But by no means are we excluding students who will be addressing shorter, more discrete problems and projects. Our discussions are set up in such a way that different emphases can be fully accommodated, both for students and for teachers.

Solving problems in the professional world takes imagination, dedication, and energy. It is hard work, but its benefits are many. We hope we demonstrate this fact effectively in the following pages.

We want to thank our colleagues who graciously gave us helpful and encouraging suggestions for improvements to this edition: David Strong, Winona State University; Donald Brotherton, DeVry Institute of Technology; Melanie Woods, University of Central Florida; David F. Marshall, University of North Dakota.

George E. Kennedy
Pullman, Washington

Tracy T. Montgomery
Pocatello, Idaho

ACKNOWLEDGMENTS

Pages 78–79: Excerpt from a review of the year's activities of a regional planned parenthood agency. Reprinted with permission of Planned Parenthood of the Inland Northwest.

Pages 81–82: "Germs," copyright © 1972 by the Massachusetts Medical Society, from *The Lives of a Cell* by Lewis Thomas. Used by permission of Viking Penguin, a division of Penguin Putnam, Inc.

Pages 144–146: Adapted from Martha Nichols, “Third-World Families at Work: Child Labor or Child Care?” *Harvard Business Review,* January–February 1993: 12–23.

Pages 146–148: Adapted from Alistair D. Williamson, “Is This the Right Time to Come Out?” *Harvard Business Review,* July–August 1993: 18–28.

Page 440: Photograph by George E. Kennedy.

Professionalism and Problem Solving

PREVIEW OF THE CHAPTER

CASE STUDY

SUSAN JAMES'S OFFICE

ITE Industries, Inc. ▪ Richmond, California

It may have been the lack of a lunch hour, or maybe the lousy pastrami sandwich she had delivered from the corner deli, but Susan James felt particularly out of sorts as she stared at her computer, trying to complete the outline of her latest quarterly report. As head of Purchasing, Susan had been under fire recently. The Engineering Department had been pointing to her and her staff as the main reason for missing several deadlines and not meeting some of the objectives that had been set for Engineering. This was not an uncommon position to be in, however. Purchasing fit the description of the perfect scapegoat in a production facility: as an administrative function, as a support activity rather than a production activity, Purchasing often had to make up everyone else's lost schedules.

But further complicating Purchasing's existence these days were the reams of forms to be filled out to meet regulations for a federal government project that ITE had recently landed, restrictions on placing certain percentages of company business with the most reliable suppliers in an effort to bring in untried minority-owned suppliers, and what seemed to be an unusual lack of engineering planning—or pre-planning, as they now call it in the latest management jargon. In addition, Susan was currently shorthanded, since a couple of her key buyers defected to other departments in the company, giving as their reasons for leaving "undue pressure."

Despite all this difficulty, Susan had to report progress. She had to have something to say for the category "Extraordinary Achievement," and preferably, it should have something to do with Purchasing's support of Engineering. So she wrote this: "Purchasing supported Engineering's crane replacement project by finding an in-state supplier within 24 hours of Engineering's request." An impressive accomplishment and true as far as it went. But what Susan knew, as others did not yet, was that the in-state crane was actually inadequate for ITE's needs and that, furthermore, no firm decision to buy the crane could result without long negotiations with the supplier.

Susan also knew that if management read her report quickly, her statement would gain her points. But she suspected that when the rest of the story came out, her reports would never be read so casually again. She decided after some thought that she simply couldn't report an "extraordinary achievement" this quarter, so she left the category blank. She further decided that until she actually had something extraordinary to report, she would continue documenting all ongoing efforts to support Engineering for her own file. Small consolation in the short run, but a sound strategy in the long run.

Having made this decision, Susan felt a little better and finished her outline rather easily. But she made one decision before going on to her next chore: on days like this, instead of pastrami, she'd order roast beef.

PROFESSIONALS AND PROFESSIONALISM

The word *professional* has come to mean several things. People talk, for example, about the difference between amateur and professional sports. Amateur athletes may be the best in the world, winners of gold medals in the Olympics and in world championships, but they are not considered professionals until they accept their first checks for what they do.

On the other hand, people often talk about amateurs as being truly professional, that is, acting in a way that commands the respect of others. They perform with singular effort, with practiced discipline and skill, and they win or lose with dignity and grace. But, then, some performances are called unprofessional. The participants may be highly skilled and perform with a singular effort, but they violate the conventions of professional behavior by disputing, arguing, and fighting about calls on their performance.

Sports is only one area where we hear the word *professional* used. Indeed, for many, it also brings to mind the traditional professions—theology, education, medicine, law. And those have been extended to include other areas of science, technology, and industry. To most people, however, professionals are those, in any field, who have special training or education to do specialized work that involves making decisions and solving problems. This work may require degrees ranging from the associate of arts to the Ph.D.

Susan James, in the situation described here, is just such a professional. In college, she majored in business with an emphasis in management systems. She worked for a couple of years as a purchasing agent for a small manufacturer of electrical novelty items—revolving pen displays, signs advertising beer, and the like. She has worked for ITE Industries for three years, the last two as head of Purchasing. The way she should report her department's quarterly activities is just one example of the problems she has to solve and the decisions she has to make. We will say more later about how professionally she solves those problems and makes those decisions.

Professionals, in general, are usually held in high esteem by other members of society, and much is expected of them. It is they who keep the society running well—effectively, efficiently, humanely—and who move it forward under the popular notion of progress.

Talking about professional versus unprofessional behavior touches on the idea of what it means to be a professional. Being a professional is a matter of attitude, or the way people think of themselves and their responsibilities in the context of their occupations. Professionalism is, then, a social notion. It is determined by the members of a group or, more loosely termed, the profession, and it has to do with the conventions of behavior within the group. Those conventions may differ from profession to profession, but the differences are less important than the similarities. Most people will agree that professionalism means maintaining high standards of performance, accuracy, thoroughness, honesty, and integrity.

Examples of Professionalism

Although the definition of *professionalism* contains abstract terms, it can be made more concrete by looking back at Susan James's situation. Susan was under pressure from both the Engineering Department and upper management to support effectively and efficiently some of the vital operations of the company. But she had to do this under adverse circumstances: increased bureaucratic paperwork, restrictions on the kinds of suppliers she could use, and a decrease in the number of key departmental personnel. As she saw the need to report on her support activities, including a category that would cite "extraordinary achievement," she was tempted to report as extraordinary the fact that her group found a crane supplier within twenty-four hours of Engineering's request. She was also tempted to obscure the facts—that this crane would not actually serve Engineering's needs and that an agreement to buy the crane could not be reached until long negotiations with the supplier had taken place. To say that she had found a supplier in twenty-four hours would be true and accurate. To say, however, that the crane could do what it was needed to do and that it could be on site in a short time would simply be a lie—but one that she could embed in language that would not sound like a lie.

Susan's decision not to mention the possibility of buying the crane under the category of "extraordinary achievement" is an example of her intelligence, honesty, and high standards of professionalism. Tempted as she was to make something of almost nothing, she resisted, knew that the truth of the situation would finally surface anyway, and decided in the end that no claim for "extraordinary achievement" would be better for all concerned than a bogus claim. Susan acted professionally.

The same abstract terms just used (high standards of performance, accuracy, thoroughness, honesty, and integrity) can be made more concrete by considering some instances in which high standards were not maintained. The Wall Street insider trading scandals of several years ago are a good case. The people involved in those scandals, such as Ivan Boesky, were certainly thought of as professionals. They had performed for years with singular effort and were highly trained or educated. They did specialized work with good results, making decisions and solving problems, all supposedly for the benefit of their firms and clients. When it was discovered that they had made decisions on the basis of confidential information, which would benefit only themselves and a few others, it was clear they had violated the trust that people had placed in them. They in turn had to forfeit their claims to professionalism, since they had ignored its standards of honesty and integrity.

Other examples are more dramatic, since they have to do with the loss of life, not just of money. Take the case of the space shuttle *Challenger* disaster in 1985. As the voluminous reports on the event show, Morton Thiokol, the engineering firm in charge of the design of the booster rockets, had been pressured by NASA to give the go-ahead to the launch even during weather that technically was too cold for certain vital functions to operate correctly. When the O-rings, part of the fuel supply system, failed, the booster rockets ignited, completely destroying the shuttle and killing all who were aboard.

This tragic event was also the result of breaches in professionalism. The launch, since it was to take the first schoolteacher into space, was a much-touted public relations venture, and America was waiting. Morton Thiokol felt pressure from NASA in two ways: to keep to a schedule that already had fallen behind and to win again the contract it had had with NASA, which had just gone up for a new bid. Bowing to these pressures, executives at Morton Thiokol overrode the objections that their engineers had to the launch and permitted it to happen. We all know the result of that decision: the loss of seven lives and a setback of three years in the space shuttle program—certainly a costly conclusion to breaches in professional integrity and ethics.

Professional Activity

Solving Technical and Rhetorical Problems. Professional activity—the kinds of things professionals do regularly—is broad and diverse. It can mean teaching, lawyering, preaching, curing. Depending on the field, it can involve basic research, research and development, production, quality control, marketing, sales, public relations, or any combination of these. But whatever is done, and in whatever field, a common activity has to take place: solving problems.

And the problems that professionals face are many and varied. They may reflect the needs of a particular job (how to prevent failures in a computer system); or the needs of a professional-social situation (how to get Research and Development people to work more closely with Production people); or the needs of the problem solver (how to put oneself forward without seeming pushy). Throughout this text we will consider these problems as operational problems, political problems, and ethical problems. But it is also important to remember that we use the word *problem* in a broad sense to mean any unresolved situation. An unresolved situation may appear placid on the surface. For example, communications may be proceeding quite smoothly in an organization despite the lack of a style guide. Nevertheless, the quality of those communications is inconsistent and, therefore, much less efficient than it might be. So the lack of a style guide is still a problem, although it doesn't suggest the urgency of the more dramatic problems that catch the headlines every day.

It is also necessary to see that professional problems worthy of anyone's attention are composed of two basic elements: the technical and the rhetorical. The *technical* elements have to do with the fundamental workings of one's profession or discipline. The *rhetorical* elements have to do with the communication of those workings—what has been done or will be done about them—to various people interested in them. The two elements are always present in any problem.

How to prevent failures in a computer system, for instance, will take a good deal of technical expertise and attention. Understanding the hardware and the software and how the two interface, and deciding how either needs to be altered to make the interface work better to prevent failures in the future, clearly require that computer engineers draw on their education, training, and experience in computer science.

But that process doesn't mean that they can ignore telling someone else about what they have found out. Their superiors will need to know what they have done to address or solve the problem so that they can make decisions about the conduct of future operations. Their peers will need to know what they have done in order to anticipate similar problems in their own work and be able to apply reasonable remedies. Certainly, technicians will need instructions about how the system will be changed and operated in the future so that they can keep it on-line and functioning as it should.

All these situations call for written communications that require different rhetorical approaches. The superiors should get at least a memo report, summarizing for them why the failures in the system occurred and what will be done to correct them. They need to know how long it will take and how much it will cost to get the system back in good operating order. They also need to know what other decisions they will have to make to ensure that troubles with the system won't occur again.

The peers should get a copy of the report to the superiors, with a technical addendum that discusses in detail the causes of the failures and the remedies applied. They need this information to keep current, to know what troubles could turn up again and how to deal with them.

The technicians need new instructions. Although they may have some interest in the basics of the failures from a theoretical point of view, they are most interested in the mechanics of correcting the system and running it well in the future. If the changes will require new training for them, they will need to know how and when that will happen.

Thus it is clear that even though computer engineers in this situation may spend much of their time dealing with the technical element of the problem, they still have compelling needs for rhetorical analysis and writing.

Writing in Response to Problems. In fact, attempts to solve or at least to address problems almost always require writing of some sort. The writing may include proposals, reports on data collection in the field or in the lab, explanations of designs for new production methods, suggestions for finding new markets, correspondence with present and future clients or customers, reports on projects in progress, or memoranda on possible implementation methods.

Simply scanning the ranks of professionals shows how much writing results from attempting to solve problems. Engineering firms, for instance, write proposals to prospective clients, both private and public, stating what they know about flood control, or traffic control, or the ways of automating inventory control, to suggest specifically how they might solve problems in those areas.

Professional researchers in the hard sciences or in the social sciences collect data, either in the field or in the lab, and write proposals to study and prove some hypothesis, for example, that drug abuse is increasing among those who profess religious affiliation or that the presence of the AIDS virus is increasing and will continue to increase in all sectors of the society.

Production experts report on new ways of making production more efficient, and marketing experts construct surveys of consumers in different parts of the country and the world to see what consumers' tastes are and how they might be affected by new products. Teachers report to their school districts about how a change in curriculum has affected students' learning of a particular subject, and nutritionists report on how dietary habits have changed among people who have taken part in state-supported nutritional education programs.

PROFILES OF PROFESSIONALS

The Problems They Solve and the Documents They Write

Following are profiles of six professionals: an engineer, a chemist, a social worker, a museum director, a computer specialist, and a businessperson. The profiles describe the responsibilities these people have for solving problems in their day-to-day work, but especially they describe the kinds of documents these people write in response to those problems. The writing tasks are divided into three kinds:

- Technical and professional writing tasks are those that require the specialized knowledge of professionals, depending on their fields.
- Administrative writing tasks are those that grow out of needs and demands of the positions these people have within their respective organizations.
- Review and/or editing tasks are those that require all professionals at some time to review and perhaps edit other people's writing. This process includes anything that they have specifically assigned others to write; it also includes writing that others are normally responsible for, but that because of its importance to the work these people do, should be reviewed and/or edited before it is sent out.

The profiles that follow are typical, but they are not necessarily inclusive and do not suggest all problems these professionals might face or all the documents they may have to write. They do suggest, however, a realistic and reasonable range of professional activity and the amount of time that each person usually spends on writing or on tasks related to writing.

PROFESSIONAL PROFILE 1

Joseph T. Brown
Operations Engineer
(Working for a nuclear waste disposal site)

PRIMARY AREAS OF RESPONSIBILITY FOR SOLVING PROBLEMS

Design of the facilities for storage of nuclear waste. Supervision of other personnel, including other engineers within his operations group. Liaison to other operations groups and to upper management.

TECHNICAL WRITING TASKS

1. **Policy statements,** e.g., on training required to operate special equipment.
2. **Facility descriptions,** e.g., a full description of the low-level waste-handling room to be used for federal officials to plan and get authorization for funds.
3. **Technical procedures,** e.g., unplugging the borehole (storage sleeve) for high-level waste retrieval.
4. **Process descriptions,** e.g., a description for a lay audience of the facility's plan for handling low-level waste from receipt to emplacement.
5. **Instructions,** e.g., extremely detailed descriptions for operating unfamiliar equipment.
6. **Formal technical reports,** e.g., a report on the inadequacies in the waste transporter.
7. **Lesson plans** for the Training Department to use to train waste-handling operators.
8. **Proposals,** most likely as part of a team, adapting standardized (boilerplate) materials for specific purposes and audiences.
9. **Accident reports,** as part of a team, of significant accidents (such as a fire in a silicon-controlled rectifier motor that drives the waste hoist).
10. **Responses** to quality assurance audits that detail fixes for complicated technical problems.
11. **Safety procedures** (particularly for Brown's own group), e.g., procedure for entering a hot cell (area where high-level waste is located).
12. **Papers and articles** on theoretical and practical issues and discoveries for professional meetings and journals.

ADMINISTRATIVE WRITING TASKS

1. **Periodic progress reports,** which are prepared weekly, monthly, quarterly, semiannually, and annually, for various audiences (including the general public) who require separate reports.
2. **Justifications** written to upper management for more staff, more money, more equipment, more overtime. (Budget justifications are also often written to show the effects of proposed cuts on Brown's operation.)
3. **Personnel documents,** such as evaluations, e.g., of the manager who reports directly to him; and justifications, e.g., to hire a particular person or for raises, promotions, leaves, and so on. (Included here would be letters of recommendation for people in his group applying for other jobs either inside or outside the organization.)
4. **Letters and memos** to upper management on particular problems, e.g., how the procedures of another group are slowing down the operation of his group.
5. **Reports** to upper management on the annual goals of his group (particularly required by management-by-objective organizations).
6. **Articles** for the in-house newspaper on the activities of his group.
7. **Documentation** for his own files on troublesome or potentially troublesome situations, e.g., plans for instituting new procedures over which he has little control.

REVIEW AND/OR EDITING TASKS

1. **Manual materials,** which are lower-level procedures for performing a number of tasks within Brown's operation.
2. **Specifications** for particular operations within his operation, or from other operations that may involve his.
3. **Minor accident reports.**
4. **Procedures** from other departments that will affect his operation.
5. **Engineering drawings** and schematics.
6. **Planning documents** written by superiors that require his technical review before they can be published.

Total time spent on writing or on writing-related tasks: 70%

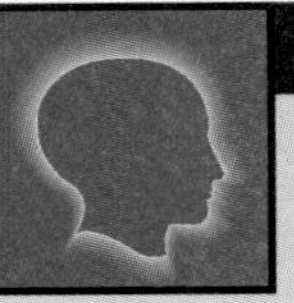

PROFESSIONAL PROFILE 2

Elizabeth J. Frederickson
Chief Chemist, Assistant Manager; Department of Engineering Research
(Working for a Pacific Coast utilities company)

PRIMARY AREAS OF RESPONSIBILITY FOR SOLVING PROBLEMS

Direction of the management of hazardous materials, including supervision of laboratory analysis, field work (soil sampling), training of new personnel, and continuing review of pertinent federal and state legislation. Liaison to the Legal Department and occasionally to lobbyists at the state legislature.

TECHNICAL WRITING TASKS

1. **Policy statements,** e.g., on special training necessary for work with hazardous materials.
2. **Technical procedures,** e.g., methods of analysis of hazardous materials; methods of soil sampling.
3. **Safety procedures** for handling hazardous materials.
4. **Instructions,** e.g., extremely detailed descriptions for operating unfamiliar equipment.
5. **General training manuals** for new personnel.
6. **Formal technical reports** written about the progress of certain established projects.
7. **Field logs or notebooks**.
8. **Site reports** (emphasizing soil sampling and analysis) on new projects just getting started.
9. **Proposals,** most likely prepared as part of a team, for changes in procedures.
10. **Critical analyses** of existing, new, and potential legislation pertinent to the use, handling, and disposal of hazardous wastes.
11. **Papers and articles** on theoretical and practical issues and discoveries for professional meetings and journals.

ADMINISTRATIVE WRITING TASKS

1. **Periodic progress reports,** which are prepared weekly, monthly, quarterly, semiannually, and annually, for various audiences, depending on their needs for periodic reporting.

2. **Justifications** written to upper management for more staff, more money, more equipment, more overtime. (These would be passed through her manager to upper management.)
3. **Personnel documents,** e.g., first-level evaluations of staff written for her manager's review, and new personnel training reports.
4. **Articles** for the in-house newspaper on the activities of the department.
5. **Documentation** for her own files on troublesome or potentially troublesome situations, e.g., plans for instituting new procedures over which she has little control.

REVIEW AND/OR EDITING TASKS

1. **Existing, new, and proposed laws** dealing with the use, handling, and disposal of hazardous materials.
2. **Site and laboratory reports** from staff personnel.
3. **Segments of proposals** written by staff personnel. (Frederickson would review these, but they would be discussed more fully in staff meetings.)
4. **Reports and position papers** written by her manager, for which Frederickson's particular knowledge and expertise are needed before they are published.

Total time spent on writing or on writing-related tasks: 60%

PROFESSIONAL PROFILE 3

Leah S. Feldstein
Director of Client Relations
(Working in a drug education and rehabilitation program for youth ages 12 to 17)

PRIMARY AREAS OF RESPONSIBILITY FOR SOLVING PROBLEMS

Direction of general drug education program in the public schools; direction of outreach work to identify adolescents who are using drugs and who should benefit from specific education and rehabilitation. Supervision of other social workers who work as counselors of adolescents inside and outside the public schools. Liaison to the director of the program, to the board of directors, and occasionally to the government agencies funding the project.

TECHNICAL WRITING TASKS

1. **Informative pamphlets** written for teachers in the public schools on the kinds of drugs available, the psychological and physical effects they can have on people who use them, and how their use is viewed generally by their students.
2. **Informative manuals** written for outreach workers on how to identify adolescents who are using drugs, and on what drugs they are most likely using.
3. **Instructions,** which amount to detailed lesson plans, for teachers on what should be covered in classroom units on the use and abuse of drugs.
4. **Instructions to outreach workers** on the kinds of counseling that can take place both on and off the counseling site.
5. **Informative articles** written for the PTAs and local newspapers explaining exactly what drug education and rehabilitation have come to mean and how they are generally conducted.
6. **Proposals,** usually written with others, to various agencies—local, state, and national—for initial or continued funding.
7. **Persuasive/informative papers and articles** written for professional meetings and journals in the areas of social work and education.

ADMINISTRATIVE WRITING TASKS

1. **Periodic progress reports,** which are prepared monthly, quarterly, and annually, to three audiences: the director, the board of directors, and the funding agencies, who require such reports at different times, depending on their needs.
2. **Requests to the director** for more staff and expansions of physical facilities for counseling.
3. **Numerous memos to outreach staff** answering particular queries, commenting on reports they have submitted on individual clients and suggesting new approaches in counseling. (These memos would probably be written in confirmation of suggestions discussed at staff meetings.)
4. **Letters to other agencies or programs** doing similar work in other places, to request information on the structure of those programs and on their approaches to drug education and rehabilitation.
5. **Personnel documents,** e.g., evaluations of existing outreach personnel and of potential new hires for outreach work; and informal evaluations, i.e., without direct administrative impact, of teachers in the

public schools. (Also included here are letters of recommendation for people looking for jobs elsewhere, and letters of commendation of teachers who have done particularly well in discussing the units on drug use and abuse in their classrooms.)

6. **Formal annual report** to the director, but ultimately intended for the board of directors, on the workings of the Client Relations unit. This report would discuss how well the education and outreach efforts have worked, and how they have been evaluated, e.g., how many students were given information in school, how many students were identified as potential clients for rehabilitation, how many potential clients were identified outside the schools for education and/or rehabilitation. It would also discuss the results, as they could be determined, of educational and rehabilitation efforts: how many students cut down on or stopped taking drugs, when in the process of education and/or rehabilitation they did so, what general effects of those two conditions have been realized.

REVIEW AND/OR EDITING TASKS

1. **Reports from outreach staff** on initial client contacts, continued or ongoing client contact and counseling, and plans for new outreach efforts. (These reports come in monthly, more often if a situation demands immediate attention.)

2. **Proposals** from outreach staff for new procedures in the bureaucratic dealings with clients, or for new approaches to counseling of existing and prospective clients.

3. **Segments of proposals** written for funding agencies by various members of the outreach staff on areas in which they have particular interest or expertise. (Feldstein would review these, make editing suggestions, but would regularly bring together the contributors to discuss the questions or changes she has in mind.)

4. **Reports and position papers** written by the director, for which Feldstein's professional knowledge and expertise are needed before they are published.

Total time spent on writing or on writing-related tasks: 65%

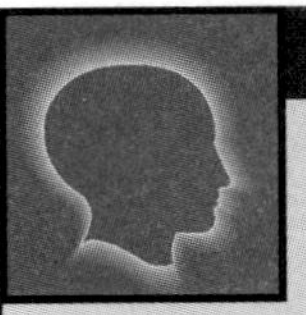

PROFESSIONAL PROFILE 4

Victoria P. Wilson
Director, University Art Museum
(Working in a small art museum in a public university in the Northwest)

PRIMARY AREAS OF RESPONSIBILITY FOR SOLVING PROBLEMS

General planning and supervision of all museum activities, including conduct and timing of all exhibits (both permanent and temporary), tours, and other uses of the museum for activities related to the forwarding and preservation of the arts. Liaison to the university's Campus Arts Committee and Museum Advisory Subcommittee, to the Friends of the Museum, and to the office of the Dean of Humanities and Social Sciences, College of Arts and Sciences.

PROFESSIONAL WRITING TASKS

1. **Correspondence** with museums, galleries, private lenders, exhibit touring services, and individual artists to arrange exhibits.
2. **Loan requests and loan agreements** with museums, galleries, private lenders, exhibit touring services, and individual artists.
3. **Requests for permission** to publish images, e.g., for posters if the museum does not have the copyright.
4. **Correspondence with cocurators,** i.e., people elsewhere who share curatorial duties for exhibits being planned either on or off the museum premises, but nevertheless in cooperation with the museum.
5. **Invitations to prospective exhibit jurors,** i.e., in a competitive, or juried, exhibit, people (jurors) are needed to assess the quality of the work to go into the exhibit.
6. **Proposals,** usually written with others, to state and federal agencies (such as the National Endowment for the Arts) for aid in developing the permanent collection, the showing of temporary exhibits, and generally for furthering the work of the museum.
7. **Correspondence with the Dean's office,** including proposals for art programs, branch campus exhibits, security for paintings, plans for exhibit programming, plans for exhibit publications.
8. **Program planning,** including exhibition schedules and budgets.
9. **Replies to inquiries** from museums, galleries, exhibit touring services, and private individuals about certain works or works by certain

artists in the museum's permanent collection; about traveling exhibits offered to the museum; about art in general, artists, or certain works of art.

10. **Liability documents,** such as correspondence with lenders, insurance companies, lawyers, shipping companies about damage to works in transit or while at the museum.
11. **Development documents,** such as annual report to the Friends of the Museum of Art, letters to various committees of the Friends, letters to prospective donors of works to the museum.
12. **Thank-you letters** to lenders for exhibits, board members, museum documents, volunteers, donors of art, donors of funds, lecturers, people who host events for the museum, and musicians at receptions. (Wilson writes many of these letters herself, but she also reviews some that she has asked others to write for her; see the "Review and/or Editing Tasks" section that follows.)

ADMINISTRATIVE WRITING TASKS

1. **Funding requests** to the Dean of Humanities and Social Sciences for more equipment, minor capital improvement, transfer of funds, visiting lecturer funds. The museum's detailed annual budget is also included here.
2. **Personnel documents,** e.g., requests for sabbaticals or leaves without pay; annual reviews of faculty and staff; requests for additional faculty or staff positions; letters of recommendation for employees, faculty members, artists, museum professionals at other museums.
3. **Police services documents,** including security procedures, reports on breaches of security, requests for police-record checks on security guards, requests for special police coverage.
4. **Periodic progress reports** on the use of private, state, or federal grant funds.
5. **Correspondence with the Department of Physical Plant** about fire protection, repairs, and general maintenance of the museum.
6. **Administrative planning documents,** including biennial review, long-term (five-year) planning, statements of purpose for future operations.
7. **Correspondence with professional organizations,** including various documents related to membership in professional museum organizations.

REVIEW AND/OR EDITING TASKS

1. **Thank-you letters** (see "Professional Writing Tasks," item 12, earlier).
2. **Exhibition handouts.**
3. **Museum newsletter.**
4. **Press releases.**
5. **Postcards and exhibit posters.**
6. **Brochures** about museum programs.
7. **Segments of grant proposals** written by others.

Total time spent on writing or on writing-related tasks: 75%

PROFESSIONAL PROFILE 5

James R. Brookings
Director of Marketing
(Working for a third-party developer for a major computer company)

PRIMARY AREAS OF RESPONSIBILITY FOR SOLVING PROBLEMS

Determines corporate marketing policy and oversees domestic and international distribution of company products. Designs annual advertising campaigns, coordinates company participation in national and international industry trade shows. Oversees the creation of product packaging and the writing of hardware and software documentation; arranges cooperative sales ventures with compatible hardware and software manufacturers. Trains new marketing personnel and acts as liaison among sales personnel, technical support, and upper management.

TECHNICAL WRITING TASKS

1. **Policy statements,** e.g., on special training necessary for work with on-site computer systems.
2. **Press releases** regarding the marketing of new products and product upgrades for circulation in industry magazines, newsletters, and other related literature.

3. **Promotional brochures** written for end-users describing company products and possible applications.
4. **Informational pamphlets** to be used in-house for customer service sales representatives, which describe company products, possible applications, and compatibility information.
5. **Product documentation,** including technical specifications for company products, and hardware and software users' manuals.

ADMINISTRATIVE WRITING TASKS

1. **Periodic progress reports** to the president and vice president assessing the effectiveness of existing marketing strategies and advertising campaigns.
2. **Long-range projections** of sales figures and marketing projects, submitted to the president and vice president on a biannual basis.
3. **Monthly finance reports** to the Business Officer accounting for expenditures in the marketing department.
4. **Requests to the General Manager** for more funding for additional staff and for the advertising budget.
5. **Letters** to other hardware and software computer companies, negotiating cooperative promotional strategies and compatibility features for new products; and to group end-users (such as churches, schools, and governmental agencies requesting discount bulk pricing).
6. **Personnel documents,** e.g., evaluations of marketing, customer service, and technical support staff members that inform in-house decisions regarding promotions and salaries.

REVIEW AND/OR EDITING TASKS

1. **Proofs from advertising agencies** of all print ads. (These proofs generally include graphics as well as text.)
2. **Drafts of hardware and software manuals** from in- or out-of-house technical writers.
3. **Promotional slide shows** on disks created by marketing staff members.
4. **Reports or memos** written by the Director of Sales that affect dealings with domestic and overseas distributors.

Total time spent on writing or on writing-related tasks: 75%

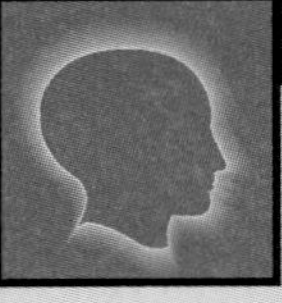

PROFESSIONAL PROFILE 6

Bruce K. Kennewick
Director of Catering
(Currently working for a medium-size resort hotel in the Midwest)

PRIMARY AREAS OF RESPONSIBILITY FOR SOLVING PROBLEMS

Scheduling and supervision of all catering services: temporary workers for meetings, including refreshments, lunches, dinners, full-scale banquets; temporary workers for private parties, including weddings, bar mitzvahs, anniversaries, retirements, etc.—all food and beverage planning and presentation. Some responsibility for sales of catering services to local organizations and individuals. Liaison to upper management, to other lateral department heads, and to clients (both existing and potential).

PROFESSIONAL WRITING TASKS

1. **Local Standard Operating Procedures (LSOPs),** e.g., changes in Home Office SOPs for physical setup of meeting rooms or for presentation of food and beverages, as determined by local needs.
2. **Letters to clients** explaining and confirming arrangements for a meeting or party; frequent follow-up letters until all arrangements are understood and set.
3. **Sales letters** to prospective clients explaining the facilities and services of the Catering Department.
4. **Descriptions** of the Catering Department's services written for general sales brochures produced by the hotel.
5. **Brief proposals** written to upper management (usually Director of Sales) on new sales approaches to be attempted locally.

ADMINISTRATIVE WRITING TASKS

1. **Monthly memos** to the Director of Food and Beverages confirming arrangements and needs for catering services already booked.
2. **Memos to the Director of Sales** reporting on the results of local sales.
3. **Proposals to the Director of Food and Beverages** for changes in LSOPs.
4. **Personnel documents,** e.g., evaluations of the various managers who report to Kennewick, letters of recommendation for his managers.

5. **Formal annual report** on the performance of the Catering Department.
6. **Formal annual statement** outlining goals for the future as required by management-by-objectives organizations.

REVIEW AND/OR EDITING TASKS

1. **Reports on various operations** that Kennewick has asked his managers to write.
2. **Segments of the formal annual reports** on performance and on goals his managers write.
3. **Evaluations of staff** reporting to his managers.
4. **Reports or memos** written by lateral heads of departments (directors) that speak of catering services and that need his check for accuracy.
5. **Reports or memos** written by the Director of Food and Beverages that speak of catering services and that need his check for accuracy.

Total time spent on writing or on writing-related tasks: 40%

Highlights of the Professional Profiles

These profiles, which represent a reasonable cross section of the professional world, come from the pure and applied sciences, the social sciences, the humanities, and business. They show that professional life presents a series of problems that should be solved and that require communication of some kind, either speaking or writing. We also chose these six people because they are in management or supervisory positions, which demand even more writing to help solve administrative problems. Joseph Brown, for instance, had no idea, when he was promoted to Operations Engineer, that he would have to spend up to 70 percent of his time on writing or on tasks related to writing. As a staff engineer, he had spent a good deal of time writing—maybe 50 percent of his time—but with the added responsibilities of management, he assumed added responsibilities for writing.

The same is true of Bruce Kennewick, who is on the low end of the spectrum for time spent on writing. When we asked to interview him, he shrugged and said, "Oh, I don't really write very much. I mostly solve problems by talking—a lot of talking." But when we asked him to catalog the kinds of problems he had to solve, he realized that his position as Director of Catering—

having to deal with managers under him, as well as those above him—meant a good deal more writing than he had first thought. Forty percent of his time spent on writing, he acknowledged, is substantial, and he admitted that with a promotion to Director of Food and Beverages, the percentage of time would increase.

The consensus among these six professionals was that as they progress upward through the middle ranks of management, more writing is required of them. They also agreed that the time spent on technical and professional writing will decrease, whereas the time spent on administrative and review and/or editing others' writing tasks will increase. Writing descriptions, procedures, instructions, accident reports, proposals, and the like will be assigned increasingly to people who work for them, and their own responsibilities will involve more statements of purpose and policy and review and the editing of work they have assigned to others.

They agreed, finally, that if they were to reach the upper ranks of management, the time they would spend on writing would decrease overall. All the basic writing and much of the review and editing would be done by others, even though they would sign many of the final documents. The actual writing responsibilities of upper management are limited generally to major policy and planning statements, correspondence with lateral upper management, and reports to boards of directors or boards of trustees, and even these generally are written by others. However, there was no question in anyone's mind that progress through the ranks of any organization is realized only through effective problem solving, which in turn is expressed only through effective communication, both oral and written.

The end of the matter is that *good professionals solve problems effectively, and effective problem solving happens only through good communication.* There is no other way.

SOLVING PROBLEMS BY ASKING AND ANSWERING QUESTIONS

Although the problems that professionals have to solve are many and diverse, in both their technical and their rhetorical elements, the process of solving them always involves asking questions and attempting to find answers to them. Each step in problem solving can be developed fully only when the professional responds to the questions implied there. (Chapters 2 and 3 present detailed analyses of the questions that must be dealt with at each stage of technical and rhetorical problem solving.)

Before attempting to proceed along the stages of the problem-solving process, professionals must first determine whether a problem truly exists. The answer to this question can be determined by asking four basic questions:

1. What is wrong?
2. What is lacking?

3. What is unknown?
4. Who feels the need for and will benefit from a solution?

Some problems will present answers to all four questions; some will present answers to three; but all problems must present answers to at least two, the one constant being, Who feels the need for and will benefit from a solution? If no one can be identified as needing or benefiting from a solution, no problem exists.

To demonstrate, we suggest the problem of cancer, which provides answers to all four questions:

- *What is wrong?* People are dying in increasing numbers and in all segments of the population.
- *What is lacking?* A cure for the disease in its various forms is lacking.
- *What is unknown?* A clear cause of most forms of the disease is unknown.
- *Who feels the need for and will benefit from a solution?* The population in general, but in particular, those who are at risk or who already suffer from cancer feel the need for and will benefit from a solution.

If, however, we take the case of geologists studying the history of glacial movement in Antarctica, the problem is defined adequately by answering only two of the questions:

- *What is unknown?* The direction and speed of centuries-old glacial movements are unknown.
- *Who feels the need for and will benefit from a solution?* Geologists and other environmental scientists interested in reconstructing the natural history of the earth in order to predict future environmental changes and conditions feel the need for and will benefit from a solution.

It is important to remember that a problem must be defined at least in terms of a felt need; that is, some segment of the population must feel the need for a solution. Cancer, in its many forms, is a problem felt by a much larger segment of the population, and by its nature is felt much more forcefully than is the problem of determining and understanding the history of glacial movements. Although perhaps only a small segment of the population feels the need of a solution in the latter case, the need is felt nevertheless, and therefore, understanding the history of glacial movements qualifies as a problem. This process of defining a problem is the first stage of technical problem solving, which is discussed in detail in Chapter 2.

The process of asking and answering questions suggests that all problems can be solved. But that is not true. Lack of knowledge and resources at times simply may render a problem insoluble—at least at the time it is first confronted, and perhaps for some time after that. That does not mean, however, that the problem should not be addressed or that professionals should not continue to ask and attempt to answer questions to try to find out as much about the problem as they possibly can. For example, although no actual solution to the problem of cancer—a cure—has been found, procedures and treatments have been discovered that can control the spread of the disease or that can alleviate the symptoms. These steps along the way—the results of asking and attempting to answer questions—are important because they may lead to an ultimate solution.

Problem solving is a continuing process. Some problems are solved, some are not, and some are in the state of being solved. The process cannot be denied, ignored, or lightly undertaken: it is essential to the work of professionals. In Chapters 2 and 3 we present detailed analyses of both the technical and rhetorical aspects of problem solving.

EXERCISES

1. Conduct an interview with a professional in your major field of interest—an academic (on campus) or someone working in the field (off campus). Ask questions that will provide information about what this person does, how he or she works, what kinds of problems this person confronts and attempts to solve, and what he or she has to write in response to those problems. Ask the person you interview for some samples of the writing that he or she has to do, and then return to class with your interview notes and the samples of writing for small group and general discussion.

2. Consider whether the following problems can be defended as problems, given the criteria on page 21 of this chapter.

   a. The wolf is losing habitat in the Yukon.
   b. Hearing aids aren't covered anymore by insurance plans.
   c. Ten large elms now surround the old mine shaft.
   d. Work in the office gets done with a typewriter.
   e. The local cable company wants to upgrade service to its customers.

ASSIGNMENT

1. Write a three-page memo report to the director of undergraduate studies of the department in which you are majoring, telling her or him what you have learned from your interview (in preceding Exercise 1) and suggesting that some kind of newsletter highlighting this kind of information be published for undergraduate majors. In your memo, emphasize the kinds of problems that professionals in your field face, and the kinds of writing that result from addressing or solving those problems. You should refer specifically to the samples of writing you have obtained and attach them as exhibits to your memo.

TERM PROJECT OPTION

Students who decide to pursue a term project will be able to produce original research from their individual fields of study or from fields they have an interest in. These projects will take the form required by the problem to be solved—depending on students' judgments. For example, students completing these projects can produce several kinds of products, such as technical reports, feasibility studies, articles for publication, and computer manuals. Students will end the term with an oral presentation based on the completed written project.

The first thing to do is make a short list of problems you would consider exploring throughout the term. When asked, be sure to defend them as problems on the basis of the criteria discussed in this chapter.

Technical Problem Solving

CASE STUDY

SUSAN JAMES'S OFFICE

ITE Industries, Inc. ▪ Richmond, California

It was time to go home: Susan James was tired. The outline of her latest quarterly report, which she had worried about the day before, was harder to fill than she had anticipated. Although she was still pleased with her decision not to report an "extraordinary achievement," she nevertheless had to show that her group had indeed made progress, especially in meeting the Engineering Department's needs. And that progress would still depend largely on Engineering's request for a new crane and Jones and Son's attractive bid—especially in terms of time—since this company could have the crane in place and operating in forty-eight hours.

Susan decided, therefore, to define the problem from a new perspective, the fact that Jones and Son's bid was unacceptable for two basic reasons: the available crane couldn't do the work required, and Jones and Son wouldn't qualify under the new push (initiated by the recent contract with the government) to hire

PREVIEW OF THE CHAPTER

minority contractors. In writing the report, she would have to keep in mind Engineering's priorities—the need for good equipment delivered in a timely manner—and upper management's priorities—efficient, timely, cost-effective delivery, preferably from a minority contractor.

When writing the report, she knew that she would have to depend on some research she had already done—indeed, research that went on in her department every day—on suppliers, the kinds and qualities of products they can supply, and their records of timely delivery and of acceptable performance. The only trouble was choosing the best, most useful, and pertinent information from that research. In many ways, doing that took more defining, but it also took analysis—breaking up what she knew about the suppliers into its parts and seeing the relationships among those parts—in order to understand what ITE Industries ultimately needed to do to stay competitive.

Susan eventually figured this all out. She defined what her department's problems were, how they were affected by Engineering's problems, and how they were affected by ITE's problems in general. After she had analyzed the various connections, she resolved to write the report. She decided that it would follow generally the outline she had written the day before but that it would contain more information to indicate that she had done her homework—that she knew what the problems were and had some idea of how to solve them.

AN OVERVIEW OF TECHNICAL PROBLEM SOLVING

The process of technical problem solving can be divided into five basic stages: definition, research, analysis, resolution and synthesis, and implementation.

In the first stage, *definition* (from Latin *definire,* meaning "to set limits"), problem solvers establish what the nature of the problem is, what its scope (or limits) is, who or what is affected by the problem, and what benefits would result from a solution to it.

The next stage, *research* (from French *rechercher,* meaning "to look for" or "to search again"), involves collecting data, or information, pertinent to the problem. The research can be primary, that is, firsthand, when problem solvers collect their own data through lab experiments, fieldwork, or surveys; or it can be secondary, when they collect data from other researchers who have published their findings for public review. Or it can be, and most often is, a combination of both.

Analysis (from Greek *ana-* plus *lyein,* meaning "to break up") is the point at which problem solvers look at the data from various points of view to determine, even tentatively, what the data mean, given the problem as it has been defined.

Questions That Technical Problem Solving Answers

Definition of the Problem: What is the felt need?

- What is wrong?
- What is lacking?
- What is unknown?
- Who feels the need for and will benefit from a solution?

Research of the Problem: What information exists?

- What sources are available to provide data?
- Are they primary sources? Of what kind?
- Are they secondary sources? Of what kind?
- How can data be collected from these sources?

Analysis of the Problem: What do the data mean?

- Do the data address the problem as defined?
- Are there relationships among the data?
- Do they show cause and effect? How so?
- Are there patterns? trends? anomalies?

Resolution and Synthesis of the Problem: Are there solutions to the problem?

- What significance emerges from analysis?
- Does the significance suggest solutions?
- Are the possible solutions reasonable?
- Are some more reasonable than others? Why?

Implementation of the Solution: What is necessary to carry out the solutions (or steps toward dissolution)?

- What steps are needed to carry out the solution(s)?
- How will the steps be performed?
- Who will perform them?
- What equipment or materials are needed?
- How much will the process cost?
- How long will the process take?

FIGURE 2.1 **Questions to ask and answer in the technical problem-solving process**

Resolution and synthesis (from Latin *resolvere,* meaning "to loosen" and from Greek *syn-* plus *tithenai,* meaning "to place or to put together") happen when problem solvers assemble all they know about a problem—from definition, research, and analysis—and suggest a solution or alternative solutions to the problem.

Finally, *implementation* (from Latin *implere,* meaning "to fill up" or "to finish") happens only when the other stages of the process are complete and when a solution, or steps toward achieving a solution, has been determined.

As noted in Chapter 1, each stage in problem solving can be developed fully only when the professional responds to the questions implied there. Note that for each stage, there is one general question that can be answered by asking other related questions. The importance of asking good questions cannot be overstated. Figure 2.1 suggests the kinds of questions that need to be asked and answered within each stage.

DEFINITION

This first stage is a crucial part of the process. Without a clear definition of the problem, problem solvers cannot know the right directions to take in researching, analyzing, resolving and synthesizing, or finally implementing a solution to the problem. Although clear definition at the outset is essential, a problem may also be redefined on the basis of what happens in the research and analysis stages. Problem solving, then, at least in the first three stages, can be a recursive process, doubling back on itself, as information is uncovered and understood.

For example, computer engineers working for a large manufacturer concerned with protecting the firm's computer system from computer viruses may discover in the research stage and then again in the analysis stage that viruses attack with different magnitudes, depending on the level of exposure to other computer systems beyond the firm's own. The engineers would do well, then, to return to the definition stage and redefine the problem as having to do with the level of outside contact rather than with the simple fact of outside contact, something that may not have occurred to them when they first set out to define the problem. Such refining of the definition would help in further research and analysis, and ultimately, of course, in the resolution and synthesis stage. When defining and redefining a problem, problem solvers must remember that a problem must be able to be defined in terms of a felt need (see Chapter 1).

If the problem has been well defined, the research and analysis can be more specifically designed and carried out, and a reasonable, feasible resolution or synthesis is more apt to follow.

Defining Multiple Problems

Professionals often face situations having multiple problems to be defined. Susan James had three problems. The first, though for ITE in general not the most important, combined her own and her department's problems.

She had to show progress in meeting other departments' needs, in particular, Engineering's needs, since it most frequently complained about Purchasing's performance, often blaming its own missed deadlines on Purchasing's failure to provide necessary equipment on time. The definition of that problem, therefore, had to do with only Susan and her department. She knew w*hat was wrong:* Engineering was always pushed to produce and was likely to suggest Purchasing's performance as a reason for many of its failings, because Purchasing couldn't or didn't *supply* what Engineering needed on time. And since Purchasing is a support department, not a production department, upper management was likely to be sympathetic to Engineering's complaints. She knew also who would benefit from a solution to the problem: immediately, the Purchasing Department, and then, ITE in general.

The second problem needing definition was fulfilling quickly Engineering's request for a crane. *What was wrong* and *what was lacking* were essentially the same: the need of a new crane in place and operating in a short time. And it was ITE's production operation that immediately felt *the need* for and would benefit from a solution.

Embedded in this second problem was a third, and in some ways the most important, problem. Susan knew that Jones and Son could get ITE a crane in forty-eight hours, but she also knew they did not have the kind of equipment that ITE eventually would need. She knew further that the contractual agreements would not meet the government's regulations for the minority contractors program, and so she had to define what the obligations for program participants were and how they would affect Purchasing's activities, which in turn would affect ITE's activities, not only in this particular case but also in the long run.

This embedded problem points to something important in the whole process of defining and solving problems. Problems that on the surface seem simply and straightforwardly defined often contain other problems that also need to be defined, researched, and analyzed. The embedded problems ultimately may be more important for those who need to have them solved. If so, the writer needs to get this greater importance across to persuade the reader.

In this case, Susan knew the difference between getting a crane expediently and getting one that wouldn't meet specifications for its use or regulations of the minority contractors program. It was much more important that Engineering have a crane that fulfilled specifications than that it have the first one available. Likewise, it was more important for upper management to understand how the process of purchasing in general, which clearly affects production, would be affected by the minority contractors program than that Engineering's requests be fulfilled rapidly, or even worse, only for the short term or illegally.

In this case, the quality of the equipment and the need for understanding and meeting contractual obligations should supersede speed of delivery and installation.

Once Susan had decided that she could actually report some progress, the definition phase was crucial. She would have to say specifically why the crane supplier that she had found in twenty-four hours and who could deliver the crane in another twenty-four was inadequate to serve the needs of the Engineering Department, and that further investigation into other suppliers' capabilities most likely would suggest a better supplier and a better product.

She decided she would give a physical description of the crane Jones and Son could *supply*—for example, the tonnage it could lift, the ranges of movement from one lift spot to another, and the time required for lifts and placements. These elements would then have to be compared with or contrasted to the capabilities of other cranes produced by other suppliers and related to the needs of the Engineering Department to show that Jones and Son, although it could have the crane in place in a very short time, nevertheless would not be supplying ITE's needs appropriately.

The additional problem Susan had to define was having to seek minority contractors. Jones and Son did not qualify. Its owners are Anglo-Americans; the founder and still Chairman of the Board, Hershel T. Jones, and his son, Raymond L. Jones, Chief Executive Officer (CEO), run the family business. Though they hire minority workers, none has been hired in managerial positions. There are no women, Hispanic Americans, African Americans, Asian Americans, or Native Americans in any policy-making or policy-reviewing positions. The government's minority contractors program requires that at least 22 percent of those positions, in lieu of direct ownership, be held by minorities. The government also requires that minorities in a range of 12 percent to 24 percent be employed in all other areas. And finally, the new guidelines require a more complicated process than would normally be required of an agreement with an old, trusted supplier like Jones and Son.

Though Susan knew that the Sales and Development Department was aware of these stipulations in its dealings with the government and that upper management would have some idea of the resulting necessary contractual obligations between ITE and its suppliers, she was not sure that upper management knew how those obligations would affect purchasing agreements made by her department. She decided that she would define in the main text of her report the basics of how the minority contractors program affected purchasing, especially with regard to doing business with Jones and Son. She then would include as an appendix more detailed information about how purchasing from other suppliers would be affected.

In the appendix, she decided she would include some of the history of the minority contractors program and the way that other firms have dealt with the problems of purchasing. She could also use the experience of some firms she knew well, or could get information about, as an example of how purchasing is affected: what kinds of restrictions that purchasing agreements imposed, especially as to timely delivery. Finally, she could include a more detailed description of the process required in bidding for contracts: what the requests for bids (RFBs) or requests for quotes (RFQs) should include and how long the bidding process is supposed to take—something that presented a problem in the case of Jones and Son. (For more information on RFBs and RFQs and on proposals in general, see Chapter 6.)

Techniques of the Definition Process

Although it is essential to understand the process of definition, knowing how to ask and answer the right questions so that the problem at hand is framed effectively and knowing how to fill in the frame—the techniques of defining—are also important.

Techniques to Use for a Well-Informed Audience. Because Susan's audience was relatively well-informed—used to reading general descriptions of equipment used regularly in their operations and familiar with the bidding process—she used the following techniques to define her problems:

- Physical description
- Comparison and contrast
- Process description
- Example

Physical Description. Susan needed to describe specifically the various physical characteristics of the crane that Jones and Son could supply—lift capability; range of movement; speed of lift, movement, and placement; and the like.

Comparison and Contrast. Susan needed to compare and contrast the qualities of Jones and Son's crane and other suppliers' cranes not only according to physical characteristics but also according to contractual obligations for delivery and service. She had to point out how other cranes would better serve the physical needs of the Engineering Department and also how other crane suppliers might meet ITE's contractual obligations having to do with minority contractors.

Process Description. In preparing the appendix to her report on the minority contractors program and the way that it would affect purchasing, Susan had to write what amounted to a process description. A description of a process differs from a description of an object, or physical description, in that there is no tangible matter to be described—no height, weight, volume, energy consumed, and so forth. Rather, a series of related steps, considered both individually and in the ways they relate to one another, needs to be described. What needs to happen to make an individual step take place (cause)? How has that step caused or allowed another, subsequent step to take place (effect)? And finally, how does the process as a whole fit together to produce the intended result?

Susan needed to describe the purchasing or procuring process, which required three steps:

STEP 1 The first step was writing and sending to potential suppliers the request for quote (RFQ), which had to include exact specifications of the object to be purchased (all physical characteristics and operating capabilities); allowable times for delivery and installation (twenty-one days was specified for the minority contractors program); and all stipulations and obligations on the parts of both parties for maintenance, service, and replacement. The minority contractors program also required that the RFQ be sent to all minority contractors qualifying to supply the necessary materials (Susan had a list of them) and that no more than the same number of RFQs be sent to nonminority contractors.

STEP 2 The second step was to receive and evaluate the quotes. The minority contractors program required that ITE allow ten working days from the day that suppliers received the RFQ for them to respond. No more could Susan rely on calling a well-known, trusted supplier and have something delivered in a few days. This method was not allowed under the minority contractors program, except under emergency conditions, which had to be defined and documented fully. In fact, all this activity had to be defined—what the time is that it takes to review quotes, how minority and nonminority contractors should be compared and contrasted, and finally, how to decide on a contractor.

STEP 3 The last step was awarding the contract. If, after the review of quotes, a minority contractor were awarded the contract, nothing more than a copy of the contract, including the delivery date and all stipulations regarding maintenance, service, and replacement, need be supplied. But if a nonminority contractor were chosen, then extensive documentation as to why it was chosen over a minority contractor had to be supplied. Susan would need to define more carefully the process of taking part in the program and thereby remind upper management of the difficulties that will inevitably arise—difficulties that shouldn't be blamed solely on her or her department.

Example. The last method of definition Susan used is an example. She knew that Norgaard, Inc., was competitive in only some areas of ITE's operations and that it had a similar corporate structure. She also knew through contacts with people at her own managerial level that Norgaard had agreed to take part in the minority contractors program and that Norgaard was generally encouraged by the possibilities of doing work with minority contractors.

But Susan also found out that Norgaard realized that the contractual obligations, starting with the bidding process, were fairly cumbersome and restricting. Although it had thought that this situation would probably be true when considering the change, Norgaard decided that participation in the minority contractors program ultimately would be more beneficial (in large part because of an increase in the number of government contracts it was apt to get) than troublesome. Susan's contact, Henry Caswell, voiced these concerns and problems, but he did not suggest that Norgaard, and certainly not ITE, pull out of the program.

Susan was able to use Norgaard's experience as an example of how participation in the minority contractors program actually works, something a thorough job of defining should include if at all possible.

Given Susan's audience—the upper management of ITE—all these techniques were appropriate, and depending on the amount of detail she included and where she included it, the techniques were also effective. Her audience certainly needed to know why the crane she had decided not to buy was inadequate to serve the needs of ITE (physical description); how it was similar to or different from other suppliers' cranes (comparison and contrast); how the minority contractors program worked and therefore influenced her decisions and activities (process description); and finally, how other firms like ITE had dealt with, or were dealing with, the stipulations of the minority contractors program (example).

Techniques to Use for a Lay Audience. When an audience is not well-informed on the subject, other techniques of definition may be used. Techniques more appropriate for a lay audience include the following:

- Etymology
- Synonyms and antonyms
- Similes and metaphors
- Analogies
- History of discovery or development

Etymology. Etymology lays out the linguistic origins of the term in question. It tells what language(s) the term is derived from and how the origins make the derived term easier to understand. Determining the etymology of a word should operate in two ways. The *analytical* process breaks up the whole into parts to see what the whole means. Earlier in this chapter, for example, the etymology of the word *analysis* showed that the word is made up of two Greek words, *ana-*, a suffix meaning "up," and *lyein,* a verb meaning "to break." It is easier, then, to understand and remember that *analysis* is the process of breaking up a whole into its parts so that they can be considered and understood by themselves, and then in their relation to other parts.

The other way is *analogical,* that is, likening the essence, meaning, or function of a term to the same characteristics of another term. Thus the term *crane,* denoting a machine that does heavy lifting, moving, and placing by means of a long, swinging projection, gets its name analogically from the bird of the same name, which is noted for its long, swinging, projecting neck.

Etymology is useful only when the derived term is in fact more clearly defined by either analysis or analog. The origins of some terms in use today are either unknown or so far removed from present meaning that they are useless in extending a definition; also, there may be no analogs in present knowledge or experience that can be used meaningfully or effectively. Using etymology in these cases is basically a waste of time.

In sum, using etymology is a good technique for extending a definition only under these general circumstances: the linguistic origins, either analytic or analogic, are relevant and vivid enough to enhance definition and understanding; and the audience for whom the definition is written needs such relevance and vividness to further its understanding of the term.

Synonyms and Antonyms. Synonyms and antonyms are words that mean either the same thing (synonyms) or the opposite thing (antonyms) as the term being defined. Thus for the word *hydroponics,* a choice of synonym might be *soilless agriculture,* or *water gardening,* whereas its antonym might be *traditional horticulture,* or *agronomy.* Again, the whole point of extending a definition by using synonyms or antonyms is to enhance the understanding of the term being defined by suggesting other terms that might be more familiar to the intended audience. Such a technique would not normally be useful or necessary for an audience educated in the field of the problem.

Similes and Metaphors. Both are figures of speech that compare the term being defined with one that is already understood. Similes use the words *like* or *as* to form the comparison. Howard Ensign Evans, for instance, in his "In Defense of Magic: The Story of Fireflies" from *Life on a Little-Known Planet* (New York: Dutton, 1968) uses these similes to help define the flashing behavior of fireflies: "Not only would the synchrony of the flashes increase the brightness but the alteration of light and dark would also be eye-catching, *like . . . flashing neon signs*" [italics added] and "brilliant, synchronous flashes serve *as a beacon* to attract females from the surrounding forest" [italics added].

Metaphors, however, compare the known with the unknown by equating them, without signals such as *like* or *as.* In describing the South American "railroad worm" with its "eleven greenish lights down each side of the body [and] two red lights on the head," Evans suggests that the insect becomes *"a fully lighted railroad train"* [italics added] as it moves along the ground.

Although similes and metaphors are more often used to extend definitions for uninitiated or lay audiences, their use is by no means limited to such audiences. Writers for professional audiences often use similes and metaphors to define and explain problems at hand. Thus scientists talk about inexplicable voids in space as "black holes" or the structure of the DNA molecule in terms of "chains," "ribbons," and "backbones." In most scientific and technological discourse, there are times when situations and phenomena cannot be defined easily in anything other than familiar terms—terms that everyone, including experts, can understand immediately, because they share the common structures of language and culture.

Analogies. Analogies are similar to similes and metaphors, in that they compare unknowns to knowns, but they deal with larger concepts and situations, not with specific words. In creating analogies, writers may also use similes and metaphors, but this practice usually happens in cases in which the writer is writing for uninitiated or lay audiences, not for expert audiences. Consider, for example, the following, an excerpt from Lewis Thomas's article "Germs" in his collection *Lives of a Cell* (New York: Viking Penguin, Inc., and London: Garnstone Press, 1974):

> The microorganisms that seem to have it in for us in the worst way—the ones that really appear to wish us ill—turn out on close examination to be rather more like bystanders, strays, strangers in from the cold. They will invade and replicate if given the chance, and some of them will get into our deepest tissues and set forth in the blood, but it is our response to the presence that makes the disease. Our arsenals for fighting off bacteria are so powerful, and involve so many different defense mechanisms, that we are in more danger from them than from the invaders. We live in the midst of explosive devices; we are mined.

Thomas constructs two analogies here. One of them likens microorganisms (which he calls germs, also an appeal to a lay audience) to the homeless. He anthropomorphizes them, giving them the human characteristic of willfulness. Once the germs find a place to live, they burrow in deep, they "invade." In keeping with that notion, the other analogy likens the body's reaction to a war, talking about "arsenals," "fighting," the "danger" from "invaders," and the fact that we "live in the midst of explosive devices; we are mined." Both analogies trigger recognition in readers: the activities of bacteria become better understood more quickly. Most people know about the homeless; they certainly know about invading, fighting, and living among explosive devices (being mined) in wartimes, because they hear about them regularly or have to live with them.

Though we find analogies useful in many situations for the expert audience, writers by and large limit their use to uninitiated or lay audiences.

History of Discovery or Development. When using this technique, the writer or speaker notes the background of the term, concept, or problem in question, and the way that it came into being. Often it is useful to know what has gone on in the past to see why a problem is pertinent today. In Susan James's case, for instance, she could have reviewed briefly the beginnings of the minority contractors program, as well as what the government had in mind in promoting minority contractors, why it was necessary to do so, what the government thought the private sector could gain from the program, and finally, how that possible gain might meet the notions of governmental and social mandates.

But again, the use of this kind of information is generally reserved for lay audiences, who require more basic information. If Susan were to include this information, she would do so only briefly in the appendix to her report as a reminder of what the purposes and goals of the program are and how ITE's participation helps meet them. For a less-informed audience, her treatment would be longer and more detailed, and might well need to appear in the body of the report instead of only in an appendix.

Summary of the Definition Process

Professionals must first clearly define a problem before they can attempt to solve it. If they don't know what the components and the limits of a problem are, attempts to research, analyze, and solve it will be simply hit or miss. That kind of approach is expensive in all terms: time, money, and effort. And even if they have spent much in these areas, there is never any guarantee that the problem ultimately will be solved. Although that outcome may be true of any problem and of any attempt to solve it, it is most surely true of one that has not been clearly defined. Thus at this definition stage, three things should be kept in mind:

1. The felt need must be clearly known and expressed.
2. The audience who will benefit from an address of, or a solution to, the problem must be clearly known and analyzed.

3. The techniques of definition to accommodate the felt need and the audience who will benefit from an ultimate address of, or solution to, the problem need to be understood and used appropriately.

RESEARCH

Once a problem has been defined, problem solvers need to collect information or data. They need to know what has already happened to create the problem, what is going on in the present, and what is likely to happen in the future. They need to know what sources of data are available and whether they are primary in nature, that is, providing data that the researcher needs to collect firsthand, or secondary, that is, data provided through publications of research already performed. Problem solvers also need to know how best to collect the data—which research design is most effective for the primary work, which sources are most useful and reputable for the secondary work. Most research is conducted using both methods, but depending on the nature of the problem, one method may dominate.

Primary research dominates when the problem has received little attention in the past, only recently has been defined, or perhaps needs to be redefined by discovering new elements through laboratory experimentation or fieldwork. Such problems center on the phenomenal world; they are tangible. Several examples are relevant: research in genetic engineering, nuclear fusion, drug therapies for the relief or cure of cancer or AIDS, or the research of social scientists about human behavior.

Cases in which secondary research dominates usually involve possibly significant changes in existing theory about intangible matters, such as philosophy, history, and the humanities in general. For example, a philosopher might reassess Thomas Kuhn's notion of scientific revolution, or a literature theorist might take on Derrida's idea of textual deconstruction.

Straightforward Primary Research

Susan James's case provides, as many problems do, a need for fairly straightforward primary research. To report on the problem of the inadequate Jones and Son crane, Susan would have to collect specifications on the proposed crane from Jones and Son, as well as specifications from other suppliers. Specifications, although published by individual suppliers, are not analyzed for the reader; they simply list capabilities or ranges of performance of the product in question. Researchers who collect the specifications have to analyze them themselves and must also determine how well they meet the needs of the problem in question. Culling that information takes what has been defined here as primary research.

Susan had to determine exactly which specifications were pertinent for purposes of comparison and contrast, for example, how the elements of lifting, shifting, lowering, and placement would meet the needs of the Engineering Department. Thus she had to review Engineering's specifications and perhaps clarify them by talking to members of that department. If time permitted, she might also have to observe the crane on site: watching it in actual operation to determine how the work progressed in terms of time and space provided, although she would probably have to have an engineer along to help her determine (and interpret) pertinent events. In other words, any collection of data that has not already been collected and analyzed qualifies as primary research.

Complex Primary Research with Fieldwork

As noted earlier, Susan James's need for primary research is fairly simple and straightforward. Other problems that professionals address require more involved research designs and methods of collecting data. The following two examples require fieldwork, something that Susan would not have to do unless she did site observations.

The first example is that of Elizabeth Frederickson, chief chemist for a Pacific Coast utilities company, whose professional profile appears in Chapter 1. Unlike the problems Susan James has to solve, those that Elizabeth Frederickson has to solve require a great deal more primary research. Determining how to manage hazardous materials that her company produces requires two kinds of primary research: fieldwork, to perform soil sampling, and laboratory work, to analyze those soil samples.

When designing the fieldwork, Elizabeth must decide on the following:

- The sites for soil sampling, and the reasons they are necessary or appropriate
- The methods and specifications for sampling: the size of areas, the amount of soil, and the depths from which soil should be sampled
- The necessary equipment and personnel to do the sampling

When designing the lab work, Elizabeth must make similar decisions, such as the following:

- How many soil samples need to be analyzed
- Which tests they will be subjected to, and why
- The amount of time needed for doing the analysis, which of course must be coordinated with time constraints that the company has identified for optimum production

Good research design is the first and an essential step in carrying out research, but design should not be confused with performance. Like any problem solver, Elizabeth must be sure that the mechanics, or performance, of the design are carried out well—that contingencies and variables are accounted for—and that the research will indeed provide the kind of data necessary for meaningful analysis and later synthesis. This undertaking requires, then, careful supervision of the methods of field and laboratory work, not only of others who may be working on the project but also of oneself as a part of the research team.

A second example of primary research is the case of Patricia Wynjenek, a cultural anthropologist who wanted to study the economic survival strategies of low-income Mexican Americans living in an American-Mexican border city. (This case is discussed again in Chapters 4 and 5.) The problem, simply stated, was posed in this question: How do intact families whose incomes are far below the officially designated national poverty level manage to survive economically and to stay together? (The question presents an implicit contrast to other urban poor, whose families are not intact and who survive by way of social economic programs like Aid to Dependent Children.) The answer to the question might suggest different approaches to and definitions of social and economic programs that all urban poor could benefit from.

The approach to this problem requires a good design for primary research, one that assures that all the necessary primary evidence will be collected. The first part of the design would require participant observation, whereby Patricia takes up residence in one of the poorer neighborhoods of the town, establishes contacts with some of the key individuals (those whom everyone seems to know and who take some kind of leadership role) in the neighborhood, and generally takes part in the life of the community when it is possible and appropriate. By living this way, she can take notes—observations on how the community gets along—in the following areas:

- What and how social events, like parties, dances, and sports events, take place
- What and how religious or other cultural events, like holidays, elections, dedications of new civic facilities or programs, take place
- How the foregoing events are influenced by the interactions within, between, and among families
- How families deal with their private occasions of change: births, deaths, initiations, graduations, marriages, and the like

This kind of information is essential for research anthropologists to collect, but in this case, it is just part of the work of primary research, because it relies on the ability of the researcher to get to places and situations where she can observe. The other part of the research means gathering information on a one-to-one basis, talking to specific people about how they live and how they get along economically.

This approach requires work that most anthropologists cannot do on their own. Because they are foreign to the culture or subculture they are studying, specific people they need to interview are often shy about giving information about their lives. So the first task is to find members of the community who would be willing to conduct interviews that pose specific questions about family employment, education, medical care, and family life in general.

Patricia then needs to decide on several other essential elements of her research design:

- How many households should be interviewed to get a reasonable sample
- How many interviewers she will need
- How the households will be chosen—at random or by some means of stratification—and the reasons for choosing either method

- How the interviews will take place—whether questions and responses should be recorded by hand or be taped
- What kinds of questions and how many of them interviewers should ask
- How long the interviews should last

It is essential, then, that plans for primary research be made and justified before the work proceeds; without planning and clear ideas about the logic of the plans, the work will stumble and falter.

Acquiring and Evaluating Secondary Research Materials

As we suggested earlier, most problems require a mix of primary and secondary research. The preceding examples help illustrate that need. In Susan James's case, most of the information she can get on the minority contractors program will come in the form of secondary materials: government publications on the history of the program, the provisions it has established for its operation, and reports of how the program has worked in specific instances. The source of these publications is usually the government itself, which will send information of this kind to anyone who asks for it. But such information can also be found in full-service libraries, either public, corporate, or academic, that have large holdings in government documents.

Though this kind of information may be fairly easy to acquire, Susan, or any researcher like her, must remember that it is written with a certain amount of inevitable bias. The government, from whatever mandate it feels the most pressure, will publish documents that reflect well on its efforts. It may not knowingly alter information to suit its needs or the audience for which the information is intended, but it will certainly present, in the most positive light, the intent and accomplishments of its efforts. In other words, Susan, in reading government materials, would be likely to get only part of the story—and the most positive part at that.

In order to get a more impartial picture of how the minority contractors program is working, Susan might turn to reports from other firms participating in the program. These reports generally can be found in trade journals—publications written for particular audiences in various areas of interest, in Susan's case, those having to do with heavy industrial production or with purchasing in particular. Since publications of this kind are designed to present up-to-date information on the workings of the trade—new ideas, new trends, recent accomplishments—they are apt to present fairly full and realistic assessments of what is going on. It is true that much of their reporting tends to be upbeat, but in fairness to their readers, they do not avoid reporting on the downside of things—and certainly would report it if the minority contractors program was not working well.

The important thing for Susan to remember is that because secondary materials often have been written with bias, more than one source of information should be consulted to get a fuller picture.

Elizabeth Frederickson's need for secondary research is slightly different. Since her primary research is based on an established tradition and methods of soil sampling and laboratory analysis, she usually does not need to do secondary research to explain or substantiate those methods. Through her education and training, she has already consulted those sources; she knows how the methods she uses are supported by theory and testing. Only when new methods are introduced would she do well to consider the published material that explains and analyzes them.

The real job of secondary research for Elizabeth lies in the review of federal and state legislation having to do with hazardous materials. She must be aware not only of the laws but also of how they affect her firm's activities; she must also know about ideas and movements for future legislation. Her basic sources for information would be the *Congressional Record,* California's and Oregon's state legislative records, and of course, copies of the laws that have already been enacted. But beyond these, she would want to look at analyses of the laws in force and of those proposed, which are written by several groups: offices of legislators and members of Congress, lawyers representing industries in question, and environmental protectionists. She would find these various reports in a full-service library, but also, and more immediately, directly from the various offices that write and produce them.

Most of what she would have to read will be rather straightforwardly written. The records of Congress and legislatures are themselves not interpretations: they are direct transcriptions of what went on and are biased only to the extent to which the authors of testimony are biased. The analyses of legislation written by other interested parties will inevitably bear the stamp of the interest that produces them. When the overall viewpoint of the interest group is known, there is little need to ferret out subtleties of analysis and presentation. For example, an analysis of an abortion bill by a right-to-life group would reflect in a very straightforward way the group's stand against abortion.

The case of Patricia Wynjenek involves more of what is thought of as traditional academic secondary research. In designing her primary research, she would have to draw upon the methods others have used in similar situations. She would find those described in ethnographies published by other anthropologists. She would also have to reconsider anthropological theory, making sure that she understands how it has evolved and how it informs current practice. In addition to recognizing differences in schools of thought (or theoretical bias), she would want to be concerned with how current the secondary material is and whether it reflects the best that is known presently.

The sources of these materials are almost exclusively academic libraries, the depositories of ethnographies, monographs, and journal articles that discuss a variety of issues and problems that anthropologists want to address and to solve. And since almost all large libraries, academic or otherwise, have automated access to their holdings, it is relatively simple and rapid to search the database or to search other databases that a library may be connected to cooperatively, in order to locate pertinent materials. This kind of service is not limited to the anthropologist's needs; both Susan James and Elizabeth Frederickson could use similar search services in completing their secondary research.

The Annotated Bibliography: A Helpful Tool. Researchers doing secondary research often find the annotated bibliography a very helpful tool. The annotated bibliography (AB) is a list of pertinent secondary sources that includes the researcher's commentary (annotation) about each source. The AB can be adapted for any researcher's needs, but customarily includes brief statements about the thesis of a source, its scope, its particularly useful sections and the reasons why they are useful, and its strengths and weaknesses. The length of annotations can range from a couple of sentences to a paragraph, although it is clear that the thoroughness of an annotation can save much rework down the line. More complete ABs might even include significant quotations from the source, with page number noted, of course.

To generate the material for an AB, researchers need only peruse and skim the source. If it is a book, they may look at the preface to determine the author's theoretical or rhetorical position. They can skim the table of contents to see what topics are covered. They may want to read chapter summaries, introductions, or concluding paragraphs.

If the source is an article, researchers may use the article's abstract or summary, which if it is well written, will state what the article's thesis is, what the author's findings were, and, perhaps, why they are significant. In other words, the summary will be conclusive.

It is a good idea at this point to distinguish between abstracts and summaries (often called executive summaries). Abstracts are short, usually only a paragraph long, and appear separately at the beginning of articles. The purpose of an abstract is to identify the main point of the discussion, plus any other information required, such as topics covered or key words. But abstracts are not conclusive. They don't reveal important secondary points, conclusions, or recommendations. They don't tell the significance of the author's findings. That's the job of a summary. Obviously, in writing a useful annotated bibliography, researchers will find the job easier if they can read useful, conclusive summaries.

Armed with a conclusive AB, researchers can build a resource base that will communicate clearly to them even after they've not seen the original source for some time. A conclusive annotation will reveal the main point of the source; it will not simply list the topics within the source without telling what the author thinks about those topics. In other words, a conclusive annotation will tell "who done it"; it will give away the answer.

Although it is tempting to resist being so explicit, explicitness is a trait to be cultivated in professionals' lives because it keeps them closer to the truth. It's *good* to remember that people don't read professional material for fun; they read it for answers.

A short but conclusive annotation of *Cinderella* might read like this: "A much put-upon stepsister finds her destiny to wed the kingdom's prince, after her fairy godmother helps her escape her hearth-bound existence." An inconclusive annotation might say something like this: "A young girl suffers under her stepmother and undergoes many trials." The inconclusive annotation reveals the general topic, but does not reveal the final outcome—marrying the prince—and that's what the conclusion of *Cinderella* is all about.

To use another example, we could imagine the following annotation: "This report takes on the cigarette industry to examine whether safety standards have been upheld." Such an annotation gives the topic of the report but does not give the answer that the report is supposed to reveal. A conclusive annotation would provide the answer: "This report of safety practices in the cigarette industry concludes that safety practices are not only keeping up to standards, but have, in fact, raised standards over the past five years." This annotation gives the answer, the conclusion. It is a conclusive piece of writing.

Summary of the Research Process

We can see from the examples of Susan James, Elizabeth Frederickson, and Patricia Wynjenek that the research stage of problem solving is very important. Figure 2.2 provides a checklist for the research process. If the research design is good, that is, if the researchers have asked intelligent questions about what they are seeking to know and have devised the best means to collect information to answer those questions, then they will have supplied the requirements for the next stage of problem solving—analysis. The quality of the analysis depends to a large degree on the quality of the research. No matter how good the researchers' analytical skills are, if the body of data gleaned from research is incomplete, biased, irrelevant, out-of-date, or some combination of any of these, the analysis will yield very little of use in solving the problem.

Primary Research

_____ 1. Be sure that the design for primary research includes all the areas and variables that need to be studied and considered.

_____ 2. Consider, use, adapt, and/or reject methods used by other researchers. Know why you are using, adapting, and/or rejecting them.

_____ 3. Once you decide on a course and method of research, follow it consistently and thoroughly.

Secondary Research

_____ 1. Be sure to review the literature on the problem you are researching. Use automated search capabilities of a full service library to ensure thoroughness.

_____ 2. Decide carefully, in reviewing the literature, which ideas agree or conflict with your ideas and analysis of the problem you are researching.

_____ 3. Analyze the problem as you have defined it and show what your analysis contributes to a general understanding of the problem with the field.

FIGURE 2.2 **Checklist for the research process**

ANALYSIS

As noted in Figure 2.1, the analysis stage answers the basic question, What do the data mean? Here researchers look carefully at the data to see how the various parts paint a picture or tell a story. Analysis requires discovering relationships—how one event or situation or fact brings about or affects another, how that connection may bring about another of its kind, but perhaps slightly different in its makeup. The important thing to remember is that analysis in solving problems is not a discrete process that happens only after definition and research have taken place, or that does not have to take into account resolution and synthesis (the suggestion of a solution) and possibly implementation, the last steps in solving problems. Analysis goes on all the time in solving problems, as we have illustrated in Figure 2.3.

Analysis is a part of the process of definition: very few, if any, problems are made up of only one identifiable element. Knowing what the various elements are and how they contribute to the full definition of the problem requires analysis.

Analysis is a part of research. One has to know what research is necessary and why. Therefore, deciding on a research design—the needs and parts of both primary and secondary research—requires analysis. Analysis is also a part of resolution or synthesis and of implementation. Solutions likewise have parts that depend on each other and influence each other; knowing and understanding the relationships among the parts, so that the solution as a whole seems sensible, also requires analysis.

Although analysis is not a discrete process but rather is a continuous one, it is important to understand that there are times when analysis will take center stage, and that process happens after definition and research. Thus Susan James must go beyond the analysis of her needs for primary and secondary research and must analyze fully the data she has collected. For example, from the primary data, she must actually compare and contrast, perhaps in tabular form,

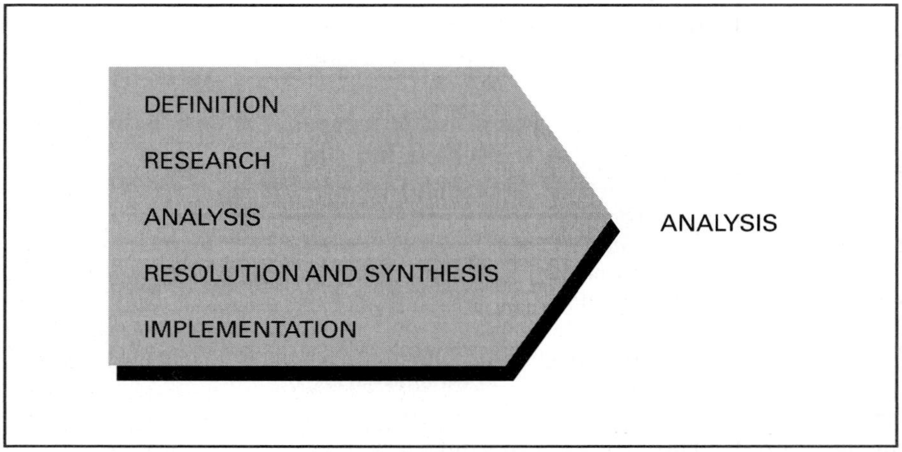

DEFINITION
RESEARCH
ANALYSIS
RESOLUTION AND SYNTHESIS
IMPLEMENTATION
ANALYSIS

FIGURE 2.3 The role of analysis in the technical problem-solving process

the specifications of cranes from other suppliers to those from Jones and Son and then to the specifications suggested by the Engineering Department. She then has to point out where the specifications differ, to what degree they differ, and how important the differences are for the proposed use and operation of the crane. For this she needs some advice from one of the staff engineers.

If Susan observed the operations on site, she would have to analyze her notes, perhaps again with some help from someone in the Engineering Department and then decide how elements of the actual operation reflect the published specifications and how that outcome might influence the needs of the Engineering Department.

Her analysis of the secondary data would be essentially the same, involving comparing and contrasting. From her reading of government documents and trade journals on the history, provisions, and operation of the minority contractors program, Susan would compare how the program is intended to operate to how it actually does operate in various situations. She would further compare or contrast those situations to her own to decide how the program most likely will affect her purchasing operation, and then how it will affect generally the operations of ITE. The result of her analysis of both primary and secondary data will form the substance of her resolution and synthesis of the problem.

Analysis of primary data for Elizabeth Frederickson is a highly technical affair. It involves exact laboratory procedure that will reveal the presence of hazardous materials and their type and concentration in the soil. It will also suggest the source of the hazardous material, since Elizabeth knows what the possible sources are, especially within the workings of her own firm.

As for analysis of secondary data, Elizabeth's job is similar to Susan's in that she has to do a certain amount of comparing and contrasting. She has to compare or contrast, for instance, how legislation on the control of hazardous materials has affected the operations of utility companies similar to the one she works for. She also has to predict the operations of her company as compared with others and the way that future legislation will affect the operations of her company.

But Elizabeth's other concern in analysis is what the impact of reviews of legislation written by environmental groups has already been or yet may be. She has to assess the importance of these groups in influencing public opinion and bringing about legislation. She can do that by noting how often environmental groups have succeeded in getting measures they want passed, and what kinds of measures they are, how restrictive, how permissive, how expensive to those who must comply with them, and so forth.

Finally, Patricia Wynjenek's job of analyzing primary data also has to do with comparing and contrasting, though in a slightly different way. She must categorize the answers to the questions of the survey she conducted, which amounts to deciding which answers are alike, which are different, and of course, how alike or different they are. For example, in response to a question on birth control practices, she could get responses from women that fall into several possible categories—yes, they consciously control the size of their families, and by what means they do that; or no, they let whatever is to happen, happen; or they practice a combination of the two.

Patricia must then decide how the answers to her survey questions compare or contrast to what she knows about the community from her own participant observation. She must take into account how the answers she gets to questions from her survey are similar to or different from her own observations. She must consider the effect her interviewers have on the people whom they interview—people who are not used to formal questioning, or maybe to any questioning at all. She must consider the following kinds of questions:

- Are those people interviewed shy or reticent? To what extent, and why?
- Do they answer questions willingly?
- How fully do they answer questions?
- Do they appear hostile? If so, in what way?
- What is the likelihood that they are giving false information?

She must decide, in general, what restraints the people she is observing feel about her presence and also decide how they feel in general about what she is trying to do, even with the help of people from the community.

The main job of analysis of secondary data comes for Patricia when she assesses the validity of the conclusions drawn in other ethnographic studies similar to her own. She must determine whether the means by which data were collected—kind and duration of participant observation, kind and method of execution of survey instruments, and so forth—can actually support the conclusions the ethnographer has drawn. She must then, once again, compare and contrast her own fieldwork methods to those of other ethnographers to see whether her methods have provided the kind of information she needs in order to draw valid conclusions. She would do this latter analysis to confirm the solidity of her research design, since in planning it initially, she had to consider the designs of other ethnographies. But it is always possible, even this late in her work, that Patricia could discover some defect in her methods and have to correct it. Even though making a correction may delay her final step, the resolution and synthesis, she knows that it is better to find and correct that error now than to risk rendering invalid conclusions.

RESOLUTION AND SYNTHESIS

Throughout the problem-solving process, problem solvers have to make decisions—decisions about the nature and scope of the problem, about research design and methods of collecting data, and about what the data collected actually mean, given the problem as defined. But it is in the resolution and synthesis stage that decisions must be made about what may or ultimately will solve the problem. These decisions mean that the significance of the analysis of the preceding stage must be determined, as well as whether that significance suggests a solution or solutions to the problem.

Decisions also have to be made about how reasonable the solutions are and why some are more reasonable than others. In other words, what are the general advantages and disadvantages of any one of the possible solutions?

An answer to this question involves general assessments of costs and time and the final effect of the solution in light of expected or possible results.

Many would think that resolution and synthesis should be the last step in technical problem solving, because it implies a solution to the problem. On the face of it, that is true, but it is important to remember that not all problems can be solved, or at least not readily solved, within the limitations of time and resources. The trouble we as a society have had in solving such problems as poverty and the plight of the homeless; in finding cures for cancer, heart disease, and AIDS; and in wiping out our international trade or national budget deficits is evidence enough that not all problems can readily be solved. It may take years, decades, even centuries before some problems are solved, and even then we have to be careful when deciding whether a solution is indeed a solution, and not just a diversion from the basics of the problem, or whether the solution creates other problems. A common example of this kind of situation is one that Rachel Carson talks about in "The Obligation to Endure" from *Silent Spring* (Boston: Houghton Mifflin, 1962). In our quest to improve agricultural production and to rid ourselves of the pests that attack our crops, we have come up with ever more powerful chemicals to spray on the pests (but also onto the crops and thus into the environment). In so doing, we have added to already staggering conditions of pollution, risking, as Carson says, "the contamination of man's total environment with substances that accumulate in the tissues of plants and animals and even penetrate the germ cells to shatter or alter the very material of heredity upon which the shape of the future depends." What we have done is to find a short-term solution (killing the present pests) that in itself creates a much longer-term problem. In Carson's words, "It is ironic to think that man might determine his own future by something so seemingly trivial as the choice of an insect spray."

This notion should not be overstated and therefore misunderstood. Most problems that professionals have to face on a day-to-day basis can be solved, at least in the short term, without the kind of damage Rachel Carson talks about. But it seems more realistic and useful to think about this step of the process as one by which the elements of the problem, after being defined, researched, and analyzed, are put back together and then some decision is made, rather than think of it as absolutely the last step, where a solution for better or for worse is declared. In this way, we can account for both those problems that actually can be solved within limitations of time and resources, and those that require repeated addresses, and resolutions and syntheses along the way, before they ultimately can be solved.

Susan James's problem fits into the category of problems resolved or synthesized, but not actually solved. Although she was initially troubled and unsure about how to report on progress that showed no extraordinary achievement, she was satisfied, after she had defined the problem and collected and analyzed the data, that the decision she had made not to purchase the crane from Jones and Son was sound. She was sure that her recommendation to go through a more extensive bidding process, which now would meet the guidelines of the minority contractors program, would eventually turn up a supplier who could best meet ITE's needs. She was also sure that her decision and recommendation were unassailable, and that if she were to be criticized for making them, that outcome would have to happen for reasons other than the good practical sense she had shown. But the fact is that she had not actually solved the original problem: finding and putting in place a crane to meet Engineering's needs in record time. The solution to that problem was yet to come.

It may seem obvious that the resolution and synthesis step involves making decisions, the kind of decisions that finally initiate action. But other decisions are usually made at various earlier steps in the process of technical problem solving, and they anticipate resolution and synthesis and the final step, if it is possible to carry out, implementation.

To illustrate this circumstance with Susan James's case, once she suspected that the Jones and Son crane probably would not serve ITE's needs, she decided to compare and contrast Jones and Son's specifications to those of other suppliers. And further, since she knew that Jones and Son would not qualify as a preferred supplier under the minority contractors program, she decided to research and explain more fully how the program affected purchasing and the other operations of ITE as she understood them. Both of these decisions were based on information Susan obtained before this resolution and synthesis step.

So we see that resolution and synthesis means putting back the pieces of the ongoing analysis, making things simpler to see as a whole, and confirming decisions that have already been made, at least tentatively, earlier in the process. The result of resolving and synthesizing a problem is seldom a surprise, and it shouldn't be. Good preparation, which means careful definition and thoughtful collection and analysis of data, will suggest how a problem is to be solved, or resolved and synthesized. Resolution and synthesis becomes, then, both a resting place and a spur to action.

For Susan James, it meant that she could see the sense in what she had decided and that there should now be no barriers to acquiring a crane through acceptable means, given the findings of her research and analysis. This problem for Susan James is solved, at least for now. In her quarterly report, she will explain why she did not contract Jones and Son; will recommend strongly that ITE seek other suppliers, in keeping with the guidelines of the minority contractors program; and will stress that in all future major acquisitions, the same process be followed.

Elizabeth Frederickson's task of resolving and synthesizing is hard. It means deciding how her firm's hazardous materials affect the environment and how the impact as defined and dealt with coincides with concerns of environmental groups and with legislation. It becomes a balancing act, one that she has to perform to meet at least three demands: the needs of her employer, the needs of the public, and the needs dictated by her own position as a professional chemist.

Each situation poses problems of ethics. Elizabeth must exhibit loyalty to her employer and try to further the company's best interests, so that she can get along in the company; she must understand and have a certain sympathy with the public, because the public has a right to be heard and protected; and she must understand and stand up for the conduct and ethics of her own profession. Her suggestions, then, based upon her laboratory work and her library work, have to take all these considerations into account. Absolute solutions, under these conditions, are not possible. But Elizabeth can suggest ways of proceeding that will address, if not exactly solve, all the problems that each group perceives in an attempt to bring all three together in some kind of reasonable and responsible package. This understanding is tricky, but not impossible.

Patricia Wynjenek's resolving and synthesizing situation is different from the preceding cases. She has to decide two things, which initially may not be seen as solutions to the problems that suggest them. First, she has to decide how the people she is studying actually survive economically, while keeping their family structures and their extrafamilial social structures intact. She also has to decide what that information means for the way those people are perceived by others.

The first decision comes directly from the analysis of her research. This decision serves as a resolution and synthesis of the kind defined earlier—one that offers no solution. Economically speaking, the people she is studying survive on very little, but they are willing and able to do so as long as they can keep their families intact. Ironically, however, it is because they keep their families together that they are worse off economically than many of their counterparts in other poor urban areas of the United States.

These families have few material goods, and those they have are old, used, and in constant need of repair. Their housing is likewise old or cheaply constructed and small. Their nutrition is below standard, relying heavily on starches and other carbohydrates and less on animal protein and fruits and vegetables. They have limited access to medical care, though they do not avoid modern, "scientific" methods; neither do they throw over completely traditional means of healing; many do not have family doctors and rely largely on a handful of free clinics or hospital emergency rooms for necessary medical attention. Nevertheless, the population has a strong sense of community, family obligation, and self-worth. Although statistically, and in real terms, they are a part of the growing underclass of American citizens, they manifest few of the commonly understood signs of alienation and disaffection associated with other American populations suffering from economic deprivation.

Deciding what these elements mean about the ways that others, such as sociologists, social workers, and government bureaucrats, perceive these problems and phenomena and deal with them, and more important, how they will deal with them in the future, is a problem Patricia Wynjenek has yet to solve. But having done her study, having resolved and synthesized many of the problems she had before doing her work, she is now in a much better position to attempt to solve that problem.

IMPLEMENTATION

Implementation grows directly out of the preceding stage, resolution and synthesis. In fact, ideas of implementation are rooted in that stage—that is where the possibilities and feasibilities of solutions are generated and first-level decisions are made.

Implementation deals with practicalities: the actual means by which solutions will be realized, who will bring them about, how they will be brought about, how much they will cost, and how long they will take. Decisions made at this level are practical rather than theoretical. If a certain route should be taken, how could it reasonably be taken, given the constraints of time and resources? Decisions made at this level may change the means of implementation, but they usually don't change the basic need for a structure of implementation.

Implementation requires four elements:

1. A clear organizational plan to realize the solution, that is, to see the solution over various structural hurdles inherent in any organization
2. A reasonable plan to employ resources, either internal or external, to bring about the solution
3. Specific knowledge of what it will cost and how long it will take to bring about the solution
4. A clear idea of what the results of implementing the solution will be

Implementation has to be systematic and careful. Once an organization has decided to take a certain path, even if it's only an interim path along the way to a final solution, the organization must take it confidently. To do otherwise means misfires, false starts, and ultimately time, money, and effort wasted.

Had Susan James actually solved her original technical problem and found a crane for the Engineering Department in twenty-four hours with delivery in another forty-eight, she would have had to implement the solution quickly. She would need approval from the Vice President of Operations to spend $250,000 to buy the crane and would therefore probably have to make a personal visit to him or to some significant subordinate to lay out the immediate need.

She would also have to make sure that Jones and Son would be able to deliver the crane in the seventy-two hours since the deal had been approved. Doing this would probably require a trip to Jones and Son, but if not that, at least several telephone conversations with the production manager—maybe even with Hershel Jones himself. And, finally, she would need some assurance from the Engineering Department that the new crane would do what needed to be done. Engineering would need to look at performance specifications and judge them appropriate and effective for its operations.

Susan's position is a good one to illustrate implementation, because the nature of the problems she solves requires her direct involvement. But many other professionals seldom implement the problems they solve, simply because their work does not call for it. Elizabeth Frederickson and Patricia Wynjenek are two examples.

Although, as we pointed out earlier, Elizabeth's job of resolving and synthesizing is difficult because of the pressures on her ethical stance, once she has solved a problem and made a decision, she passes the implementation on to her superiors. If, for example, she decides that her firm must adopt a new method of disposing of hazardous materials at a particular location, she will recommend the new method, pointing out why it is necessary and suggesting the means of implementation. However, she herself will not be directly responsible for the process. But note that in her recommending a new method and suggesting a means of implementation, she lays the necessary groundwork for someone else to do the work. Therefore, she has to be aware of what implementation involves.

The same is true of Patricia Wynjenek. Although at some time she may be able to recommend new social programs on the basis of what she has found out about the economic survival strategies of the people she was studying, she is not in a position to implement them. Others in public or private agencies would be responsible for implementing the solutions that Patricia recommends. But again, Patricia's understanding of what implementation involves will make her recommendations more specific and therefore more likely to be carried out effectively.

CONCLUSIONS

The basics of technical problem solving—definition, research, analysis, resolution and synthesis, and implementation—are dynamic and demanding tasks. They require a commitment to understanding and willing participation in professionalism that most who seek meaningful work in their lives are willing to make. But if they are to work—if they are to be used successfully to solve problems—they require close attention to the process involved in using them.

It is not enough simply to define a problem. One then has to research the problem to find out the details of what has defined it. Once the details have been gathered, they have to be analyzed—looked at as discrete entities, then as parts of the whole to see how they fit into a larger picture. Then the whole process needs to be resolved and synthesized—to be bound up to decide ultimately what it might mean. And, if possible, a decision must be made as to how a solution will be implemented. At every stage, the need for redefinition, further research, and reanalysis may present itself. Solving problems is not a clean and linear activity; it is often messy and frustrating, doubling back on itself; but once honestly and conscientiously taken on, it can render great rewards. It is not merely honorific to say that any job has been well done. The ability to say that with conviction is the ultimate test.

EXERCISES

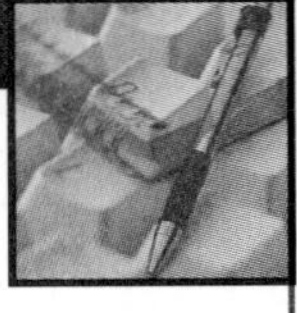

1. In small groups, use the definition techniques discussed in this chapter to try to convey the following concepts to the groups indicated:

   a. Shin splints to a nonexerciser
   b. Having a brother or sister to an only child
   c. A word processor to someone who has used only a typewriter
   d. The ocean to someone who has never been out of North Dakota
   e. An engine and generator to an English major
   f. The value of poetry to an engineering major

2. Either in small groups or as a class, consider the problems in the following list by answering the fundamental questions posed in defining a problem: what's wrong, what's lacking, what's unknown, and who will benefit from a solution to the problem? Remember that not all problems will have answers for all questions but that all problems must have some group that will benefit from a solution.

   a. Protection of the spotted owl in the Pacific Northwest
   b. Occurrence of acid rain in the industrial Midwest
   c. Maintaining the balance of international trade
   d. Growing drug abuse among teenagers
   e. Social and economic effects of an aging population
   f. Developing automobiles powered by alternative fuels
   g. Fighting the AIDS epidemic

3. In small groups, suggest the kinds of problems you already know that professionals in your various fields of study have to face. Choose one of the problems, and construct an outline of problem-solving stages around it. Be as specific as possible about the questions that would have to be answered and the ways of answering them at each stage of the outline.

ASSIGNMENTS

1. Using an article your instructor has assigned, write first an abstract of the article and then a summary. Review the differences between the two on page 40 of this chapter. Your reader should be able to see marked differences between the two.
2. Write a conclusive annotated bibliography of five sources on a topic your instructor has assigned.

TERM PROJECT OPTIONS

1. Write a memo to your instructor that fulfills the following requirements:

   a. Uses standard memo format. (Memo formats are discussed in Chapter 10.)

b. Defines in detail the research problem you want to address or attempt to solve during the remainder of this term. Keep in mind the needs and techniques of defining problems discussed in this chapter.
c. Suggests the ways you will go about gathering information, in both primary and secondary research.

2. Write an annotated bibliography of ten secondary sources that you believe will be helpful in addressing or solving your research problem. On the title page of this assignment, be sure to state what your research problem is. At this point, you may have a problem clearly in mind or perhaps only a question you would like to pursue. This annotated bibliography will help you narrow your interests and focus on a problem. Following the title page, present the bibliography, heading each annotation with the appropriate bibliographic information for the source (for example: Carson, Rachel. *Silent Spring.* New York: Houghton Mifflin, 1962.). Sources should be listed alphabetically by author's last name, or in lieu of an author's name, by first substantive word of the title of the piece.

   Your annotations should consist of two full paragraphs. In the first paragraph, write a conclusive summary of the source. In the second paragraph, explain how you think the source will be useful in your work. For example: Will the source provide introductory material? Will the source provide an interesting perspective on the problem as you understand it?

   Note: You will be able to use this annotated bibliography to help you write your proposal to research the problem (proposals are discussed in Chapter 6).

CHAPTER THREE

Rhetorical Problem Solving

CASE STUDY

MARY ANNE COX'S OFFICE

Outerware, Inc. ▪ 17 D Street ▪ Hartford, Connecticut

Mary Anne Cox got to her office at the usual time—7:45 A.M.*—early for some people, late for others, but just about right for her. She went about the usual chores of making some coffee, sorting out yesterday's mess on her desk, and turning on her desktop computer, one of three in her office. When she did so, she found a message from her immediate superior, Tim Sanford, saying that something had gone terribly wrong with the last test of the environmental control system that Outerware had installed in late February in the capitol building, and that they'd need to discuss the situation as soon as possible. Mary Anne got a little nervous: to get a message like this from Tim meant that he was upset—he usually dropped in on her when he had something to talk about, and he seldom left a message on her computer. It also meant that since she hadn't got the message until now, he'd either stayed late, later than she had stayed the day before, or that he had*

PREVIEW OF THE CHAPTER

gotten in earlier this morning. She checked the time Tim had sent the message: 6:15 P.M., March 20. She had left at 5:30 that day, having put in a full day, and she began to relax, at least about the timing.

Mary Anne had worked for Outerware, Inc., for five years. She had earned her B. S. in Computer Science from the University of Connecticut, had interviewed with several firms in the area, and was hired by Outerware as a systems engineer. Outerware was currently developing an automated environmental control system, by which the ambient temperatures of many different offices in a complex could be controlled at a central location, thus eliminating the need for individual thermostats and yet keeping offices at comfortable temperatures throughout the year. Outerware had named the system ENCONTRO. One of Mary Anne's primary responsibilities was the programming of ENCONTRO—and working out any bugs that popped up in it. Tim, her superior, maintained a liaison position between research and program development and Outerware's clients. The two worked closely together, Tim providing Mary Anne with analyses of the needs of new clients, Mary Anne providing Tim with the technical analyses and answers to those needs.

Mary Anne met with Tim as soon as they could both arrange a time. Tim told her that the capitol building's chief operations engineer, Janice Trippling, reported that ENCONTRO had rendered the capitol building basically unlivable on its last test run: in the mild late March weather, half the offices, mostly in the east wing, had been cooled to 62 degrees F (just about the outdoor ambient temperature), while most of the west wing offices had been warmed to 78 degrees F, well over the usual 70 to 72 degree range. This situation had occurred in the last couple of hours of the previous working day, and thus the late report. Mary Anne agreed to meet with Janice Trippling that day to discuss the problem and to report in writing her analysis of and proposed solution to the problem.

The meeting with Trippling did not go particularly well. Trippling was annoyed, even a bit hostile, since, as she pointed out, she had had nothing to do with the decision to install ENCONTRO. That decision had been made somewhere up the line, and she, as Chief Operations Engineer, had to implement it and live with it, even though she had seen no particular advantages to using the system. Even given the hostility, Mary Anne learned that there was nothing intrinsically wrong with the system, but that the assistant operations engineer, Sonny Baker, who had responsibility with others for monitoring it, had misunderstood crucial procedures for using the control system software. These people would have to be trained once again. The system was sound and would remain so, if only those who use it would do so properly.

Mary Anne went home that night with questions and worries. How should she write a report that Tim would read; that John Springer, Vice President of Program Development at Outerware, would read; and that Janice Trippling and several of her superiors and subordinates would also read? Mary Anne didn't sleep well, but she woke up with some ideas.

AN OVERVIEW OF RHETORICAL PROBLEM SOLVING

Mary Anne Cox's story is typical of professional activity in all areas. In fact, it is in the nature of professional problem solving that a number of people with differing interests in a problem will be affected by a solution to it. This fact speaks of the complexity of professional life and of the need for careful analysis in the work that professionals perform.

The ideas that Mary Anne woke up with had to do in part with the reality that literally hundreds of people had been affected by ENCONTRO's malfunctioning. She knew that she would have to deal with that technical problem first, which in fact she had already begun to do. In her meeting with Janice Trippling, Mary Anne discovered that Trippling, Sonny Baker, and others who were in charge of monitoring ENCONTRO did not fully understand some of the basic functions of the system. Mary Anne now realized that the training she had given them had been inadequate and that she would have to retrain them to meet their needs. She had already started to shape the contents and tone of those future training sessions, in which she was planning to emphasize the links between theory and implementation, whereas in the first training sessions, when ENCONTRO was originally installed, she had taken shortcuts and had stressed implementation only.

Mary Anne felt pretty good about her ability to solve the technical problem, but she was still a little worried about the *rhetorical* problem—the need to write a report about *this* situation. She realized that she would probably have to write two reports. The first would have to deal with both ENCONTRO's malfunctioning and the reasons for it; a second would have to evaluate how the proposed solutions had worked. Both would have to be written to multiple audiences, even though there would perhaps be only one addressee, or specifically identified audience: Tim Sanford, Mary Anne's immediate superior. Mary Anne thought rightly that if she did a good job of analyzing and solving the rhetorical problem for the first report, which would be expected in a day or two, she would be in good shape for writing the second report, since the primary audiences would be the same.

Mary Anne's task, rhetorical problem solving, was one of effective communication through texts, words, and graphics, and also through oral means. The word *rhetoric* in everyday use has come to have negative connotations. People often say that a politician's speech, for example, is "just so much rhetoric," meaning that the speaker has used nice-sounding words that really have no content but contain only empty promises. These speakers seem to give "only rhetoric"—that is, only what some may want to hear, but not necessarily the truth.

Rhetoric, however, has a more general, less negative meaning than its popular connotation suggests. Defined simply, it means "effective communication." These two words—effective communication—are loaded with implications that need to be fleshed out later. But the first point to remember about rhetorical problem solving is that it concerns effectively communicating the essentials of technical problem solving. Chapter 2 details the process of technical problem solving; this chapter details the process of rhetorical problem solving, that is, effectively communicating the pertinent matter of technical problems that already have been solved. The written and spoken products of rhetorical problem solving develop after technical problem solving has taken place. Even if the solution to a problem may not yet be clear, professionals generally will still be responsible for some written and oral communication about the technical work in progress.

So, as one concern in technical and professional writing, writers should remember two levels of problem solving:

1. **Technical**—defining, researching, analyzing, resolving and synthesizing, and implementing (if possible) feasible addresses and solutions to a problem.
2. **Rhetorical**—effectively communicating the results of that work. Technical problem solving is only as good as the ability to communicate it effectively. Many professionals get locked into a certain slot in the hierarchy of their organizations because of their inability to communicate effectively the technical problems that they have solved.

It is now time to reconsider the phrase *effective communication,* to see what it means when writers try to accomplish it. Communication refers to activities ranging from writing formal technical reports to running a meeting. It encompasses both written and oral forms. The profiles of professionals in Chapter 1 show clearly that professional work requires constant written and oral communication.

The qualifier in the phrase *effective communication* is obviously the word *effective*. Professional and technical data must be transformed from their raw state into something that an audience—any audience—can understand and deal with significantly. Upper management, peer experts, subordinates, or members of the public outside the organization all need to have the results of technical problem solving explained to them clearly and significantly so that they can act upon what they learn in ways appropriate to their positions.

It is important for professionals to realize that good technical problem solving and effective communication of the results are important and intrinsically human activities. They require human responsibility. Computers cannot generate professional and technical writing that solves problems meaningfully, because computers cannot think. It's only through careful thinking and accepting one's responsibility for good technical problem solving and effective communication that the world's work gets done well.

The elements of effective communication are purpose, audience, context, and ethical stance. Every writing or speaking task depends for its success on careful consideration of these four elements. As professionals tackle a communication task, they begin to think systematically, and that habit of mind helps establish the confidence needed to solve the rhetorical problem. A closer look at the elements shows how rhetorical problem solving works.

To establish the *purpose* of the task, writers or speakers ask themselves several questions. Why am I writing this, or why am I giving this talk? What do I hope to get accomplished? What do I want my readers or listeners to know, or to be able to do, or even to become as a result of my writing or speaking? It is helpful to think of the purpose of communicating in terms of actions that produce results, such as persuading, explaining, directing, teaching, and informing.

The *audience* of a communication is the readers or listeners. Writers or speakers ask themselves whom they will be trying to reach with this communication. What are they like? What do they know? What don't they know? What do they want to know? What do they need to know? Basically, why are they reading or listening? Sometimes the answers are obvious and specific: I'm writing to my boss, Tim, whom I've known for ten years, and who wants to know what went wrong in our service of the Cushing account.

But sometimes the audience is more difficult to define: for example, potential buyers of VCRs in the California bay area. This audience will be mixed, having in common, perhaps, only that they have discretionary money to spend and that they might like to watch films at home. It is hard, but not impossible, to appeal to such an audience; considering audience is one of those constants that has interesting twists.

Context in this discussion has to do with the environmental factors that must be taken into account. We use the word *environmental* broadly to include workplace realities that define place and time. For example, as far as place is concerned, writers and speakers need to remember organizational politics and hierarchical levels, whether they exist in two-tiered family-owned businesses or in multilevel, multinational corporations. It is important to consider the lines of communication and understand one's place in them. Are the lines direct and consecutively ordered? Or are there lateral detours that need to be taken, such as someone who should be aware of what is going on, though his or her position on the face of it wouldn't seem to demand such consideration? In other words, have writers and speakers considered all parties who, for whatever structural or political reasons, should be cognizant of the workings of the organization?

Addressing time constraints includes consideration of production schedules, past correspondence and agreements, and assignment due dates. It is important to remember to orient readers and listeners to constraints of time, especially so that contractual obligations can be met. But time constraints are also important in precontractual, planning stages so that opportunities for the effective conduct of business won't be missed. Because most readers and listeners have many demands on their attention, exigencies of time need to be clearly and directly laid out.

Ethical stance concerns the writers' or speakers' conscientiously considered ethical position on the subject of the writing or speaking task. In some cases, one's ethical stance can be diametrically opposed to the context. Whereas the context considers the phenomenal world involved in the task, the political realities, and the organization's interests, the ethical stance considers the writers' or speakers' personal and professional opinions about the task. They ask themselves questions such as the following: Is there anything about this subject that gives me a philosophical or an ethical problem? Can I live with myself for taking a position that I know my readers and listeners will agree with but that may not be my own? Am I telling the truth, or do I know what the truth is about this situation?

The important thing to remember is that it is part of professionals' responsibility to consider ethical stance just as carefully as they consider purpose, audience, and context. Although professionals usually want to act to serve their employers, they must also remember to serve themselves as well, but only in ways that coincide with the accepted notions of professionalism (as discussed in Chapter 1). By taking their ethical responsibilities seriously, professionals save themselves from becoming mere cogs in a wheel, facing work as meaningless drudgery.

The four elements of rhetorical problem solving can be summarized as follows:

1. *Purpose* considers the communication task, and answers the question of why it should be performed.
2. *Audience* considers the readers or listeners, and answers the questions of who will be concerned, and why.
3. *Context* considers the environment of the problem, and answers questions about where the message is being sent and when it needs to be acted upon.
4. *Ethical stance* considers the positions of individual writers or speakers, and answers questions of where they stand on the issue at hand.

Questions That Rhetorical Problem Solving Answers

Purpose: What do I hope to accomplish with this communication?

Primary Purposes

- Am I simply answering questions of fact? (record keeping)
- Do I need to communicate fully why something was done? (formal documentation)
- Am I simply providing information? (informing)
- Am I providing information so that others can act or make a decision? (enabling)

Secondary Purposes

- Am I creating, maintaining, or improving my professional image?
- Am I maintaining or improving interpersonal relations, either internally or externally, or both?
- Am I creating, maintaining, or improving a sense of trust or reliability?

Audience: Whom am I trying to reach with this communication?

First Level of Audience Analysis

- What are my audiences' backgrounds, levels of education, and technical expertise?
- What positions, biases, or attitudes are they likely to have on or toward my subject?
- What purposes do they have for reading or listening to what I am communicating?

FIGURE 3.1 **Questions to ask and answer in the rhetorical problem-solving process**

As they approach the task of rhetorical problem solving, professionals must continue to ask and answer questions, just as they did in the technical problem solving process. Figure 3.1 presents the types of questions that must be considered during the rhetorical aspect of problem solving.

A short example of the rhetorical problem-solving process will help introduce the more complex case of Mary Anne Cox. In honor of National Engineering Week, Max Chen is assigned by his employer to give a short talk to twelve fifth graders at Sunnyvale Bible School, a small fundamentalist Christian school, on any subject he chooses from his field of expertise, which is chemical engineering. Having received the directive, Max puts what he knows about the situation through the process of rhetorical problem solving.

1. *Purpose:* Why is Max giving this talk? There are a couple of answers to this question. First, he is to provide a public service for his company, from which both the community and the company will benefit. Second, he is to teach these students something about engineering and what

- What do I want them to know or become after reading or listening to what I have to say?
- What do I want them to be able to do after reading or listening to my communication?

Second Level of Audience Analysis

- What do they already know about the problem?
- What do they want to know?
- What do they need to know?

Context: What environmental factors must I take into account?

Considerations of Place

- What is the environment of the problem?
- Are there physical, financial, political, or philosophical constraints?
- What are they, and how do they affect what I have to communicate?

Considerations of Time

- What are the constraints of time?
- Are there things I must know or do by a certain time?
- To what extent am I dependent on others to meet deadlines?

Ethical Stance: Have I considered my ethical position?

- Does anything about this subject present an ethical problem?
- Am I writing or speaking only to please my audience, but not myself?
- Am I telling the truth as far I know and understand it?

FIGURE 3.1 (***Continued***)

engineers do, because it is National Engineering Week. But the most important purpose is to suggest to these children that engineering can be fun and interesting. He wants them to get a positive impression of chemical engineering.

2. *Audience:* What does Max know about these children? What do they already know about chemical engineering? What do they want and need to know? Why will they be listening to him? Max knows that these students are about ten years old, come from conservative backgrounds, and have some basic notions and information about science, but that they probably know little or nothing about chemical engineering. Because they are ten and have been taught to respect school and learning, they will probably want to know whatever Max can tell them. What they need to know is slightly different. They will need to know that chemical engineers help provide many of the products that make their lives easier and better—from safer, more durable plastics that their toys are

made of, to safer, more environmentally sound fertilizers and insecticides that their parents use on their lawns and gardens.

3. *Context:* What about the environment of the speaking situation does Max need to remember? In this case, *where* is the important question, not *when,* since Max doesn't expect the students to respond specifically to what he says (except perhaps by way of their approval), as he would expect of professional readers or listeners. But *where* is significant, since he's to speak at this small, not particularly prosperous church-run school. Max figures that he will have to supply any visuals, models, or other tangibles for the children, but that he should also describe these items at least to the principal and class teacher, accompanied by a brief outline of his talk, so that they will know that the topics of his talk and all supporting materials are appropriate, given the school's educational orientation.
4. *Ethical Stance:* Since Max believes in both religious and educational freedom, he decides that he has no ethical problem with talking to this group. But he also decides that any topic he chooses to talk about should be as free of philosophical controversy as possible; he doesn't want to risk contradicting any of the school's premises.

Having put his task through the four basics of rhetorical problem solving, Max returns to his primary purpose: to inspire the children. Because he wants to breathe some life into what might seem to them a hopelessly technical field, he resolves to take a "Mr. Wizard" approach. He will let the children work with him to break down several familiar substances into their chemical components. The children will enjoy the hands-on approach, and Max hopes they will get the message that chemicals are an important, but also everyday, part of life and that the work chemical engineers do enhances the quality of life for everyone.

PURPOSE

Mary Anne Cox's situation is a more complex demonstration of rhetorical problem solving. Since she had already solved the technical problem, she needed to move to solving the rhetorical problem—communicating the solution to those people who were vitally interested. She would have to ask and answer several questions about what she had to write. The first series of questions Mary Anne had to consider centered on her purposes—both primary and secondary—for writing.

Primary Purposes

In the case of primary purposes, Mary Anne could be writing simply to be able to answer questions of fact: what has happened in her work, what is happening now, or what is likely to happen? *Record keeping* is, in fact, an important part of a professional's work: the documentation of work accomplished, or work in progress, or work yet to be done, which is written in the form of notes or logs, is usually necessary to remind oneself and others of the directions a project may take, and how those involved may be affected by factors already known and predicted.

But since the malfunction of ENCONTRO demanded immediate attention and since Outerware ultimately would have to shoulder the responsibility for what went wrong at the capitol building, Mary Anne knew that the sometimes informal activity of keeping records was not enough. She would have to *document formally* what she had done and why she had done it. Although she may never have to show this documentation to anyone, she knew that at least it would be there in case questions arose, and she would also be able to use it to help her form the substance of the reports that would go to specified and directly concerned audiences.

In addition, she thought that she should be writing to impart *information.* Clearly that is a function of any document: if the document doesn't provide information, what good is it? But the difference here is whether the document is written for the sole purpose of defining, describing, or explaining—showing what ENCONTRO is, what it can do, and how it might be used—or whether it is written to exact a decision about how the system should work. In this latter case, the writer's purpose moves from simply informing to informing on matters of policy or value, so that readers can take action or make a decision. Mary Anne concluded rightly that her initial report to Tim would require some action from him, if nothing else than for him to say, yes, go ahead and retrain the operations personnel at the capitol building. Although she figured he would probably come to this decision anyway, she decided she would have to write her report in a way that would make the decision inevitable—and one that Tim, not she, had made, on the basis of her advice. Figure 3.2 summarizes graphically Mary Anne's analysis of her purposes for writing to this point.

Record Keeping	What happened?
Formal Documentation	Matters of Fact or Definition
	What is currently happening?
	What might happen in the future?
Informing	What was Mary Anne doing about any of this?
Persuading	What should be done to solve the problem?
Moving to Action (Enabling Decision Making)	Matters of Policy
	Make decisions about two necessary courses of action:
	■ Retraining personnel
	■ Changing attitudes toward use of the new system

FIGURE 3.2 Mary Anne Cox's primary purposes for writing

Having analyzed the technical problem, Mary Anne decided what she had to do to solve it: retrain operations personnel. She was then able to decide that her main purpose in writing about the problem was to bring about decisions in at least two quarters: Tim's liaison position and John Springer's program development position, both within her own firm. Decisions within the capitol building concerning this problem would certainly be influenced by the decisions that Mary Anne's superiors made.

Secondary Purposes

Because a purpose is deemed secondary, it is not necessarily of lesser importance. The distinction is simply that primary purposes are those that come immediately to the foreground in attempts to solve a problem. Secondary purposes lie perhaps in the background or are not immediately apparent. In Mary Anne's case, the primary purposes were the needs for formal documentation of the ENCONTRO malfunction and for Tim to make a decision about retraining the capitol building personnel. Her secondary purposes were to show Tim and his superior, John Springer, that she was on top of things, had analyzed the situation well, and was assuming responsibility for the initial inadequate training of capitol building personnel. In other words, she wanted to be perceived as acting in a fully professional way. This perception would be important later on when she was considered for new assignments, a promotion, or a raise in her salary.

One could argue, of course, that secondary purposes for writing will always be the same for any professional. No one wants to be perceived as unprofessional or as not caring about later promotions and raises. But these considerations only would make the notion of secondary purposes too sim-

What are the purposes for writing?

Primary Purposes

- Record Keeping
- Formal Documentation
- Supplying Information
- Changing Opinion (persuading)
- Moving to Action (persuading and enabling and decision making)

Secondary purposes

- Creating, maintaining, or improving image
- Maintaining or improving interpersonal relations, either internally or externally or both
- Creating, maintaining, or improving a sense of trust or reliability

FIGURE 3.3 **Summary of primary and secondary purposes for writing**

ple. There may be other equally legitimate and compelling secondary purposes for writing. Mary Anne may want to help direct the decision to retrain and, as we said earlier, to make the decision inevitable and one that she would have to implement in her own way. Certainly, these are important purposes for writing, although they fall within the secondary category. Generally speaking, however, we can summarize the purposes for communicating as shown in Figure 3.3.

AUDIENCE

Realizing that a problem is not solved until others know about it and understand and approve it, Mary Anne began to analyze her audiences more fully. Mary Anne thought that the question of who her audiences were seemed simple. After all, she could name her audiences easily enough: Tim Sanford, John Springer, Janice Trippling, and perhaps several other people at the capitol building who were in positions either superior or subordinate to Janice's but who were, for some reason, vitally interested. But as she began to think of who those audiences actually were—to think beyond names to the people themselves—she remembered that she had been in this situation before: she had had to solve problems that multiple audiences would be concerned with, in sometimes distinctly different or sometimes only subtly different ways. She knew that those concerns would be dependent in part on the audiences' backgrounds, education, experience, and positions in the structural setup. Moreover, she knew that she would have to take them all into account, which made the seemingly simple task something less than simple.

She began, then, the more detailed analysis of who her audiences were and what kind of information she had to give them by asking questions at two levels. The first level concerned the purposes she had for writing; the second concerned the specific knowledge that the audiences required about the situation.

First Level of Audience Analysis

To find the answers to who Mary Anne's audiences were and what information they required, she had to ask herself questions very similar to those she had asked when considering her purpose. Purpose and audience are closely related; one cannot be considered without the other. Some of the questions she asked are shown in Figure 3.1 under the heading "First Level of Audience Analysis."

This first level of audience analysis allowed Mary Anne to gain a general understanding of her readers and their needs. But this alone was not enough. The answers to these questions, although providing Mary Anne with important information on who her audiences were and what their backgrounds and roles in the rhetorical situation were, remained rather theoretical and unspecific. To understand what requirements her audiences had specifically, she had to look to a second level of analysis.

Second Level of Audience Analysis

Mary Anne was able to fill in the specifics lacking on the first level of questions by asking and answering questions on the second level:

- *What do the audiences already know about the problem?* Answers to this question will distinguish what can be assumed from what needs to be explained, laid out, or reiterated, at least at this initial stage.
- *What do the audiences want to know about the problem?* In most cases, the information is specifically requested, through either oral or written communications, but in either situation, it is clear that some specific information must be supplied.
- *What do the audiences need to know about the problem?* Answers to this question reveal what information audiences should have to make informed decisions. Sometimes a fine line is drawn between what audiences want to know and what they need to know. But it is an important part of the writer's analysis to ask the question and to determine the difference if one exists. Audiences may have asked for information that would supply an answer to a problem that they have to solve, but they may not realize that they need something else, some other kind of information, to solve the problem fully.

Once Mary Anne had asked these preliminary questions, she began to answer them for each of her audiences. And as she answered the questions, she began also to think about the problem of what her report should contain, the tone it should set, and the ethical limitations of the problem. She knew from other writing experiences that *tone*—the attitude of the writer toward her or his subject—is extremely important in producing an effective, successful document. A report, memo, or letter may contain all the technical information necessary to address the problem at hand, but if its tone is wrong, for example, aggressive or condescending, or if the writer has conveyed an attitude that is somehow inappropriate or offensive to the audiences in question, the problem will not have been adequately addressed and therefore the document will fail to do what it was intended to do. The following audience profiles show how Mary Anne analyzed her audiences individually.

PROFESSIONAL PROFILE 1

Tim Sanford
Projects Director
Outerware, Inc.

Tim was Mary Anne's primary audience, that is, the person who asked for the problem to be solved and who will be the direct addressee of some written response to the request. Tim had a B.S. in electrical and computer engineering, also from the University of Connecticut, and had worked for Outerware for the past five years. He had started out in the job Mary Anne now had and had moved into his present position as Projects Director two years after he arrived. He was proud of his advancement and was ambitious to go further.

Mary Anne had known Tim for ten years and had worked for him for the last two. They got along well. Each understood the nature and pressures of the other's job; they respected each other for the way they met their responsibilities.

Mary Anne knew that Tim, as Projects Director, had to maintain informative and congenial relations between Outerware and its clients, and so would require information on the political context of the problem. She also knew that in her position as Systems Engineer, she had to know all the technical details of any system that she had worked on and that Outerware had sold and installed. Further, she was expected to have contact with clients about any technical problems that arose, especially through any maintenance contract agreement.

WHAT DID TIM ALREADY KNOW?

Mary Anne was able to catalog the events and Tim's knowledge of them fairly simply: on March 21, ENCONTRO had malfunctioned, causing extreme temperatures (both cold and hot) in both wings of the capitol building. Tim also knew that Janice Trippling was very upset and had demanded immediate technical aid. And finally, Tim knew that aid had to come from Mary Anne, that she would have to be in touch with Janice Trippling, and that he was ultimately responsible for maintaining good relations between Outerware and the capitol building.

In determining the information that Tim already knew, Mary Anne could eliminate any mention of this kind of background information in her report. Tim knew all the background and would not want it fed back to him. Mary Anne, then, could get on to answering the next question.

WHAT DID TIM WANT TO KNOW?

The answer to that question was straightforward: Tim wanted to know *what went wrong and how she proposed to fix it.* She would have to tell him, then, that Janice Trippling, Sonny Baker, and others who monitored

ENCONTRO had misunderstood its operation on a day-to-day basis. He would also want to know both what Mary Anne planned to do to correct that problem and how she would bring the correction about. So she knew that she would have to talk about the need for more training of capitol building personnel and that the training would have to emphasize more the links between the theory and the actual operation of the system, something for which she was clearly responsible, but had not thought necessary to explain when the system was first installed.

Mary Anne turned then to the final question.

WHAT DID TIM NEED TO KNOW?

Initially, Mary Anne thought that this was one of those times when the distinction between what an audience wants to know and what it needs to know is so fine as to appear virtually nonexistent. Certainly, Tim would need to know what he also wanted to know: what went wrong and how Mary Anne proposed to fix it.

But as she gave the question more thought, she realized that Tim also *needed to know* that Janice Trippling and Sonny Baker did not presently have the theoretical understanding and technical expertise to run ENCONTRO efficiently. Mary Anne had simply misread their needs at the initial training sessions and had not given them the right kind of training. Although she thought that through retraining, she could improve the situation and bring them to a reasonable, even effective, understanding and use of the system, Tim also needed to know that future dealings with the personnel at the capitol building might be troublesome. Trippling and Baker did not seem particularly amenable to help or advice from the outside, and this tendency combined with their lack of technical knowledge might make for some rough going in the future. Again, she was willing to accept responsibility for at least part of the difficulty: had she read their needs more carefully at the beginning, perhaps the failure would not have happened, and they would not have been embarrassed and resentful.

Because of the delicate nature of this information, Mary Anne decided that it could not be included in the general report on ENCONTRO's malfunctioning. Instead, she would write a separate memo to Tim, who could then decide, perhaps with the help of John Springer, what to do about the problem. She thought that under the time constraints (Tim wanted her report "as soon as possible" after her meeting with Janice Trippling), a separate memo would be the best way to go, though she knew that she could also write two reports, one for in-house review, one for external review. The two of course would reflect the differences necessary in Mary Anne's approaches to her audiences.

Because Tim was Mary Anne's primary reader, her analysis of his needs had to be rather detailed. She knew after answering these questions that she had essentially answered the questions on the first level:

she would have brought Tim up-to-date on the nature of the problem (what do I want my audiences to know or become?) and would have given him enough information to decide what next had to be done, especially on the political front (what do I want my audiences to be able to do?). By spending time at this preliminary stage—before writing—to analyze her audience carefully, she would save time in the end because much of the content of her report and her approach to establish the appropriate tone would have been decided. Actual writing time, then, could be significantly reduced. Such detail about the primary audience usually meant that the time needed to analyze her other audiences could also be reduced.

PROFESSIONAL PROFILE 2

John Springer
Vice President, Program Development
Outerware, Inc.

As Vice President of Program Development and as Tim Sanford's immediate superior, John Springer, in this situation, became Mary Anne's most direct *secondary* audience. Springer had a B.A. in political science from Trinity College and an M.B.A. from the University of Hartford. What he knew about computers he had picked up through special courses and his seven years of experience at Outerware, the last two of which he had spent as Vice President of Program Development.

Since he had been responsible for landing the account at the capitol building, Springer took special interest in what Mary Anne had found out and was recommending, because of the following:

1. Springer was responsible for developing new business, and the capitol building, which was still a new account, was thought to be valuable not only in itself but also as a means of generating more business elsewhere. It was entirely likely that he would have to make some kind of decision about the problem that Mary Anne or Tim could not make.

2. As Tim's superior, Springer also, of course, wanted to see how well Tim approached problems and how well he attempted to solve them.

Springer would receive his own copy of the report. We should point out that not all secondary readers are this directly involved. Depending on the problem and the reader's position within the organization, secondary readers may take a more casual interest in documents written to address or solve a problem. They would read for information only, not expecting to have to take any action or make a decision. Also, they would probably not get their own copy of a document but rather would be included on a list of interested readers attached to a circulating copy.

Mary Anne knew all of this about John Springer, and therefore she took some time to analyze his needs as a direct secondary audience.

WHAT DID JOHN SPRINGER ALREADY KNOW?

Mary Anne thought rightly that what Springer knew would be about the same as what Tim knew. Springer might not know some of the details of ENCONTRO's malfunctioning that Tim knew, but she was sure that he would have the essential outline. She needn't add anything to what she would already provide Tim.

WHAT DID JOHN SPRINGER WANT TO KNOW?

The answer to this question was also similar to what Tim wanted to know: what had gone wrong, why it had done so, and whether there was something that could be done to solve the problem to keep the client's good faith—and business. The answer to the last question was for Springer the most important, maybe even more important than for Tim, so that Mary Anne realized she would have to be very clear about what the retraining program would entail; what kind of time it would take out of the capitol building personnel's daily schedules; and, in general, what efforts Outerware was willing and able to make to bring matters back into line. Since that was the kind of detail Tim would expect anyway, Mary Anne did not worry that she would be too explicit. She also knew from her experience that it was better to include more detail than less: good readers realize at once a writer's attempt to provide what is wanted. They can always skip the particular details if they don't need them, but if the document is written without sufficient detail, the writer will just have to go back and rewrite what was already written.

Mary Anne went then to the last question.

WHAT DID JOHN SPRINGER NEED TO KNOW?

This question again presented some problems for her. Springer needed to know basically what Tim needed to know—that the capitol building personnel lacked certain technical knowledge and expertise that Mary Anne had not discerned, and that she therefore had not provided adequate training. Finally, he needed to know that maybe because of this situation, the capitol employees were reluctant to take help and advice from outsiders. But since Springer had struck the contract with superiors of Janice Trippling, it was through them that he would have to deal, if he were to deal at all. He would have to let those people know that ENCONTRO would work well and that Trippling, Baker, and the others who would monitor the system would more willingly accept the system and would feel more comfortable using it, if they were agreeable to get some additional training—training provided by Mary Anne Cox.

As in Tim's case, Mary Anne decided that the nature of the information Springer needed to know was delicate enough to warrant its own internal memo; even allusions to the problems with Trippling and Baker in her general report might backfire. So she decided that the most efficient way of dealing with the problem was to write the separate memo to Tim and to show Springer as the direct recipient of his own copy. In that way, Mary Anne could keep the analysis of the problem at her level within Outerware, and Tim and Springer could decide how to approach Trippling's and Baker's superiors at the capitol building.

Again, Mary Anne was doing for Springer very much the same as she was doing for Tim: providing him with policy-related information that would enable him to make a decision.

PROFESSIONAL PROFILE 3

Janice Trippling
Chief Operations Engineer
Capitol Building, Hartford

Normally, Trippling would have been considered a *tertiary* audience, two steps removed from Tim in the direct concerns of the report. Tertiary audiences almost always read only to gain information, seldom to make decisions, and therefore require a less thorough analysis on the writer's part. The decisions a writer makes about content and tone are almost always directed by the analysis of the primary and secondary readers, seldom of the tertiary readers, as Mary Anne was aware. However, she also knew that Trippling had specifically asked her for a copy of her report, something she certainly could not refuse to provide. And thus Trippling was brought into the position of a *secondary* audience, requiring at least as thorough an analysis as she had given Springer.

So Mary Anne reviewed what she knew about Trippling. She had a B.S. in mechanical engineering technology from the University of Massachusetts and had minored in computer science. She had worked for the capitol building as a systems engineer for ten years, and only within the last six months had been promoted to Chief Operations Engineer, jumping ahead of several men who had been with the state for a combined thirty-five years. Trippling was ambitious, but also nervous in her new position, thus making her wary about anything that Mary Anne might do that could possibly throw her work into a bad light.

Once again Mary Anne asked and answered the pertinent questions.

WHAT DID JANICE TRIPPLING ALREADY KNOW?

Trippling already knew basically the same things that the other audiences knew. ENCONTRO had malfunctioned while she was in charge; the capitol building had gotten either too hot or too cold, depending on the locations of offices, and Outerware was going to have to do something to solve the problem. She also knew that the decision to install ENCONTRO had been made at levels above hers and that it was probably unlikely the decision would be reversed. This awareness, Mary Anne noted, might have a dual effect on Janice's reading of the report: she might look for reason to take offense, since she had had nothing to do with making the decision, and she would be looking for any suggestion that she was perhaps to blame for the malfunction. Though Mary Anne had already decided, through her analysis of Tim and Springer as audiences, that the content and tone of her report would have to show that Outerware, particularly Mary Anne herself, was responsible for inadequate training of the capitol building personnel, this part of her analysis of Trippling as an audience confirmed that need.

WHAT DID JANICE TRIPPLING WANT TO KNOW?

When she considered this question, Mary Anne again came up with similar answers to those she had already formed, though again colored a bit differently, given that this audience was a client and not in-house personnel. Trippling obviously wanted to know what went wrong and what it would take to fix it, but she would also want to know what part she had had in the malfunctioning, assuming correctly that she, as the direct superior of the monitors of the system, would have had to have at least some part in the problem. Mary Anne would have to mention the inadequate training, but mainly as a problem she was responsible for. And, so as not to alienate Trippling, she would have to be very careful not to mention that the personnel seemed to lack the necessary theoretical background and technical expertise to run the system without more training—or perhaps training of a more fundamental kind. That problem should be discussed between Tim and her, and maybe John Springer, at a different time and in a different context.

WHAT DID JANICE TRIPPLING NEED TO KNOW?

She needed to know just what everyone else needed to know—that Outerware had a workable solution to ENCONTRO's malfunction in the form of additional training, and that she and others who would monitor the system could come to use it comfortably and efficiently.

Whereas in Tim's and Springer's cases, Mary Anne knew that she was having to provide them with enough information about the problem to enable them to make decisions, this was not the case for Janice Trippling. Trippling would not have to make a decision on the basis of what Mary Anne had provided, but she would have to continue to use ENCONTRO, and Mary Anne knew that it was much better to have her become a willing user than an unwilling one. Unwillingness could only mean trouble for Outerware, and particularly for Tim and her, in the future. The tone and content of her report would indeed have to be carefully crafted to make all audiences into something Mary Anne wanted them to be: willing and efficient decision makers and users of the system.

CONTEXT

Having fully analyzed her audiences, Mary Anne next turned to the questions concerning the context of her problem. As for considerations of time and place, Mary Anne considered specifically a few issues. The significance of place was already somewhat analyzed in her consideration of John Springer as a secondary audience. The capitol building was a client, and a government client at that, so Mary Anne knew she couldn't assume much flexibility from this group when she scheduled her retraining sessions. Second, since the capitol building was a very big account, any of its problems took top priority. Also, because it was such a big account, the upper-level Outerware personnel needed to be involved in the solution. Mary Anne had taken care of this requirement through her analysis of audiences.

The time consideration was a bit tricky. She first needed to make absolutely clear to Tim the date by which she needed replies so that she could implement her solution. Then she needed to suggest retraining times that would work well both for Outerware and for the capitol building employees. Assuming that Janice Trippling was cooperative, Mary Anne could start almost immediately. If Janice was not cooperative, however, because she saw the ENCONTRO malfunction as Outerware's problem and didn't want to put herself out to be capable with the system, then Mary Anne would have to negotiate.

ETHICAL STANCE

Thinking about the ethics of the situation, Mary Anne asked herself whether she was being as straightforward as possible with all concerned. Was she telling the truth she had come to after the technical problem solving? Yes. She had as evenhandedly as possible traced cause and effect, responsibility, and solution, so that the capitol employees' parts were clear but not overstated. She accepted her own responsibility for underestimating the needs of the initial training program. She shaped a solution in the spirit of Outerware's maintenance agreement. Mary Anne had even extended herself to consider Janice Trippling's anxieties in her new position, and although she thought Janice's hostility was misplaced, she felt no need to retaliate.

Mary Anne's documents would acknowledge her own share of the responsibility, because it was part of her ethical code to do so and because the last impression she wanted Tim and John to have of her was of someone who ducked her responsibilities. In short, the ENCONTRO problem did not present difficult ethical dimensions to Mary Anne. She was sympathetic to company policy on such problems. But she might have had trouble if Outerware insisted on shouldering either all of the blame or none.

RHETORICAL PROBLEM SOLVING AS A RECURSIVE PROCESS

The discussion to this point has outlined an ideal process, one that works well, but also one that needs to be qualified. The process implies linearity, that the steps need to be taken sequentially. But that is not, in fact, the way most of us think. Most of us, instead of thinking in a line, jump forward to ideas that ideally would occur to us only after we had considered others. Or, while we are presently formulating an idea, we may fall back to some idea we have already had and consider how that idea might influence what we are currently thinking about. The process, then, is recursive—jumping forward, falling back—fairly fluid until it is complete. And when it is complete, it presents another question: is it complete when writers and speakers are satisfied they have analyzed the problem as fully as they can or when they are required to have some answer—when they have to meet a deadline? It can be either or both, but there is always an end to the process—it cannot go on forever—and that end is almost always determined by the deadline. There is a point at which the product of work will be called for; it is then that it must be complete, for better or for worse, and submitted.

Mary Anne's situation once again is a good example. She asked and answered all the questions that we have spoken about, but she did not do so linearly. In fact, she did not have to. As she realized, several of the questions dealing with the audiences' knowledge of the problem could probably be answered in basically the same way for each audience, with variations of course, but with variations that she could keep easily in mind.

She thought, for example, of what Tim, her primary audience, already knew about the problem and realized that her other audiences, John Springer and Janice Trippling, already knew the same. She also saw that what they wanted to know—what went wrong and how it could be fixed—was the same, though given their different positions, they would probably use that information differently. Tim could show Springer that he could reassure his contacts at the capitol building that matters were in hand, and Trippling could show that the fundamental problem had been generated by Outerware, not by her and her personnel, and that it was being corrected, with her cooperation, of course.

What happened in this case shows that thinking about audiences does not have to take place linearly. The knowledge, wants, and needs of Tim Sanford, John Springer, and Janice Trippling clearly overlap, and Mary Anne thought of them in that way. She realized that she did not think of any one part of the problem in sequence: sometimes she would think of needs, sometimes of wants, sometimes of what her audiences already knew. Mary Anne also knew that she had to consider her audiences as part of the package, along with purpose, context, and ethical stance. At times it seemed she was dealing with all levels of the analysis at once. Janice knew she would have to sort them out soon, because Tim had called for her report "as soon as possible," which she understood to mean "within the next day or two" after her meeting with Janice Trippling. It was then that she sat down in front of her computer to write the actual documents.

Before we look at what Mary Anne wrote, there is another important idea to mention here. The audiences for whom people write in practically every professional environment are chosen for them; the writers or speakers themselves do not choose them. Mary Anne did not have to conjure up the audiences to solve the problem of ENCONTRO's malfunction: they were there, concrete, not abstract, and clearly defining themselves, demanding information and expecting help. Tim Sanford, John Springer, and Janice Trippling were all directly involved in the problem for different reasons, and with slightly different expectations for a solution; but they were not strangers to each other or to Mary Anne, a fact that ultimately makes the whole process of audience analysis easier than it seems. The information is there: a problem and the people involved in the problem, and each has to be analyzed to produce a reasonable solution.

We should now look at the fruits of Mary Anne's analysis and then analyze ourselves what she wrote. Figure 3.4 shows a memo report on Mary Anne's meeting with Janice Trippling at the capitol building, and Figure 3.5 shows Mary Anne's memo to Tim on the potential problems she sees with capitol building personnel.

WRITING FOR SOMEONE ELSE'S SIGNATURE

We have been speaking all along of Mary Anne Cox as a professional who solves problems with the help of writing. That is, she has been confronted with a problem and has either been assigned to report on a proposed solution in writing or has herself seen the need to report; in the latter case, her writing comes from self-direction. In either case, however, writing comes about from direct or central involvement in the problem-solving process; the problems, in other words, pertain directly and naturally to her own work. But what we have not pointed out yet is that this kind of writing constitutes only a part of the writing that professionals do. There is another substantial amount of time spent on writing for others, or *ghostwriting,* that is, writing documents that will show others as the authors.

It is true that ghostwriting is also assigned, but the difference is that it addresses other people's (always superiors') problems—technically not those

The opening paragraph reestablishes the context of the problem, gives the specifics of what Cox did to meet with the capitol building personnel, and summarizes the general outcome of the meeting. Note the conciliatory tone in the last sentence. Note, too, its conclusiveness.

The second paragraph lays out for Sanford exactly what went wrong. Although it tells of the monitors' part in the malfunction, it does not lay blame on them; it suggests that the mistakes they made are understandable.

Cox, in the third paragraph, assumes the responsibility for faulty training and suggests that additional training will probably solve the problem. She is specific as to what the training would cover and when it might begin. She acknowledges that Sanford is the one to make the final decision. She closes in the fourth paragraph with a dignified apology and an appeal for action.

She sends copies to her secondary readers: Springer, who is Sanford's superior and has special interest in the problem, and Trippling, who asked for a copy of her report.

OUTERWARE, INC.

17 D STREET, HARTFORD, CONNECTICUT 06902
TEL: (609) 398-4701 FAX: (609) 398-4700

Memorandum

To: Tim Sanford, Projects Director
From: Mary Anne Cox. Systems Engineer
Date: March 23, 2001
Subject: Meeting with Janice Trippling, March 21, Capitol Building

Following our morning conversation on March 21, I met with Janice Trippling in her office at the capitol building from 2:00 to 3:30 P.M. to discuss ENCONTRO's malfunction of March 20. Sonny Baker and Jim Pierson, of Ms. Trippling's staff and monitors of the system when it went down, were also there. The meeting proved very helpful in establishing what actually went wrong and in suggesting what we can do to prevent similar malfunctions in the future.

From what we can determine, the malfunction was directly caused by a communications lockup at one of the modules: this caused temperatures to shift in both wings. Also, the monitors misread the data at the command location, figuring that since one of the two chillers was shown as operating, temperatures in both wings were being maintained at acceptable levels. As you know, this is an area where mistakes can easily be made: instead of looking for patterns in the data, monitors sometimes tend to concentrate on individual characters, which in turn causes them to misread the overall configuration.

We all agreed also that better training of the personnel when ENCONTRO was first installed could have prevented this from happening. Had I spent more time on the basics of the system, Ms. Trippling and her staff would have been better prepared to monitor it effectively. So the best solution I can suggest, and one that the people at the capitol building agree with, is for me to offer some basic retraining. I think we can cover the necessary material, such as a review of the system's diagnostic tools, and some dry runs of troubleshooting techniques in two or three sessions of a couple hours apiece. I propose to start the sessions as soon as I get your approval, but my schedule and those of the people at the capitol building could permit us to begin as soon as next Tuesday, March 27.

I'm sorry for the inconvenience of the event, but I hope you'll agree that we are on a reasonable course for preventing similar troubles in the future. I look forward to hearing from you.

cc: John Springer
Janice Trippling, Capitol Building

FIGURE 3.4 **Mary Anne Cox's memo report**

OUTERWARE, INC.

17 D STREET, HARTFORD, CONNECTICUT 06902
TEL: (609) 398-4701 FAX: (609) 398-4700

Memorandum

To: Tim Sanford, Projects Director
From: Mary Anne Cox. Systems Engineer
Date: March 23, 2001
Subject: Future Dealings with Janice Trippling's Office, Capitol Building
Re: Memo of March 23 on Meeting with Janice Trippling

I have referred to my memo report on my meeting with Janice Trippling on March 21 to help provide the context of a problem beyond ENCONTRO's malfunction of March 20. I said there that the meeting was very helpful in determining what had gone wrong and how we might fix it, and that's true; the meeting was helpful. But it was not particularly congenial, and what I learned from it, beyond the technical aspects of the malfunction, is that Trippling and her staff are embarrassed and, therefore, angry that the system went down, thus making hundreds of people uncomfortable. That they got it back up and running, so that the last test run was not a complete failure, shows that they have a good deal of energy and some basic technical expertise to correct that kind of malfunction.

In the first paragraph, Cox recognizes Sanford's need to know about the political context of the problem at the capitol building. She is candid and straightforward about the atmosphere of the meeting and about the attitudes of the people she met with, but she also concedes that they dealt with the malfunction better than might be expected.

I am worried, however, about two things. The first is less important, but nevertheless worthy of my mention: the staff's technical knowledge of this kind of system is not what it might be. They don't understand the kinds of problems that can occur and seem not to have been educated in some of the necessary theoretical bases of how a system like this operates, something I realize I should have known when I started training sessions in early February. This, I feel confident, I can correct with the right kind of retraining. It seems reasonable that I should concentrate on having them see patterns in the data that come across their screens and on diagnostic and troubleshooting techniques.

Cox, in the second paragraph, addresses the first of two problems that she could not have mentioned in her first memo: the client's lack of technical expertise. She again acknowledges her responsibility for not recognizing this, and suggests specifically the kind of retraining that will help correct the problem.

The other problem has to do with attitude. Trippling and her staff are put out by having to implement a system that they were never really consulted about. Their questions to me were aggressive—what went wrong? why did it go wrong? how do we deal with it? what are we supposed to do about it now? I felt defensive because I had helped implement the system and, as it turns out, had not trained them sufficiently. That certainly was a mistake, but I think that my plan of two or three retraining sessions of about two hours each will help correct it.

The third paragraph deals straightforwardly with the attitude problem, but again Cox concedes that capitol building personnel are in a difficult position and that she contributed to that by misreading their needs for training. Again she suggests how she will correct that situation.

I wanted you to know, however, that the personnel at the capitol building system may be reluctant and resistant to taking to my instruction and suggestions. I don't think we have a permanent problem, but rather one that will crop up again without some correcting on all of our parts. I intend to extend Janice Trippling every courtesy, but thought I might take her and her staff to lunch as well to improve the atmosphere. Can we budget for this? If there is anything else I can do, let me know. I look forward to hearing from you.

The last paragraph mentions the final element of the second problem—possible resistance to her training instruction—and suggests further political action.

Cc: John Springer

She again sends a copy to Springer, who also may have to take part in the corrective action.

FIGURE 3.5 Mary Anne Cox's memo

that the actual author is involved in. In Mary Anne's case, for example, Tim is vitally interested in knowing how she assesses the problem at the capitol building and realizes the need to correspond with his contacts there, people in management positions superior to Janice Trippling's. But for a couple of possible reasons, lack of time and/or lack of knowledge or confidence in technical matters, he is unable to write the memo or letter to Anthony Capello, Director of Operations at the capitol building in Hartford, to explain what went wrong with ENCONTRO and what will be done to correct it. He, therefore, assigns the task to Mary Anne, saying, of course, that he will review and perhaps edit her work before he signs and sends it. That undertaking places Mary Anne in the position of having to analyze the problem, not only from her perspective and role, but also from Tim's—an uncomfortable and difficult situation sometimes, but one that cannot be avoided: it is a simple fact of professional life. Although writing for someone else involves all the same steps of analysis that we have been speaking about here, it also requires additional understanding of the relationship between the sender and the recipient of the document in light of the problem that the document addresses. Thus, Mary Anne has to draw on her knowledge of the nature of Tim's position and the demands it puts upon him, as well as her understanding of the relationship Tim has with Capello, something she may know about through observation but may also need to strengthen and confirm through conversations with Tim.

The professional profiles in Chapter 1 show that ghostwriting is a common necessity for professionals in middle management, even though they, too, have occasion to assign writing tasks to those who work for them. This fact is evident by the number of documents they have to review and/or edit—a number of which they themselves have assigned to others to write. But again, as we indicated there, when professionals move up the ranks, the amount of writing they have to do in general decreases, but the amount of ghostwriting they have to do in particular decreases to almost nothing. Instead of being assigned to write someone else's document, they assign others to write for them. It is simply the way of any hierarchically structured organization.

CONCLUSIONS

What Mary Anne went through in preparing her documents—a report and a memo—is what should happen whenever a professional faces a rhetorical problem. It is important to remember that most rhetorical problems grow out of technical problems and that the whole problem-solving process takes a combination of technical knowledge and expertise (the ability to solve the technical problem) and knowledge and expertise in communication (the ability to solve the rhetorical problem). Professionals can solve problems not only with the help of what they know about their various fields—computers, engineering, social work, geology, or some other field—but also with the help of what they know about writing and speaking.

The discussion of Mary Anne Cox's situation demonstrates that rhetorical problem solving is a specifically human activity requiring effective communication. To do this, writers and speakers must think systematically, though not necessarily linearly, touching the four essential bases of purpose, audience, context, and ethical stance. As we pointed out in Chapter 1, problems that professionals face cannot really be called problems unless there are people who feel the need of and will benefit from a solution. Thus, the malfunction of ENCONTRO at the capitol building in Hartford is not just some abstract, theoretical breakdown to be corrected by simple tinkerings and adjustments in the technology. Because people were made uncomfortable and were inconvenienced, the problem takes on the necessary human dimension. And because people—both at Outerware and at the capitol building—were directly responsible for the technological mishap, the problem demonstrates a greater level of human involvement.

What we see, then, is technology that works only if people make it work. And thus we have Mary Anne's analysis of the human element in the breakdown: her own misreading of the needs for training capitol building personnel; Janice Trippling's embarrassment and resulting resentment of the system and of those who put it in place; and Tim Sanford's and John Springer's needs to reestablish easy, congenial, and productive relations with their client.

We see, further, the kinds of documents that must be written to reflect the nature of the problem and what should be done about it, without putting off any of the audiences who will read them. Reports to Tim Sanford, with copies to John Springer and Janice Trippling in the first instance, and a memo, with a copy to Springer, in the second instance, both dealt with the same problem but with different purposes and potential results in mind. We should also remember the memo that Tim has in mind for Mary Anne to write to Anthony Capello, but that will go out under Tim's signature.

In the end, the whole enterprise becomes basically human. People feel the effects of a technological breakdown; people are responsible for the technological breakdown; and people must clean up after themselves and must correct the technological breakdown through human interaction. Careful and thoughtful writing and speaking, which are the results of careful and thoughtful analysis, are the means by which this correction can be done. They are, as far as we know, the best answer to solving problems in professional life.

EXERCISES

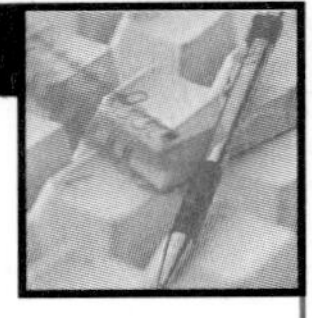

1. Following are four excerpts of technical and professional writing, taken from various fields. In small groups, read and consider at least two of the excerpts, and then answer the following questions:

   a. What purpose does the writer have for writing? You may see both primary and secondary purposes.
   b. Who is the audience for the piece? What are their levels of education and technical expertise likely to be? What positions, attitudes, and biases are they likely to have on the subject in question? How does the writer appeal to them (through emotion, authority, ethics)?
   c. What purpose does the reader have for reading?
   d. Do you pick up any clues about context—considerations of place and time?
   e. Does the excerpt suggest any ethical dilemma for the writer?

I. From a Review of the Year's Activities of a Regional Planned Parenthood Agency*

A Message from the President and Executive Director

What a year we have had! Momentous and courageous decisions were made towards accomplishing the mission of Planned Parenthood. Resounding support was received from the community for Planned Parenthood's work.

- Acting on the news of the impending retirement of a local abortion provider, and on the final report of a study commissioned by our board, the Board of Directors (after 3 board meetings in the month of August) unanimously voted to begin offering first trimester abortion services in the summer of 1990.
- In the face of ultimatums by the local Catholic Diocese, the United Way Board of Directors voted to retain Planned Parenthood as a member. The 1989 United Way campaign met its adjusted goal for funding all the remaining agencies at the prior year's level.
- During 1989, direct contributions to Planned Parenthood increased by 30% and our donor base increased by 24%. United Way contributors specifically designating Planned Parenthood increased tenfold, to over $115,000.
- Notwithstanding the withdrawal of support by key sponsors, the Smith Commission brought Dr. James Ploski to Pleasantville to speak to over 1500 appreciative teens, parents. and professionals. A private reception was held for Dr. Ploski which included Planned Parenthood staff, President Circle donors, and board members.

We . . . you . . . are part of an organization considered to be one of the most credible non-profit agencies in America. We are doing more than just

*Source: Excerpt from a review of the year's activities of a regional planned parenthood agency. Reprinted with permission of Planned Parenthood of the Inland Northwest.

talking about the need for family planning: quality, objective sexuality education; preventing unwanted pregnancies; and treatment and education for sexually transmitted diseases. Planned Parenthood is providing those needed medical services, that needed education, that needed advocacy of people's right to make their own choices. The more polls and studies we see, the more we come to realize that we do, in fact, speak for the majority of the people in this country.

. . .

We need to continue to provide medical services, to educate and to advocate. We will need your help, your contributions of time, talent, and treasure. This coming year will require at least as much work as last year for us to accomplish our mission: making every child a wanted child.

II. From a Research Memo Written by a Paralegal Working for a Large Eastern Public Power Supplier on the Subject of Unauthorized Use of Computer Programs

DATE: June 20, 1999
TO: James P. St. Pierre, Chief Counsel
FROM: Alison L. McKee, Paralegal I
SUBJECT: Copyright and Literary Property
REFERENCE: Unauthorized Use of Computer Programs

Query:

Assume that the Supply System is held liable for the unauthorized use of unlicensed software in Supply System P.C.s—explain the nature and extent of liability for punitive or exemplary damages or other damage measures beyond just payment of royalties.

Short Response:

According to 17 USCS #504, an infringer of copyright is liable for either (1) the copyright owner's actual damages and any additional profits of the infringer, or (2) statutory damages. "Punitive damages" per se are not available under the federal Copyright Act.

Discussion:

I. Background

The Supply System wants to develop agency standards for use of licensed software. The prime area of inquiry is what stance the agency should adopt when faced with situations where multiple copies are made of purchased software.

II. Copyright Protection

The Copyright Act of 1976 (17 USCS #101 et. seq.) protects the exclusive right or privilege of authors or proprietors of literary property to print or otherwise multiply, publish, and vend copies of literary, artistic or intellectual productions, when secured by compliance with the copyright statute, 17 USCS #101. The definitional provisions of the Copyright Act of 1976 encompass computer programs and data bases in the definition of "literary works."

The Act was amended by the adoption in 1980 of the Computer Software Copyright Act. This amendment replaced former section 117, added a specific definition of "computer program," and provided for the lawful copying of a program in two instances: (1) by its rightful owner for use "in conjunction with a machine"; or (2) for "archival purposes."

The discussion proceeds for three single-spaced pages, defining and talking about a variety of topics: the components of computer programs that are protected, successful court precedents, civil liability, and criminal liability. The author then concludes with the following:

V. Conclusion

The magnitude of the problem of trying to provide legal protection for computer programs is staggering: it has been estimated that some 15,000 computer programs are written each day in the United States and that the total value of this software is in the tens of billions of dollars. (See "Schmidt, Legal Proprietary Interest in Computer Programs: The American Experience" 21 Jurimetrics 345 (1981) as cited in Jostens, Inc. v. National Computers Systems, Inc., 318 NW2d 691, 214 U.S.P.O. 918, 33 UCCRS 1642, 30 ALR4th 1229.) The Copyright Act provides the owner of a copyright with a variety of remedies, including injunction, the impoundment and destruction of all reproductions, recovery of actual damages and any additional profits realized by the infringer, or a recovery of statutory damages, and attorneys fees. But despite the fact that the Supreme Court speaks of a copyright owner's "potent arsenal of remedies against an infringer," (see Sony Corp. of America v. Universal City Studios. Inc., 104 S Ct 774, 464 U.S. 417. 78 L.Ed.2d 574. 220 U.S.P.Q. 665 (1984)) a software firm which discovers that an individual has surreptitiously copied its computer program may have little to gain from filing a suit for copyright infringement since the recovery probably would not be enough to cover legal costs. This situation has been likened to piracy of phonorecords by individuals onto cassette tapes which has become so commonplace that the recording industry does not, as a practical matter, prosecute individuals. However, possible benefits from suing an individual for copyright infringement of a computer program would be publicity that might discourage potential customers from pirating programs instead of buying authorized copies.

Let me know if you need further research into this issue.

III. From a Report on the Installation of Transmission Lines in a Western Public Utility District

APPENDIX B

TRANSMISSION LINE SURVEY REQUIREMENTS
FOR
PUBLIC UTILITY DISTRICT NO. 2 OF GRANT COUNTY

I. General
   A. All work shall be completed under the direction of a professional land survey or licensed by the state of Washington.
   B. All horizontal control surveys shall be a minimum of third order, Class II accuracy. Vertical control surveys shall be a minimum of third order accuracy. Third order and third or Class II shall be defined in

CLASSIFICATION, STANDARDS OF ACCURACY AND GENERAL SPECIFICATIONS OF GEODETIC CONTROL SURVEYS as reprinted in 1979 by N.O.A.A.

C. All horizontal points shall be referenced to the Washington State Plane Coordinate System or as otherwise agreed to between the District and Contractor. When Washington State Plan Coordinates are used, provide Washington State grid factors on each drawing.

D. All field and office work required for the project surveys shall be based on a closed traverse of triangulation network.

II. Survey Requirements

A. The survey shall locate all road, highway, railroad, pipeline, transmission line, canal and other utility right-of-ways which may affect the transmission line centerline. All right-of-ways shall be referenced to section corners and State Plane Coordinate System.

B. The work shall include locating all property lines, section corners. and quarter section corners. Property lines shall be tied down to section corners and quarter section corners. Section and quarter sections shall be located in the field and referenced to the State Plane Coordinate System.

C. The survey shall extend into private property to include the transmission line right-of-way as specified by the District engineer.

D. The survey shall locate all proposed transmission line P.I.'s in relation to the section corners and State Plane Coordinate System. P.I.'s shall be staked with nominal 2. x 2. x 8. hubs and tacks and flagged with lath and surveyors flagging.

E. Point-on-Tangents (P.O.T.) shall be staked on the transmission centerline with nominal 2. x 2. x 8. hubs and tacks placed at intervals not exceeding 2,000 feet and flagged with lath and surveyors flagging. Surveyors hubs and tacks used in the control survey and online of proposed centerline shall be left intact. If removal of the control survey hubs is desired or recommended, coordinate with Jim Smithson or the District engineer assigned to the project.

F. The survey shall locate other survey monuments in the area of the transmission right-of-way such as USGS triangulation stations and benchmarks. These monuments shall be referenced to the State Plane Coordinate System.

G. When specifically requested on the request for professional services for the specific project, the survey shall include a detailed plan and profile of the transmission line right-of-way. The plan and profile shall include the information listed below in addition to that information requested in paragraphs A to E above and in Section IV below.

IV. From Lewis Thomas's Article "Germs" in *Lives of a Cell* (New York: Viking Penguin, Inc., and London: Garnstone Press, 1974)*

It is the information carried by the bacteria that we cannot abide.

The gram-negative bacteria are the best examples of this. They display lipopolysaccharide endotoxin in their walls and these macromolecules are

*Source: "Germs," copyright © 1972 by The Massachusetts Medical Society, from *The Lives of a Cell* by Lewis Thomas. Used by permission of Viking Penguin, a division of Penguin Putnam Inc.

read by our tissues as the very worst of bad news. When we sense lipopolysaccharide, we are likely to turn on every defense at our disposal; we will bomb, defoliate, blockade, seal off, and destroy all the tissues in the area. Leukocytes become more actively phagocytic, release lysosomal enzymes, turn sticky, and aggregate together in dense masses, occluding capillaries and shutting off the blood supply. Complement is switched on at the right point in the sequence to release chemotactic signals, calling in leukocytes from everywhere. Vessels become hyperactive to epinephrine so that physiologic concentrations suddenly possess necrotizing properties. Pyrogen is released from leukocytes, adding fever to hemorrhage, necrosis, and shock. It is a shambles.

All of this seems unnecessary, panic-driven. There is nothing intrinsically poisonous about endotoxin, but it must look awful, or feel awful when sensed by cells. Cells believe that it signifies the presence of gram-negative bacteria, and they will stop at nothing to avoid this threat.

I used to think that only the most highly developed, civilized animals could be fooled in this way, but it is not so. The horse-shoe crab is a primitive fossil of a beast, ancient and uncitified, but he is just as vulnerable to disorganization by endotoxin as a rabbit or a man. Bang has shown that an injection of a very small dose into the body cavity will cause the aggregation of hemocytes in ponderous, immovable masses that block the vascular channels, and gelatinous clot brings the circulation to a standstill. It is now known that a limulus clotting system, perhaps ancestral to ours, is centrally involved in the reaction. Extracts of the hemocytes can be made to jell by adding extremely small amounts of endotoxin. The self-disintegration of the whole animal that follows a systemic injection can be interpreted as a well-intentioned but lethal error. The mechanism is itself quite a good one, when used with precision and restraint, admirably designed for coping with intrusion by a single bacterium: the hemocyte would be attracted to the site, extrude the coagulable protein, the micro-organism would be entrapped and immobilized, and the thing would be finished. It is when confronted by the overwhelming signal of free molecules of endotoxin, evoking memories of vibrios in great numbers, that the limulus flies into panic, launches all his defenses at once, and destroys himself.

It is, basically, a response to propaganda, something like the panic producing pheromones that slave-taking ants release to disorganize the colonies of their prey.

I think it likely that many of our diseases work in this way. Sometimes, the mechanisms used for overkill are immunologic, but often, as in the limulus models, they are more primitive kinds of memory. We tear ourselves to pieces because of symbols, and we are more vulnerable to this than to any host of predators. We are, in effect, at the mercy of our own Pentagons, most of the time.

2. Using the questions asked earlier, analyze two of the documents obtained from your interviews of professionals in Chapter 1.

ASSIGNMENTS

1. Choose a fairly sophisticated technical article from a professional journal in your field of study or in a field that interests you. Using the

DOCUMENT ANALYSIS WORKSHEET

(Adapted from work done by John Harris and Donald Cunningham)

Name of the document or situation that requires that it be written:

Primary Purposes for Writing

1. Record Keeping/ Formal Documentation?
2. Supplying Information?
3. Changing Opinion (Persuading)?
4. Moving to Action (Persuading to enable decision making)

Secondary Purposes for Writing

1. Creating, maintaining, or improving professional image?
2. Maintaining or improving interpersonal relations, either internally or externally, or both?
3. Creating, maintaining, or improving a sense of trust or reliability?

Audience(s) Intended for This Document

1. Who are they? (Primary? Secondary?) What are their backgrounds, levels of education, and technical expertise?

   Primary:

   Secondary:
2. What positions, biases, attitudes are they likely to have on or toward the subject?
3. What purpose do they have for reading this document?
4. What should they know (or become) after reading this document?
5. What should they be able to do after reading this document?

Preliminary Plan for the Document

1. What sources are available and should be used to gather data? Primary (collected firsthand, not published)? Secondary (collected secondhand, already published)?
2. What limitations need to be set for gathering data (i.e., what is the scope of the proposed document?)?
3. How might the information of the document be organized, and why? Inductively? Deductively? Chronologically? Climactically? Reverse climactically?
4. What graphics are necessary, and why?

FIGURE 3.6 **Document analysis worksheet**

substance of the article, write an article of your own for one of the audiences in the following list. To be sure that your article will be appropriate for the new audience, fill out the Document Analysis Worksheet (see Figure 3.6) *before* you start writing.

a. You have been asked by a high school counselor to write an article on a subject related to your field of study. The purpose of the article is to acquaint high school seniors with the subject so that they might begin to think about possible majors in college.
b. *Science Digest, Time,* or *Newsweek* (choose one) has asked for an article from you on a subject of interest and importance to the general readership.
c. Your boss (middle management) has asked you for an article on a subject that is important to a current or future company project. Your boss has little technical background in the subject but needs to appear knowledgeable at the next meeting of the board of directors.

CASE STUDY

The Situation

A. You are a member of the Personnel Relations Committee of Omni International, a highly diversified firm with interests in many areas. The committee is charged with reviewing personnel problems within the firm and making recommendations to James Peterson, Vice President of Operations. The chair of the committee, Joseph Doakes, has presented you with this problem: his boss, Peter Mason, a regional sales director for the Consumer Products Division, has been showing signs of wear. His last two quarterly reports were late, sales in his region were down 5 percent for the last quarter, he missed two important meetings with sales representatives in the last two months, he has been absent seven days in the last month, and he sometimes doesn't make it back to his office from lunch.

Mason is sixty-two, has worked for Omni for twenty-five years, is eligible to retire, but doesn't want to until he's sixty-five. Doakes points out that Mason has done an excellent job until recently and has been very good to those who work for him, especially to Doakes himself, whom Mason has been grooming to take over his job when he retires.

THE TASK

A. Doakes wants the committee to recommend a course of action to Peterson, but because of the sensitive nature of his relationship to Mason, wants to withdraw from the deliberations. He claims that he will go along with anything you recommend to Peterson, who is a no-nonsense person and who is well respected for his efficiency and fairness.

B. In this case, you have two audiences to analyze and write for. Your prime audience will be Peterson, because he wants your recommendations and will, in part, base a decision on them. But you have a secondary audience in Doakes, who is both the chair of the committee and your colleague, who has entrusted you with recommending a solution.

C. Specifically, you are to do two things:
   i. Analyze your audiences and the situation according to the considerations and methods discussed in this chapter.
   ii. Write a memo to Peterson recommending a plan of action to deal with Mason. Remember that Doakes will review the memo before he sends it along.

Some Tips

Although your ultimate recommendations should be your own (and you will have to invent some of the details to round them out), here are some possibilities to start you thinking:

i. Force Mason to retire in December. Doing this, of course, would reduce his pension, but it will save the company money and would probably elevate Doakes to Mason's position.

ii. Force Mason to retire in December, but pay him his pension as if he had retired at age sixty-five.

iii. Reorganize the regional pecking order so that Doakes takes over many of Mason's responsibilities until he retires at age sixty-five. Raise Doakes's salary; keep Mason's as it is.

iv. Make Mason a consultant to the company at a reduced salary, but pay him his normal pension when he reaches age sixty-five. Elevate Doakes to Mason's position.

Note: Firing Mason outright is not an option.

TERM PROJECT OPTION

Assuming that you have chosen a problem from your field to work on, researched it, and written an annotated bibliography, you are now ready to interview an expert to help you put your research into perspective and to clarify any questions you might have about your project.

First, conduct your interview, having prepared good, interesting questions, not basic questions you should be able to answer yourself from the research. Then write up the essence of your interview for your instructor. This is a basic rhetorical problem, because your instructor was not at the interview and will have to rely totally on your report for a sense of what was said and why it is important. Your report on the interview should cover the following topics:

1. Context of the interview
2. Introduction to the interviewee (who the person is, his or her position and qualifications as an expert)
3. Interview goal
4. Conclusion about the effectiveness of the interview
5. Summary of the three most important questions asked and their answers
6. Attitude of the interviewee
7. Plans for using the interview material

You should end up with about a page and a half of typed or word processed material. This material can then constitute another source for your proposal and your final written project.

Solving Problems Through Advanced Research

CASE STUDY

PATRICIA WYNJENEK'S OFFICE

Department of Anthropology ▪ Hastings University
Las Palmas, New Mexico

Patricia Wynjenek found herself thinking again about Indigo, Texas, a place she was first introduced to a little more than thirty years before when she was a VISTA volunteer, working on problems of economic development among the substantial poverty community. The city, which is on the border between the United States and Mexico, was at the time 93 percent Mexican American, politically and commercially run by Mexican Americans (not Anglos, as was the case with most other border cities of its size), and had the dubious distinction of having one of the lowest per capita incomes in the country. Patricia spent fourteen months there at that time, got to know numerous residents of the poor barrios, and helped establish a cooperative grocery store that was intended to encourage community economic empowerment.

PREVIEW OF THE CHAPTER

After Patricia left Indigo, she went back to school for advanced degrees in anthropology that required field work at one point in sociolinguistics. She returned to Indigo some six years later to study the effects of English on Spanish that was spoken essentially universally in Indigo. For most of the people in the poorer barrio, Spanish was not only their first language but also their only language. This circumstance was especially true of older residents, whereas their children and grandchildren learned and spoke English in school. Even though English was the language of school and often of the marketplace, Spanish remained the language of the home. Patricia's research question was, In what ways had English affected the structures and vocabulary of Spanish as it was spoken by the residents of Indigo? Through an ethnographic study involving recorded interviews and recordings of meetings conducted in Spanish, Patricia was able to determine that indeed there were some significant differences between the Spanish spoken in Indigo and that spoken in the interior of Mexico that could be explained only by the influences of English in this American border city.

Although Patricia had visited Indigo a couple of times after her linguistics research, she did so to renew friendships and to cement contacts for a time when she might do more research there. That time had come. In the almost fifteen years since she had been in Indigo, things had changed significantly. Since the passage of the North American Free Trade Agreement (NAFTA), Indigo had become one of the busiest commercial ports of entry to and from Mexico. This change brought tremendous growth in many sectors of the economy, in development, both commercial and housing, and of course in population. But the gap between the rich and the poor only widened from the days that Patricia first knew Indigo. A significant part of the population still remained below the national poverty level and yet, remarkably enough, continued to maintain a stability of community unknown in most other urban centers in the country. Patricia decided that this phenomenon was worthy of study and thus for both a sabbatical proposal to Hastings University and proposals to various foun-

dations for financial support, she formulated a research question: which strategies for economic survival do the residents of the poor barrios employ that enable them to maintain strong families and to remain viable as a community? Patricia's proposed research plan and techniques are explained in some detail in the discussion that follows.

INTRODUCTION TO SOME ESSENTIALS OF RESEARCH

Although we have been talking about research of some kind or other throughout our discussions, we thought that a separate discussion would highlight some of the primary features and answer some common questions about doing research. The first thing to remember is that research comes in all shapes and sizes and is absolutely fundamental to solving problems in the professional world. The problem can be as relatively minor as deciding on a different floor plan for an office that appears to be having difficulties routing information efficiently, to a large-scale problem such as a national retailer's attempts to regain lost portions of market share in retail trade. In Patricia Wynjenek's case, the problem is not necessarily a question of what needs to be fixed but rather is a question of what cultural practice allows for the economic survival and family stability in the poor barrios of Indigo.

All problems will require first a clear grasp of the nature of the problem and then, in those cases that call for some kind of solution, at the very least a review of past practices, in light of what is presently being done, should be performed. This in most cases will suggest a solution to the problem that, if well implemented, will bring about productive changes in the future. In Patricia's case, since a solution is not being sought, and past practices are more than likely to be very similar to present practices, the focus is on those strategies for economic survival and encouraging family and community stability that are happening presently. Understanding cultural practices in Indigo may lead Patricia to suggest changes in social programs for the people of Indigo's poor barrios or for other communities that are similar demographically and culturally. Following are the reviews and possible directions that researchers of these problems might take:

The Need for a Different Floor Plan

1. A search through internal files on office procedures, original plans for the use of space within the office, documentation of changes in and acquisition of machinery, especially those employing new technologies.
2. Interviews with long-term employees about how the office had been arranged in the past and whether changes over time have actually improved routing of information. Interviews could also be done of all

employees to see what troubles they have with the present arrangement of space. If there are many employees, a short questionnaire in place of interviews could be useful in gaining information.

3. A turn to external sources, such as libraries and the Internet, for information on the effects that arrangements of work stations, machinery, location of private offices, and the like have on the productivity of employees. By considering studies in professional journals, researchers can adapt the findings to their own situations, which can help significantly in making decisions for change.

The Need to Increase Retail Market Share

4. A thorough analysis of financial statements of past fiscal years, especially those years just before the decrease in market share began. This will help document the specific kinds of merchandise and geographical areas that have been hurt the most.
5. A separate analysis of plans for expansion, how and to what extent the plans have been carried out, and the resulting effects on profit and loss or cost/benefit ratios.
6. A third analysis of management structures to reveal the ratios of managers to workers, how management relates to workers, and the techniques used to train workers and to encourage and reward productivity. Important information on how management and workers view the established structures can be gained by interviews and surveys, perhaps performed by outside consultants to promote objectivity.
7. A review of external publications on marketing and managing techniques, including discussions on changing national demographics and their effect on the conduct of retail trade. An abundance of books and articles in professional journals offers analysis of all manner of business concerns and practices, and again can be found through libraries and the Internet.

These examples show some of the essentials of doing research:

1. A *clear problem* or *research question* that demonstrates the need for something to be done:
   a. Inefficient routing of information resulting in diminished productivity
   b. A significant decrease in market share resulting in decreased revenues and profits
   c. The need to determine what kinds of economic survival strategies exist in Indigo, Texas, that enable significant family and community stability in the poor barrios of the city
2. A *clear purpose* for doing research that reflects the problem or the research question and that suggests directions to take to collect information on the problem:
   a. Search for a new method of arranging office furniture, machinery, and spaces to bring about greater efficiency of communication and therefore productivity

b. Search for the causes of decreased market share to provide possible changes in marketing and management practices that would reverse the trend

c. Search for an understanding of economic survival strategies in Indigo, Texas ultimately to suggest changes in social programs

3. A *plan* for doing the research. The plan doesn't have to be worked out in complete detail, but researchers should have some reasonable idea of what can be done to collect needed information and how that collection will take place. In the examples we have been considering that call for some kind of solution, we see that both would benefit from reviews of *internal* and *external print sources* and from *interviews* and/or *surveys* of employees (the differences between interviews and surveys). In the case of the diminished market share problem, an ambitious but potentially very useful device would be surveys of customers to see what their buying preferences are and in which retail outlets they can fulfill those preferences.

   Patricia's plan generally would involve two major components: (1) *fieldwork,* also known as *participant observation,* that would call for Patricia to take up residence in one of the poor barrios in Indigo and take as much a part of the life of the community as possible; and (2) *interviews* of a selected number of residents using what is called an *interview schedule* consisting of open-ended questions about numerous features of family and working life.

With this foundation, we should now look at the process of research in more detail. The discussion will include a review of some past discussions and a look at some more recent developments in methods of research.

REVIEW OF KINDS OF RESEARCH

Research is first divided into two generic types: primary and secondary. From there, many varieties of approaches and techniques can be employed, especially under the heading of primary research, but we will reserve discussion of those until later.

Primary Research

Primary research is simply defined as the first-hand collection of data on the research problem at hand. The researcher is responsible for collecting the data, analyzing them, and then coming to some conclusions about or possible solutions to the problem. The definition of primary research hinges on what and how the research is to be done; it is not dependent on the extent to which it is used in any research plan. In the preceding cases, for instance, the interviews or surveys of employees, or in the case of the decline in market share, interviews or surveys of customers are clearly primary methods of research. But the need for a different floor plan might tilt the balance in favor of primary research, since the experiences and opinions of employees are key to determining the problems of the flow of information within the office environment.

Although the interviews or surveys of management and workers on management structures in the market share case are important, reviews and analyses of external documents, which amount to *secondary research*, on the practices of the larger market place would probably take on greater prominence in the research plan. These reviews and analyses allow for comparisons of those firms with greater market share with those with a lesser share to see what practices appear to account for the difference. Primary research in this case would take a lesser role than secondary research.

Secondary Research

As opposed to primary research, secondary research is simply defined as the review of data that have already been collected, analyzed, and synthesized by other researchers to produce some conclusions and perhaps solutions to the research problem at hand. These data and their analyses are most often published in books (monographs), articles in professional journals, government reports, and reports from organizations or firms, both public and private, intended for public consumption, and finally in electronic form on the Internet. Once again, the term secondary does not imply any necessary lesser importance or prominence in any particular research plan. As in the case of the market share problem, researchers would be compelled to study published reports on trends of sales in the retail market, plans of some of the bigger players in the market for expansion or contraction, changes in demographics that might affect future sales, and the like. In fact, the bulk of useful, even crucial, information will be gained from secondary sources, not from primary sources, as would be the case with the office floor plan problem.

VARIETIES OF PRIMARY RESEARCH

Experimental/Empirical/Quantifiable Research

Primary research can be generally divided into categories that reflect the purpose and scope of the research being performed. Studies intended to determine the efficacy of a new drug in a chemotherapy aggregate for treating cancer, for example, would first be tested in a laboratory setting. The tests are referred to as *experimental* and the results are termed *empirical* or *quantifiable*. Researchers must perform a series of tests designed to see how the drug will perform in reducing the number of cancer cells under different conditions, such as the reactions of other drugs in the conglomerate to introduction of the new drug, the level of dosage, and the time of application in the course of the disease. The accumulation of findings will help lead to conclusions about the safety and efficacy of the drug and thus to decisions by such oversight organizations as the Food and Drug Administration (FDA) about whether the drug can be approved for use in treating people with cancer.

Other different situations also can call for quantifiable research. The current national interest in levels of achievement in education is one area that could employ empirical research. An example would be comparing different approaches to teaching reading to elementary schoolchildren: the whole language approach that attempts to immerse students in the cultural context of their reading and to deemphasize drill exercises of recognizing and memorizing phonetic structures versus the more traditional approach that indeed places primary emphasis on teaching and learning phonetic and grammatical structures, although the social contexts of students' reading texts are not ignored. The difference between the two is more a matter of degree of emphasis than it is one of exclusion. But the question before us is, How does this situation lend itself to empirical research?

The answer has to do with the purpose that researchers have for conducting research. If it is to determine the comparative difference in students' reading abilities after having been taught in either one of the approaches, then an empirical study could be set up. The first thing to do would be to establish the size and nature of the study population, perhaps all fourth graders of a school district, and then to divide the population into those who would be learning in the whole language approach and those learning in the traditional approach. The division could present a problem in that numbers of fourth-grade classes may not be evenly divided and the number of teachers capable of and willing to teach in either approach again may not be evenly divided. Accounting for these differences may reduce the size of the study's population, but once it is determined, designating functions of the groups follows easily. Since the whole language approach is newer and less well tried, students learning in it would become the *experimental group;* students learning in the traditional approach would become the *control group.*

The job then becomes a matter of comparing the results of testing reading abilities of both groups in such areas as comprehension, fluency (speed), and vocabulary. This type of comparison will attempt to show the differences between the experimental and control groups, but since the results are subject to statistical assessment, it is entirely possible that certain measures, such as comprehension, will show no statistically significant difference. In these cases, researchers will either have to accept that both approaches in the area of comprehension are equally effective or may decide to look more closely at the method for measuring comprehension to see whether they have overlooked some flaw in the method. Memory of what someone has read is not necessarily the same as understanding what someone has read, and yet the questions asked about the reading may be constructed in such a way that tests memory primarily, not comprehension.

The comparison of the experimental groups and the control groups can also include other variables that can be revealing. The differences in the responses to questions about reading according to gender, for example, can suggest further differences in learning styles, maturation levels, or cognitive development between fourth-grade girls and boys. Although it would require more involved investigation, the socioeconomic backgrounds of students could be considered in comparing results of testing, but the testing instrument itself would become unwieldy if it sought that kind of information in addition to the basic information on the success of the whole language approach versus the traditional approach. Demographic information would have to be collected separately from other sources, such as student records that are usually updated at the beginning of each academic year.

Nonexperimental/Qualifiable Research

The basic difference between experimental and nonexperimental research is that in nonexperimental studies, there are no variables that need special attention, manipulation, or treatment. In terms of the earlier discussion, there are no new drugs to be observed under varying conditions and no new paradigms in curricula to be tried, tested, and compared with existing paradigms. In nonexperimental studies, researchers observe existing conditions in order to come to some conclusions about how the conditions came about, how durable they are, whether they are on the verge of change, what kinds of forces may bring about change, and so on. One of the most common tools of nonexperimental research is the *interview schedule*, which is a set of open-ended questions administered by interviewers to selected respondents, designed to determine respondents' attitudes, beliefs, and behaviors as held and practiced in the course of life in a family and community. This is where Patricia Wynjenek and her interest in the people of the poor barrios of Indigo, Texas, come in.

The Case Of Patricia Wynjenek

In attempting to define meaningfully the economic survival strategies of the poor residents of Indigo, Patricia actually uses two tools to collect the data necessary to her analysis and eventual conclusions: one tool is the *interview schedule*, but at the same time, she is using *participant observation*, another tool utilized particularly by anthropologists.

Participant observation calls for Patricia to take up residence in a barrio of Indigo that she wants particularly to study. She will find housing typical of the barrio, will shop as far as possible in local stores, and otherwise will take part in the life of the community, attending parties, baptisms, weddings, and funerals of the people she comes to know. She will also attend meetings of the community on civic issues, where much can be learned about the nature of debate and compromise in cultural context of the poor barrios in Indigo. These meetings of the city at large will confront issues of zoning, housing, commercial development, education, crime, and all manner of concerns that confront cities like Indigo that are experiencing tremendous growth in population.

It is important to realize that Patricia's research plan must include participant observation. Although it is not so immediate or intimate as the interview schedule that is administered on a one-to-one basis within the home environment, Patricia's participation in the life of the community on neighborhood and citywide levels is essential to provide a full picture of the cultural realities of the smaller area and population she wants to study.

The *interview schedule* is the other essential partner in Patricia's marriage of research tools. Patricia's proposed construction of the interview schedule included several steps:

1. **Choosing Respondents.** A select group of respondents is also known as a sample, a smaller number of the whole population whose responses will serve as representative of the whole. If a researcher wants to gain

responses from an inclusive population, all members of the group will be surveyed, and the sample becomes a *census.* Since questioning an entire population, unless it is very small, is both time-consuming and expensive, most researchers plan on other sampling techniques to reduce the size of the group without losing the reliability of the study. Following are the more widely used techniques:

a. *Stratified Sample.* Using a stratified sample often depends on the nature of the research problem or research question. If at election time, a researcher wants to find out what new voters think about the issues and about the impact that new voters may have on the election, he or she could limit the sample to all people within a chosen population between ages eighteen and twenty-two. The effect of stratifying is to form the sample with people with the same characteristics, thus reducing the possible number of respondents to some fraction of the whole population and refining the focus of questions and answers.

   The same could be done with people with the same occupation, for example, all lawyers within the whole population or a more refined sample of all lawyers with ten or more years in practice. The possibilities for stratification are almost limitless: people with the same religion, ethnicity, income range, marital status, and so on. But the choice shouldn't be arbitrary; it should reflect significant elements of the research problem or research question.

b. *Stratified Random Sample.* Randomness in sampling should be defined first, although it is easier to define than it is sometimes to achieve. Constructing a random sample means that everyone within a certain population has an equal chance to be chosen for responses. The ways of attempting randomness are discussed in the next section.

   A stratified random sample allows researchers to control the size of the sample more completely. A stratified sample reduces the size of the sample because it includes only respondents with certain characteristics, but the size of the sample could still be larger than the researchers want to deal with. By deciding ahead of time the optimum size of the sample, researchers can then choose the respondents randomly within the limits of size and stratification.

c. *Random Sample.* Achieving true randomness in constructing a sample of respondents is difficult. Some suggest that choosing names arbitrarily from a telephone book would provide everyone in the population an equal chance of being chosen, but that is the case only if it is agreed that the population should be made up of those who have telephones or who have telephones and want their numbers listed. Those who don't have telephones or who have unlisted numbers will have no chance of being chosen. The same would be true of any list made up of names of people with some kind of exclusive characteristic: choosing randomly from a voter registration list or a list of property owners, for example, would clearly exclude those who are not registered to vote or do not own property.

Randomness can be achieved if numbers are arbitrarily assigned to every individual in a population and then the numbers are thoroughly mixed and drawn to the limit of the desired size of the sample. The old numbers in a hat, box, or bushel basket method has been replaced by the use of computer software that will randomly mix and match names and numbers; the only remaining problem is deciding the source(s) from which names of prospective respondents will be chosen.

d. *Patricia Wynjenek's Method of Choosing Respondents.* Patricia decided to reduce the problem of choosing respondents by leaving names out of the mix. She would choose one particular barrio that she had had experience with in the past, rent a small detached apartment, and decide to interview only the residents of that barrio. She proposed to map the barrio and record the house numbers of all houses in the area. Since there are 480 houses and she would need to restrict the number of interviews to 100 or so to save time and money, she proposed to choose 100 house numbers at random and interview the residents of those houses. If no one answers the door at one of the houses chosen, then she would choose the house next door or the one next door to that until she finds someone home. This method amounts to a stratified random sample. It is stratified by virtue of the fact that anyone chosen has to live in a house within a particular limited area, and it is random since all houses in the area have an equal chance of being chosen for the sample of 100 houses.

2. **Choosing Interviewers.** Realizing that it would be beneficial to the interview process to have residents of the research barrio conduct interviews, Patricia proposed to hire three women to conduct three-quarters of the interviews. She would hire women because the delicate nature of some questions (those having to do with birth control practices, for instance) would better be asked by women than by men. And since Patricia knew that most of her respondents would be women, it made good cultural sense that women would be more likely to be candid and forthcoming to other women. But it should be noted that Patricia would reserve a quarter of the interviews for her to perform. She decided on this step for two reasons: she would need to keep within the proposed budget for interviews, but of more importance she would want to demonstrate that she has legitimate reasons as an educated Anglo woman to become a member of the community that she would be living in. It is very important in doing anthropological fieldwork that researchers gain the acceptance and trust of the members of the community, and taking part in interviewing is one good way of doing that.

   One other important feature of interviewing is payment. Patricia proposed to pay the interviewers $10 per interview, which ultimately amounted to about $10 an hour, a respectable wage, not exploitive, but not extraordinary either. She also proposed to pay interviewees $10 for their interviews, which recognizes that they should be paid for their time and the intrusion into their homes, but at the same time the sum itself is not great enough to foster envy among neighbors who had not been chosen for an interview. If such envy were to surface, it would do so quickly, early in the process, and could be very damaging to the community and especially to the chosen respondents. In Mexican-American culture, to be the subject of envy can be damning; the possibilities of being shunned socially or experiencing physical maladies rise in the minds of people who are the potential subjects of envy.

3. **Creating the Interview Schedule or Questionnaire.** Before looking at Patricia's interview schedule in particular, we should look at some of the common characteristics of interview schedules and the differences between them and other survey questionnaires. A primary difference occurs in the construction of the questions as either *close-ended* (or forced response) and *open-ended*. Close-ended questions require respondents to choose one answer of several possibilities that best fits how they feel about a subject or the experience they have had in the realm of the subject. A couple of examples could come from a questionnaire on alcohol use on college campuses. After establishing that the respondent does in fact drink alcohol, this instruction might be given:

   Check below one answer that best describes the frequency of your drinking:

   _____ Once a week
   _____ Twice a week
   _____ Three to six times a week
   _____ Every day

   After establishing the frequency of drinking, the researcher could want to know the places that the respondent drinks by giving the following instruction:

   Check below one answer that best describes the place you drink most often:

   _____ At home (dormitory room or apartment)
   _____ At parties
   _____ At bars

   This question forces the respondent to decide on what sometimes may be a fine distinction, forcing an answer that people may not feel comfortable in making. Since the researcher is seeking as much information as possible and not trying to restrain the respondent unnecessarily, the following possible answers, which might come closer to the respondent's experience, could be added:

   _____ Evenly divided between home and parties
   _____ Evenly divided between home and bars
   _____ Evenly divided between parties and bars

   Clearly, there are many possibilities for questions that could or should be asked about drinking habits in a close-ended format, such as the number and kind of drinks consumed on any one occasion or series of occasions; legal problems that have arisen (minor in possession, driving violations, malicious mischief, disorderly conduct, to name a few); negative effects on academic achievement; and alienation of family and/or friends. But researchers need to be sure that answers to all questions asked will actually add significant information. Yes or no questions are of limited value if there is no follow-up question to specify what the answer means. If a respondent is asked, "Have you had legal problems brought about by your use of alcohol?" Yes ____ No ____, a yes answer is of little use unless the kinds of possible legal problems are also known. A follow-up checklist of possible legal problems is essential to adding significant information.

Advantages and Disadvantages of Close-Ended Questionnaires. The *advantages* of using close-ended questionnaires are several:

1. *Ease of administration*—If the questionnaire is mailed or administered by telephone from some central location, the number of personnel needed for administration can be significantly reduced. This method results in a savings in costs.
2. *Ease of participation*—If the questionnaire is logically and efficiently constructed, that is, questions build sensibly on each other, offer clear and limited choices for answers, and do not ask for open-ended, written, or oral responses, respondents can complete the questionnaire in a short time. This brevity is especially necessary for telephone questionnaires, since most people don't like to spend a lot of time answering unsolicited calls (see later for an expansion of this idea).
3. *Ease of assessment*—Most close-ended questionnaires can be subjected to electronic reading. Computer software programs can count the number of particular responses to questions and perform whatever statistical test of reliability the researchers need to perform to meet the purposes of the research. Since small amounts of time are required on the computer and only a few personnel are required to run the program, costs of assessment are reduced significantly.

The *disadvantages* of using close-ended questionnaires follow:

1. Mail questionnaires present a couple of sampling problems. As Sproull (1995) points out in *Handbook of Research Methods,* if researchers use mailing lists bought from other organizations that do direct mailing, they may lose "the precision required in defining a research population. . . . It is hazardous to use mailing lists if the researcher wishes to generalize research results to a specified population" (196). If preexisting mailing lists are used, at the very least researchers should know the criteria and circumstances under which the names were chosen and the list was constructed. The closer the criteria come to those of the desired research population, the more likely it is that the lists will be useful.

   Mail questionnaires also are notorious for having low response rates. Consider how many times you have received unsolicited questionnaires and have ignored them, forgotten them, or thrown them away. If you multiply your actions by a factor of six, you will arrive at the rough percentage of questionnaires that are not returned: "some researchers consider a 35% to 40% response rate as acceptable" (Sproull:193). This likelihood suggests two actions that need to be taken: (1) Mail out enough questionnaires so that a return of 35 to 40 percent will meet the desired research sample size (N). If, for example, researchers are looking for a total N of 100 (returned and completed questionnaires) in the sample, they will have to mail out at least 250 questionnaires. (2) Attempt to increase the response rate through the following selected methods (Sproull: 195):

a. In the cover letter that is required to solicit willing participation, appeal to respondents' affiliations, such as, "Since you are a graduate of XYZ University, we think you will want to __________." Of course this method will work easily only if the sample is stratified and restricted to members of a certain group. All cover letters, however, need to explain the purpose of the research, and need to appeal to respondents' best intentions to be of help. Also, mention that the questionnaire will take only a short time to complete.

b. Be sure to write clear directions. There should be no ambiguity about what respondents need to do and how they should do it.

c. Construct individual choices for response so that they can be chosen easily and quickly. Also, construct the entire questionnaire so that the placement of questions is logical and easy to follow.

d. Avoid mixing close-ended and open-ended questions. Since close-ended questionnaires are generally easy to complete, respondents are likely to avoid answering open-ended questions, which require both a different kind of thinking and time to write. You should be the first to avoid open-ended questions, not your respondents.

e. After about ten days, mail reminders to those who have not responded. This mailing can be done inexpensively by postcard, but some effort should be made to be sure that you are not reminding respondents who have already returned completed questionnaires. This treatment is annoying to respondents, and common courtesy, at least, would call for some effort to keep good records of returned questionnaires.

f. If the postcard reminder doesn't work, a second questionnaire can be mailed or a telephone call be made to remind the respondents and to find out whether they have misplaced the questionnaire and need another.

2. Telephone questionnaires also present a sampling problem. As we mentioned before, it is impossible to get a true random sample (if one is being sought) by using telephone books or other telephone listings because they do not account for people who do not have telephones or whose numbers are unlisted. Again Sproull helps us out here: "People who do not own phones . . . tend to differ from those people who are listed in the phone book on a number of significant variables such as income level, type of housing and degree of mobility" (196). So if some degree of diversity in the makeup of the sample is being sought, using telephone listings is limiting.

The greatest disadvantage of telephone questionnaires is the growing unwillingness of people to respond to unsolicited calls. The incredible growth of telemarketing in the last decade has had a negative effect on many people who are often deluged with unwanted sales calls, so that they refuse to answer their phones, let answering machines take over, simply dismiss the caller with "No thanks, I'm not interested," or more abruptly with a hang-up. For researchers who are not selling anything and who believe fully in the legitimacy of their work, this experience can be very frustrating. At the very least, it means that they must make two to three times more calls than the desired N just to ensure that the N can be formed. We have no suggestions for a solution to the problem except perhaps for the caller to say as early as possible that she or he is not making a sales call. But with the growing impatience and cynicism on the part of those who are called, it is not at all certain that such early disclaimers will work.

Patricia Wynjenek's Interview Schedule. We have been spending a lot of time on the close-ended questionnaire largely because it is so widely used by so many professionals: sociologists, government officials, economists, businesspeople. But this focus ignores another important primary research tool, the *open-ended* questionnaire, or *interview schedule,* which is frequently used by anthropologists like Patricia and has gained popularity in other academic areas interested in ethnographic research, such as schools of education and departments of writing and rhetoric.

The interview schedule relies heavily on the respondents' willingness and ability to write or to speak in answer to the questions posed to them. That factor means in turn that the researcher must be careful to construct the questions so as to avoid ambiguity and to encourage answers that are forthcoming and truly useful. It also means that interview schedules take longer to administer, and in Patricia's case, as we have seen, require interviewers, who need to be trained and paid for their time in training and in interviewing. Also, Patricia's decision to tape-record the interviews, which would relieve both interviewers and interviewees of having to write (frequently an awkward, if not indeed an impossible, task for many residents of Indigo) would mean that tape recorders and tapes would have to be supplied and perhaps a person hired to transcribe the tapes. All of these factors add costs to the research, which generally make the interview schedule (interviewee for interviewee) more expensive to administer than the close-ended questionnaire. Given the amount of funding available, additional costs such as these can be seen as a distinct disadvantage.

We need to point out, however, that it is not our purpose to make qualitative, comparative judgments about the close-ended and open-ended methods of surveying. Both methods can provide information not collected and analyzed before that can help researchers to further the purposes and goals of their research. Researchers who use the close-ended questionnaire are generally looking to collect large amounts of data from a large sample, which will meet tests of statistical significance. Those who use the open-ended interview schedule are generally interested in smaller samples and in a limited range of data that reveals more of the texture of the lives and experiences of the respondents. Researchers may indeed want to quantify some data, but they tend to do so in percentages of the sample and not be concerned with statistical significance. Instead, they look to the innuendoes and embellishments of respondents' responses to increase their knowledge of the respondents' lives.

Looking more closely at Patricia's case, we see that she proposed a set of forty-five questions that would deal with employment, including family income, education, medical care, relations with the community, and family life. Numerous questions would ask for short answers that need no particular embellishment, such as, "How many people live in your house?" "How are they related to you?" These can be answered quite straightforwardly with perhaps a minimum of explanation. But those questions that deal with issues other than basic demographics, such as how the family's income is constituted (who works, for how long, at what wage) may need some follow-up and expansion, especially if the respondents indicate that at some point, they had to leave Indigo for work (because of a long history of high unemployment) but then returned. Interviewers would want to know where the respondents went to find work, how long they were away, how much money they made, and why they decided to return to Indigo. These are all questions that could naturally grow out of some basic questions about employment but that could not be easily anticipated or accounted for in a close-ended format.

Another interesting area of query is family life, especially birth control practices. Before proposing her work in Indigo, Patricia had to look into some of the demographic data on birth rates provided by the U.S. Bureau of the Census and by state and local social agencies and hospitals. She found that even in the traditional and largely Catholic culture of Indigo, the birthrate nevertheless had declined significantly in the last twenty years. This decline could be explained only by the use of some kind of birth control, so Patricia decided to ask about it in this way: "Do you think it is a good idea to limit the size of a family or to have as many children as come naturally?" In neutralizing the question by replacing "to limit the size of your family" with "to limit the size of a family," Patricia would be more apt to receive candid answers. Respondents would feel less constrained by the more general question, since it doesn't suggest any necessary answer about personal practices. Once again, this flexibility and chance for greater expansion in response makes the open-ended format a better instrument than the close-ended format.

Some discussion about the area of medical care will also show some of the advantages of the open-ended interview schedule. A significant number of the population of Indigo, as is true of many cities along the Mexican border, mix their calls for medical care between the scientific medical establishment and traditional (or folk) practitioners, known as *curanderas*. These practitioners can be found in some numbers across the border in the neighboring Mexican city of Pollo Blanco. Although Patricia had heard of and knew something about the mix of uses of medical care, she did not know specifically under what circumstances people would choose one over another. She suspected that a primary reason for choosing traditional care is its relative low cost; for people who by and large have no health insurance, the cost of the medical establishment's treatment is prohibitive. But in order to be sure that she would get representative responses about the use of traditional care, Patricia proposed a series of questions that would probe the subject more thoroughly. The open-ended format, once again, allows for this kind of probing more easily; anecdotal responses encourage both general conclusions and individual insights into important subjects and behaviors regarding health care.

A Word About Self-Reporting. Both close-ended questionnaires and open-ended interview schedules rely on respondents to report on their own beliefs, experiences, and/or behaviors. The practice is called self-reporting and is always held with some degree of suspicion as being unreliable. Theory and practice come together here: no matter what, some respondents will lie about beliefs, experiences, and/or behaviors because they do not want to admit to something that may be viewed as wrong, antisocial, dangerous, and/or illegal in the eyes of the researchers or indeed in the eyes of the larger society. Some researchers claim that they can nullify the lying effect by constructing questions that ask about the same basic issues but are reworded in such a way that inconsistencies in responses will show up. But even this attempt can be foiled by people who want to foil the attempt and who have enough experience and information to know how to do it. Nothing can be done about such respondents, even if they can be identified, and researchers have come to accept that there will always be some degree of unreliability. The more that respondents can be convinced that they will remain anonymous and that no one will attempt to connect their responses to them personally, the greater the possibilities for honesty and candor.

Another feature of self-reporting that puts the lying factor in a different light is respondent interpretation. If we take the earlier example of questions about the frequency of drinking on college campuses, respondents' interpretations of what constitutes a drinking event may differ significantly from one to another with no intention to deceive. If someone responds that she or he drinks only twice a week, for example, Friday and Saturday nights, the respondent may be thinking that multiple drinks, three or four or more, are necessary to constitute a true drinking event. The occasional drink, one or two, on other nights of the week, don't actually amount to a drinking event to this way of thinking. The problem, of course, is how differences in interpretation can be determined. In the close-ended format, additional questions could be asked about drinking events, but it might take numerous questions: Is a drinking event defined by the environment? Do drinking events happen only at parties or in bars and not at home when you are alone? Is a drinking event defined by the quantity of drinks consumed? How many drinks constitute a drinking event? A combination of yes or no responses and the forced choice of one of several alternative responses could eventually get to an understanding of the respondent's interpretation, but the additional questions required may make the questionnaire uncomfortably long. On the other hand, the open-ended interview schedule could handle this matter of interpretation with one question that asks the respondent to describe her or his drinking habits, including times, places, and quantities.

But once again, researchers' intentions and purposes for conducting the research will dictate the method to be used. The thing to remember ultimately is that self-reporting, like any other human activity, is subject to the vagaries of deception and interpretation.

VARIETIES OF SECONDARY RESEARCH

It used to be that secondary research had no varieties. Researchers understood it to require a certain series of steps, but all were seen as parts of one process. Before the advent of electronic search programs, researchers would consult print indexes, such as *Social Science Abstract,* under topic headings where listings of publications would open up the possibilities of finding out what other researchers were thinking about, analyzing, and writing about with regard to the research problem at hand. The sources that researchers found (monographs, journal articles, reports from government and industry) not only provided reviews and analyses of research but also provided additional sources to be investigated through their citations and bibliographies.

Assembling a body of information was in itself not difficult, although it was time-consuming, but the more important chore was deciding on the usefulness and credibility of the sources. Given the area of investigation, the recency of secondary sources can be very important. In certain fields of science, engineering, and certainly technology, changes and progress happen so rapidly that it becomes imperative for researchers to consult the most current publications. This necessity does not mean, however, that publications gain credibility simply by their currency. It is also necessary that the credentials of the authors of publications be reviewed. Their reputations for responsible, intelligent, and useful research are essential to the place in the research community they are a part of. These measures to assess the value of secondary sources have remained essential to the productive conduct of research, but with the advent of the Internet and the Web, new pressures to assess sources adequately have arisen. The sheer burgeoning volume of Web sites on literally thousands of topics of inquiry has made the assessment of the quality of those sites more difficult, especially for beginning researchers. We will return to this problem later in this discussion.

Searching Libraries On-line

The days of the library card catalog are over. No more standing in the midst of those monolithic structures with hundreds of small drawers containing thousands of small cards with the information necessary to find the source profiled. Today, virtually all libraries of any size have converted to on-line catalogs that can be accessed easily from multiple computer units in the library or from your office or home through various remote options: dial-in connections, telnet connections, or on the Web (Metter, 1999). The once laborious chore of searching the card catalog for information through author, title, or subject headings can now be reduced dramatically by asking the on-line system to search through holdings that contain or deal with certain key concepts that are approached through *key words*. We will discuss later this remarkable timesaving device in some detail.

On-line catalogs also provide additional search possibilities beyond the author, title, and subject typical of the card catalog. These possibilities include "limiting your search by date; confining your search to a particular material format, like maps or videos; finding materials in a foreign language; searching for resources from a particular publisher; and combining the author's name and subject words to pull up information" (23). Besides providing the standard information for a book, for example, of call number, title, author, publisher, place and date of publication, number of pages, presence and number of illustrations, summary or synopsis of the topics covered (more often now by chapter titles), or a listing of key words, and so on, many catalog systems will also tell searchers whether the item is in the library or has been checked out, and if it has, the date it's due back. The system will also tell searchers that a book is missing, not on the shelves but not checked out either. This information clearly saves time, and searchers can initiate a return call if the book is checked out, or a search if it is missing, although usually a search is initiated as soon as it is discovered that an item is missing.

Boolean Searching. Named for George Boole, an English mathematician, Boolean logic is a system of pairing of concepts to gain either inclusive or exclusive lists of sources that deal with a particular research topic. The three "logical operators" in the Boolean system are AND, OR, and NOT. The OR operator increases the number of documents that contain two terms of interest; the AND and NOT operators reduce the number of documents that contain the two terms. Using Patricia Wynjenek's research problem as an example, the Boolean operators work in the following way (Henniger, 1999):

Border cities **OR** Mexican-American

This operator increases the number of documents since it calls for either or both terms to be present in the document. Following is a graphic representation of this OR operator:

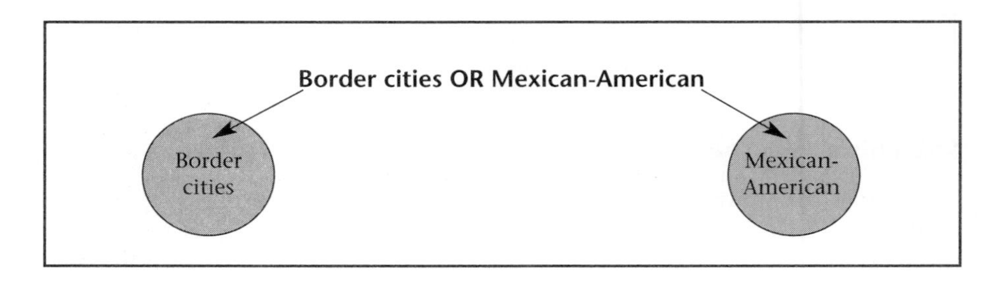

Border cities **AND** Mexican-American

This call reduces the number of documents, since both terms have to be present in the document. It is no longer an either-or situation that clearly raises the possible number of documents. Following is a graphic representation of the AND operator:

Border cities **NOT** Mexican-American

It is most unlikely that Patricia would use the NOT operator in this way, since only documents dealing with Canadian border cities would be called up. She could call border cities NOT Canadian, but that search would simply duplicate the border cities AND Mexican-American call.

Since Patricia is most interested in reading what others have to say about problems of poverty in Mexican-American border cities, her search using Boolean operators could be more usefully focused to produce the following:

(border cities and Mexican-American) **AND** poverty

The parenthetical item would be searched first and then poverty would be added for a list of documents that deal with poverty in the cities along the Mexican-American border, but this search would also include Mexican cities on the border, not just American cities. This search could be altered to exclude Mexican cities, if Patricia wants to, by calling up this combination:

U.S. Mexican-American border cities **AND** poverty

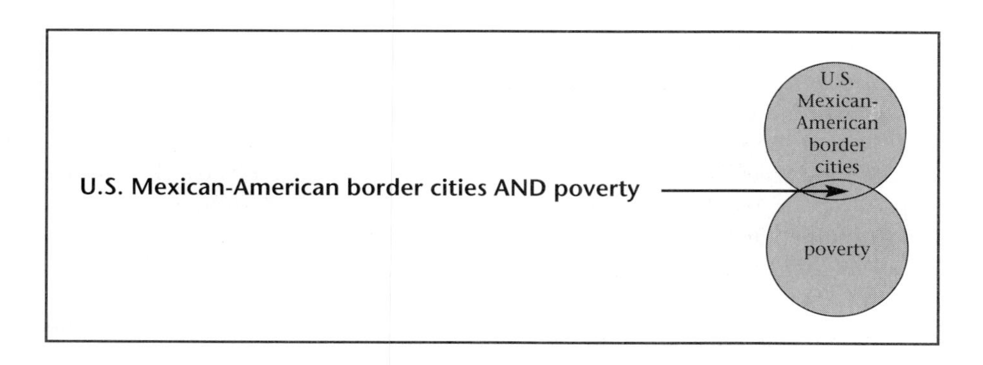

To sum up Boolean searching, Metter (1999) provides an interesting note and a bit of advice:

> A note concerning AND, OR, and NOT: Unfortunately, there's no widespread standard insisting that all publishers of search software use the same Boolean search terms. Thus, any number of commands will perform the operations [described previously]. I've seen the term ANY used for OR, and the symbol + for AND. As long as you understand the Boolean searching concepts, it doesn't really matter what they're called: you'll be able to do a precise search for what you need. Read the instructions for the system you're using. (26)

Keyword Searching. Keyword searching can be largely divided into two methods: free searching of a full text index or controlled vocabulary searching of a subject index. As Henninger (1999) points out, in the full text environment,

> Every word of the document becomes an index term and thus is searchable. . . . For the searcher, however, there is no indication of synonymous or equivalent relationships between terms: for example between *cats, tigers* or *felines.* Nor are there any contextual relationships (semantic or syntactic associations) embedded in the index: for example "blind Venetian" is not distinguishable from "Venetian blind." The searcher can only rely on the presence (or absence) of the requested word in the document. Searching this type of index can result in a lack of documents because the index terms are too precise, or too many documents, many of which are irrelevant. (35)

The controlled vocabulary environment is quite different from the full text environment. Whereas full text indexing is generated by computer programs and is thus inexpensive to create and maintain, controlled vocabulary indexing, which produces specific headings, requires human indexers, thus adding significantly to costs, and also "constrains the number of documents indexed and the depth of that indexing" (Henninger, 1999: 36). The advantages, however, are that "the context of the index terms is defined, providing semantic associations, and synonymous relationships can be established by the use of a thesaurus. . . ." (36). The important thing to remember in using controlled vocabulary searches is that although highly relevant documents can be found, the searcher must know the contents of the database and the controlled vocabulary (essentially the possibilities of the thesaurus); otherwise, the search may turn up disappointingly few documents (36). This situation can be helped significantly, however, if the searcher pays special attention to the portion of records of the documents that lists subject headings or descriptors. Metter (1999) offers a useful example of how this simple step can help expand a controlled vocabulary, subject classification search. On the subject of quitting smoking, she asks for a search under the keyword SMOKING QUITTING but gets only one record for a book called *The Enlightened Smokers Guide to Quitting* (Rockport, MA: Element, 1998). A check of the descriptors section of the record shows four phrases that describe the book: SMOKING—CESSATION PROGRAMS, CIGARETTE HABIT, TOBACCO HABIT, and MIND AND BODY. This is an example of referring to terms from a standardized list of terms called the Library of Congress Subject Headings, or the LC Headings. One can use the LC Headings under the general heading of SMOKING, but a useful shortcut is using any one of the previous subject headings to yield additional sources that in turn will suggest other headings (28). Eventually, the subject headings will begin to repeat themselves, but by then a fairly rich list of sources will have been generated. Also, it's good to remember that most of the sources will provide bibliographies or lists of works cited that can help in expanding a search.

Before we leave the subject of on-line searching, Metter provides two more useful tips: "When searching online, it's often wiser to use just a few keywords. Putting in too many keywords may inadvertently lose relevant records as the system tries to find records that match every word you've submitted" (28). Another easy method of searching on-line is by call number. If in a search, one has already found a potentially very useful source, and knowing that books on similar topics are shelved in close proximity, "by searching by the call number of the promising book, the system lists all books that show up in order before and after that book, in effect, allowing you to browse the shelves of a distant library on your screen" (26).

Searching the Internet

The Internet (the "Net") and its precocious child, the World Wide Web (the "Web"), are so widely known and written about they need no introduction. When our children return from school to tutor their parents on how to use the Net or to set up a Web page and site, we get a graphic idea of the extent to which this technology has become a part of our lives. In fact, we turn to that palpable experience to get us closer to understanding the abstractions about its size presented by those who study the Web. Although the figures change almost daily, recent estimates put the number of documents on the Web at over 350 million (Henninger: 49). That figure would mean slightly fewer than 1.5 documents for every woman, man, and child in the United States. For the researcher, the problem then becomes how to determine which documents are truly relevant to someone's research problem and the searches that must proceed from it.

Before we go into some of the details of Web searching, we need to acknowledge that we rely heavily for this discussion on two very helpful works: Maureen Henninger's *Don't Just Surf: Effective Research Strategies for the Net*, Second Edition (University of South Wales Press, 1999) and Ellen Metter's *Facts in a Flash: A Research Guide for Writers* (Writer's Digest Books, 1999). Henninger and Metter offer much good, approachable, and readable information that helps to dispel some of the myths that have grown up since the advent of the Web.

One such myth, especially for beginning researchers and users of the Web, is that the Web is the be-all and end-all of research resources. This is an easy and understandable mistake to make, since the sheer volume of documents is so impressive, even unimaginable to some of us. With all of this material available, people ask, How could anything else be necessary? But Henninger points out that "to select wisely you have to know what is contained in the information source, how it is arranged and how it can be efficiently accessed. You should treat the Web as just another information source. This means going through the same process of selection you normally do when you use a reference book, a CD-ROM or an online database via a vendor such as ***Dialog*** or ***Ovid Technologies***" (48).

As a librarian, Henninger is understandably interested primarily in providing information on how to search intelligently and efficiently for other useful information. She rightfully points to the Web as a tremendously rich resource but also implies that it can be used in ways that are inspired by a false sense of security, the idea that with the Web as a reference tool, one can stop searching; or worse than that, with the millions of documents that get published on the Web, one need never again crack the covers of a book or a periodic issue of a professional journal. But Henninger reminds us, "The Web is a communication tool and as such may be used by anyone to 'publish' a document or opinion. Therefore many of the sources are not authoritative or lack clues to their authenticity" (49). An antidote to this problem can often be the use of Web sites of organizations known to have comprehensive and authoritative data, such as the U.S. Center for Disease Control (www.cdc.gov), the government agency responsible for collecting statistics on communicable and infectious diseases. The CDC provides epidemiological data not only as they exist within the United States but also as they exist in other parts of the world. Researchers can get both historical and current information on the cases of HIV+, for instance, among different populations in the United States, but also among the populations of sub-Saharan Africa that are experiencing an alarming increase in the incidence of the disease.

Another problem, perhaps even more debilitating, is also the end result of the immense volume of documents on the Web. This tremendous proliferation of documents has caused articles to be brief, superficial, and largely undocumented. The larger culture of Web users have come to expect a more synoptic approach to items they access and read, which is another clue to beginning researchers to be wary of using the Web as an inclusive critical resource.

One feature of the Web presents both pluses and minuses on a scale of intelligent usability. Although the Web is invaluable for finding the latest updated information on areas that are always changing, such as the stock market, weather, political polls, and new software releases, the very currency of the Web makes it all too easy to ignore older material generally, in some cases valuable information that offers a necessary historical perspective but that has sometimes not been converted to electronic format for reasons of expense or contrary priorities (Henninger: 48–49). Again, this drawback results from a misunderstanding of the importance of a full foundation in doing research on the part of those with limited experience. One thing should be clear by now: as valuable and versatile a tool as the Web is, it should never be considered as some kind of convenient one-stop shopping.

Other Problems and Moves to Solve Them

Keeping in mind the issue of efficient use (in addition to intelligent use) of the Web, we turn to some of the problems hindering efficiency that Henninger discusses (50–51):

1. Being unable to detect identical documents easily is difficult, but some search tools can now identify duplicate documents as "alternate sites."
2. To solve the problem of a lack of metadata (talk about the document itself, such as author, date of publication, maintainer of the Web site, etc.) that would make retrieving information more efficient, work is being done by the Dublin Core that will help standardize and apply metadata to Web documents. There are also governmental initiatives afoot that will require all government web-based documents to be described by metadata. In the United States, for example, the Government Information Locator Service (GILS) is being developed.
3. The difficulty of rejecting irrelevant documents is connected to "word spamming"—the practice of including text that will make a search engine rank a document as more relevant than it actually is. This is sometimes done by vendors who want to sell their products and therefore want their home pages to appear near the top of a list of search results. To correct this situation, search engines have indexed terms in the metatags, a way also of raising the relevancy of a document by favoring documents whose metatags contain the requested search term.
4. In a proprietary move to increase access to their Web sites, many organizations are providing search engine indexing to augment or even replace established search engine indexing that may point to only some of an organization's documents. An organization's search engine may index the entire site, thus allowing access to all of its documents.
5. The difficulty of being unable to find documents that an index search has suggested is one of the more frustrating problems of efficient use of the Web. Since Web sites change so frequently it is easy to understand why so many users get error messages saying that the requested document cannot be found. To alleviate this problem, there are two initiatives in the works. There is a renaming project headed by the Internet Engineering Task Force (IETF) that is producing URNs. A URN (Uniform Resource Name) "identifies a resource or unit of information independent of its location (URL). URNs are globally unique, persistent and accessible over the network" (Henninger: 145). PURLs (Persistent Uniform Resource Locators) are the other means for retrieving documents that seem to have disappeared. PURLs add to URLs a "*resolver address.* Thus, instead of pointing directly to the location of an Internet resource, the PURL points to an intermediate redirection service" (145). This process permits access to the "disappeared" document and looks forward to the time that URNs will be a permanent part of the searching apparatus of the Web.

Using Search Engines. We have been talking off and on about search engines throughout this discussion of research, but we should spend some specific time on understanding what search engines are and what they can and cannot do. It seems hardly necessary to define search engines, since they have become so much a part of Web users' on-line lives, but a functional definition of what they can and cannot do is important, especially for those who have limited experience. Like library on-line catalogs and databases and commercial databases like DIALOG, search engines use prompts—words or phrases—that describe the kind of information desired to "retrieve lists of hypertext links representing different Web sites" (Metter: 43), or they provide categories of subject headings that can be used in a more general way to retrieve links to various sites. But it should be kept in mind that no search engine searches the Web completely; some cast a broader net resulting in hundreds of thousands of links, sometimes millions, whereas others are more selective, resulting in smaller numbers. The difference in the results hinges on the method of selection, whether it is in part formed by human efforts or is entirely driven by computer software programs, known as robots. These robots will respond to requests for specific keywords in the URL, the title, the text, or metatags that contain keywords and descriptions of the Web site that the author chooses (43). All search engines use robots to some extent, but some attempt to refine their selection process through the work of live indexers:

> At Yahoo! And Magellan, live indexers evaluate Web sites and place them in a hierarchical subject tree, going from the general to the specific. Such subject trees may begin with a broad subject such as art and then branch out to a subcategory like photography, then offer a link to photography galleries and collections, and so on. Such structures may lead the researcher to excellent pages—perhaps even the cream of the crop. But great sites may also be missed. (43)

Most search engines offer the more flexible, robot-driven search methods that can be activated by typing in words or phrases that describe the topic for which Web sites are being sought. But flexibility comes with a price: depending on the search engine used and the topic as described by keywords or phrases, a searcher can retrieve literally millions of hits, many of which will be useful, but others will not. Ellen Metter gives a wonderful example of a "quick-and-dirty" search of *Alta Vista* for sites that would give her ideas about a vacation in Miami. She typed in only two words, MIAMI TOURISM, and got 1,825,040 hits: "The first site was a useful Miami tourism page sponsored by Air France. The second site was a link to a story about a Miami tourism coordinator receiving an award. Not useful. Others on the opening page linked to tourism in different areas of Florida and a Miami tourism page in French" (44–45). Metter also looked at three other engines: *Excite, WebCrawler,* and *Yahoo! Excite* and *WebCrawler* rendered 424,655 and 24,648 hits respectively, and each had very useful sites at the beginning, from *Excite* "a little map of Miami and pointers to all types of tourism information from http://city.net," and from *WebCrawler Miami Information Access* at http://miami.info-access.com/ but *WebCrawler* also provided Miami information "in a more roundabout way, such as the link for the Miami Project to Cure Paralysis." The *Yahoo!* engine produced "two Yahoo categories, the first a bull's eye: Recreation: Travel: Convention & Visitors Bureaus: U.S. Cities: Miami" (45). Metter reports that by using this simple search, she was able to draw useful results from each engine, but of more importance, "there was virtually no overlap between top-ranked results for each engine" (45). The fact that each engine produced useful results and that there were no overlaps encourages searchers to try more than one engine just to get a richer supply of sites. But it also speaks to those who may have differing ideas of what is useful: finding engines that meet particular needs and tastes for doing research is no big deal as long as people are willing to investigate the structures of various engines. Metter recommends using the "help" documentation of each engine to understand the syntax and other search features and concludes, "Become extremely familiar with at least two [engines]; learning the most advanced capabilities of search engines you choose will increase your chances of retrieving useful results" (44).

Common Searching Methods on Search Engines. In general, searching methods are divided into simple and advanced types. Simple searching calls for nothing more than words or phrases in the general search box on the homepage of the search engine. But given the possibility for retrieving multiple sites that have either marginal or no relevancy to the searcher's needs through simple searching, most users will have occasion to access the engine's advanced searching capabilities. These are identified in some way as being advanced, and they will define the engine's searching options.

Boolean Searching. Although natural language is increasingly being encouraged by search engines, it is not a universally viable alternative, and thus "artificial" search methods, like Boolean searching are still essential to know and understand. Boolean searching on the Web is virtually the same as it is when using library catalogs or databases. The same word connectors, AND, OR, and NOT are used by many engines either to expand or to narrow the field of possible relevant sites, but in some engines' advanced structures, & can substituted for AND, | for OR, and ! for NOT (Metter: 46). In the general structure of many engines, "putting a plus sign (+) in front of a word indicates that the word *must* be retrieved, so the search +FATS + EXERCISE in the general search would be equivalent to FATS & EXERCISE in the advanced search. Putting a minus sign (–) in front of a word indicates that the word must not appear in a document.

Phrase Searching. Metter is quite sure about the care that needs to be taken in phrase searching: "This concept can make or break your search: You need to be sure a search engine is searching a phrase *as* a phrase. If you don't take that precaution, it often ruins the search" (47). She then goes on to suggest an example, although perhaps a bit extreme, that helps to illustrate the problems with phrase searching. If someone enters the phrase BILL GATES with no specification that the words appear in that order only, the user may retrieve sites that have to do with someone named Bill who is a good carpenter and is expert at building fences and gates. A widely used way to prevent this problem and to protect the integrity of the phrase as intended is to enclose the phrase in quotation marks, "BILL GATES." But be sure to check the phrase searching protocol for the engine that you are using because quotations may be used differently in different engines.

Relevance Ranking

We have talked elsewhere in this discussion about the relevance of sites in the results of a Web search but have said little about the ways that relevance is determined by research engines. Metter helps out here: "Each engine follows a ranking algorithm determined by the engine's developers, playing varying degrees of importance on whether keywords requested fall in a page's title, body, URL, or metatags" (48). Ranking generally happens automatically, although some engines may require a ranking search, and the user can assume that the sites that head the list of sites are the most relevant. But beyond that, many engines will list a score, usually expressed as a percentage, next to the sites that have been retrieved. A 100 percent score "indicates that one or both of the keywords appeared everywhere that the algorithm deemed most important" (48). In an engine like *Yahoo!,* for example, the presence of keywords in the title is ranked higher than their presence in the body of the site or in its URL.

Selected Search Engines. Metter (50–51) has been helpful suggesting a selection of search engines that can be used in seeking a selection of search engines and directories, such as *All-in-One Search Page.* www.allonesearch.com and *All Known Search Engines.* www.primenet.com/~rickj/index.html, or searchers, known as meta engines, that look for sites in multiple search engines through one request; the names and URLs of some of these engines follow:

DogPile. www.dogpile.com
Mamma. www.mamma.com
MetaCrawler. www.go2net.com
MetaFind. www.metafind.com
Profusion. www.profusion.com

Following is a list of selected specific engines and directories:

AltaVista. www.altavista.digital.com
Argus Clearinghouse. www.clearinghouse.net
Excite. www.excite.com
Google. www.google.com
GoTo. www.goto.com
HotBot. www.hotbot.com
Infoseek. www.infoseek.com
Looksmart. www.looksmart.com
Lycos. www.lycos.com
Magellan Internet Guide. www.magellan.excite.com

MSN Web Search. www.home.microsoft.com
Northern Light. www.nlsearch.com
Snap! www.snap.com
WebCrawler. www.Webcrawler.com
Yahoo! www.yahoo.com

The Ethics of Performing Research

It is probably safe to say that anyone reading these words has some idea of what ethics is. The term is certainly widely used, and maybe too loosely, but one can't help being exposed to some talk about ethics—in business practice, politics, research, education, professional life of all areas, family and community life—almost every day in some way. There is also probably a fairly universal definition of ethics, but in case one doesn't come immediately to mind, Beach (1996) helps us: "Ethics is defined as the discipline related to what is good and bad or right and wrong behavior, including moral duty and obligation, values and beliefs, and the use of critical thinking about human problems" (2). The general breadth of this definition is useful because it applies, as we have suggested earlier in our discussions of ethical stance, to every area of human endeavor. More specifically for our purposes, ethics is a concern to every profession and to every professional; professionals may make poor ethical decisions or may decide not to factor ethics into decision making at all, but both moves are conscious and point to the fact that ethics cannot be avoided.

Kinds of Ethical Theories. Beach breaks ethical theories into two kinds: those based on *principles* and those based on *virtues* (3). Principle-based theories, also known as normative theories, can be seen from two schools of thought: "**deontological,** the theory of *obligation* or duties, or rules and rights. The second is known as **consequentialist,** the theory that links the rightness of an act to the goodness of the state of affairs it brings about. This theory is also referred to as *utilitarianism*" (3). Principle-based theories are "all normative in that they affirm or apply norms or standards to making decisions" (3).What we have here is the conscious attempt to formulate principles of ethical behavior, to evaluate how well ethical decisions work according to how closely they match those principles, and to universalize not only the principles but also the results of applying them to similar situations. In this way, the principles and their applications must be seen objectively.

The main difference between principle-based theories and virtue-based theories is the emphasis on the situational (virtue-based) versus the objective or universal (principle-based). In the case of virtue-based theories, Beach also speaks of two schools: *communitarianism* and *relationalism* (4). "Communitarianism is based on shared community values or closed societies in which there are collective values shared by all. Relationalism emphasizes the values of love, family, and friendship inherent to the situation at hand" (4). But in either case, we need to acknowledge the importance of the individual to both schools, all those elements of an individual's life that come into play in making ethical decisions: personal history, affections, values, beliefs, and family and community obligations. Virtue-based theories focus on how the individual becomes ethically virtuous and how well her or his ethical choices reflect the deliberations of an ethically virtuous person. Virtue-based theories operate in the context of a concrete situation, not in a generalizable or strictly objective environment.

Are There Common Grounds Between Theories? If we look at the two kinds of ethical theories strictly according to their elements as previously laid out, the differences between the two approaches and the likely resulting behaviors on ethical issues appear fairly clear. In the case of principle-based theory, establishing standards of obligation (duties, rules, or rights) or standards for determining the "goodness" or rightness of an act by the "goodness" or rightness of conditions that result from that act, both within a strictly objective context, could result in an ethical system pretty much bereft of what most of us have come to believe is necessary to an ethical system. It is not hard to conjure up a societal system that protects and benefits only a social and an economic elite, that disenfranchises and disables large numbers of those who reside outside the circle of the elite, that persecutes, incarcerates, and denies basic human rights either to vocal dissidents or to the silent disaffected, all in the name of consistently applied rules, rights, or duties of those who construct the ethical system. Racial supremacist groups, some of which reach the pinnacles of power, in both our past and our present, are an easy example of elite and oppressive and murderous rule that is justified by standards of beliefs and behavior consistently and objectively maintained.

We can see the case of the virtue-based theory in a similar way. Although it is based on the actions of the individual in a unique, not generalizable, ethical situation, it is possible for individuals to define virtue in ways that are not normally associated with the notion of virtue. Even in those situations in which the individual's actions reflect the shared community values, clearly there is nothing to say that the community values must be benign, fair, and sensitive to basic human needs and rights. The limits of the community are defined by the community itself, and if they extend no farther than those suggested before—those of the social and economic elite—then the idea of ethical behavior will take on a very different hue from that many have painted.

Having said all this, we should also consider Beach's emphatic position that there is indeed a common ground between theories and that it is established by certain imperatives that serve to benefit the human condition:

> Basic to all ethical theories are three *imperatives*, moral obligations or commands that are *unconditionally and universally binding*. These imperatives form the foundation and criteria for evaluation of every moral system authored by human beings. They are the source of every moral dilemma, but, though we can explain correct judgments in terms of these imperatives, we have difficulty deciding which takes priority over which. (5)

The three imperatives that Beach presents as the bedrock from which Western values, moral duties, and choices evolve are as follows:

1. *Human Welfare—Beneficence.* This relates to the idea of helping others—protecting them from harm, healing their illnesses, or saving their lives—and the duty to promote good, prevent harm (nonmaleficence), and use of the maximization of human happiness for the greatest number of individuals as the criterion of right action.
2. *Human Justice—Fairness.* This imperative requires one to set fairness for all above benefit for some. It encompasses the responsibility to apply fairness in all dealings with others, particularly fairness in the law, which is justice. Our form of law, distributive justice, requires that we seek the morally correct distribution of benefits and burdens in society.
3. *Human Dignity—Autonomy.* This relates to the idea of respect for a person, including the person's right to choose freedom, privacy, and protection of those with diminished autonomy. The autonomy requires us to respect the choices of others and to allow them the space to live their lives the way they see fit. (5)

Clearly these "imperatives" are ideals and as such are often applied with qualifications, or in some cases, not at all. Rarely can we say without hesitation that ethical practices in all areas of professional life, including especially the creation of laws, are informed by these imperatives in a pure or unadulterated form. A familiar example might be the dramatic increase in the number of citizen initiatives or propositions on state and local levels that can alter already existing laws or practices to benefit ultimately the interests of the few over those of the many. The rollback of property taxes or the alteration of the levels of taxation, which has happened in several states, to improve the financial standing of the property owner, appears beneficent on the surface, but when such a move deprives many others (including property owners, by the way) of financial support for affordable public education, housing, and transportation, then the notions of fairness and the greater welfare of the community are seriously brought into question or compromised.

In the field of bioethics, which had its beginnings in the early 1970s, we see the sometimes problematic mix of ideal ethical theory and the complexities of moral questions resulting from developments in science and technology, especially as they have emerged in the life medical sciences. Just one example is the ability of medical researchers through in vitro fertilization to determine the perfect genetic match of an embryo as a potential donor of healthy cells to an ill recipient after the embryo is brought through a full-term pregnancy. Few would argue that where perfect genetic matches occur, donation of healthy cells to an unhealthy recipient should take place as long as the procedure in no way puts the donor at risk. But ethical questions arise with the original analysis and choice of one embryo among others for implantation and development in utero. Ethicists ask whether the next logical step is choosing embryos for future development with genetic maps for certain kinds of desired traits: hair and eye color, complexion, height, and so forth. Is this not a move toward acceptance of a latter day, more sophisticated version of the eugenics research practiced by the Nazis in the years before and during World War II?

It is developments such as these that prompt Beach to say, "The necessity of finding solutions for these complex dilemmas has forced the philosopher to move from the abstract and theoretically ideal situation to applying ethical reasoning to real situations that researchers in laboratories, hospitals, and society in general live with every day" (7). Underlying this shift to applied ethics is the notion of rights, both human and nonhuman. Making sure that rights are protected, not violated or redefined in some way that would diminish their importance, has become central to all research involving humans and nonhuman animals.

The Ethics of Patricia Wynjenek's Research

Sproull defines research ethics as those practices and procedures that help to ensure "(1) protection of human and non-human subjects, (2) appropriate methodology, (3) inferences, conclusions and recommendations based on the actual findings and (4) complete and accurate research reports" (9). The purposes of maintaining these practices are to protect subjects of research from physical and psychological harm and to ensure that the planned method of research will produce reliable results that will be reported honestly, accurately, and meaningfully. Without these basic guarantees, the value of conducting research is seriously compromised or in some cases is rendered damaging and useless.

To prevent this kind of failure from occurring, researchers such as Patricia must submit their plans for the use of human or animal subjects to institutional review committees that determine whether the proposed methodology is sound and will do no harm to subjects. Only after the plans are approved can the research be funded and conducted, so it is imperative that researchers like Patricia, whose research will involve human subjects, demonstrate convincingly that they will protect the following subjects' rights:

1. *The Right to Free Consent.* Research subjects have the right to consent to participation completely free of pressure or coercion. If the subjects being interviewed or surveyed are all employees of the same organization, for example, they must be guaranteed that their continued employment will not depend on their participation. To do this, researchers often ask the managers at whatever level is most pertinent to write and sign letters to potential subjects assuring them that their participation is entirely voluntary, confidential, and completely disconnected from their employment.

   In Patricia's case, her potential subjects are not part of any common structured group, so she has to rely on her own statement that guarantees there is no constraining connection, no coercive relationship that subjects would have to worry about. There is some question, however, about Patricia's promise to pay subjects $10 for their interviews. Especially given her subjects' depressed financial situations, some would see payment as a definite form of coercion. But if Patricia made no payment, she, at least, would see that as exploitive, so the best she can do is not to mention payment during the consent introduction and then to make the payment unannounced at the end of the interview. Although word will certainly get around to the potential subjects that payments are being made, the effort is an honest attempt to make consent truly free.

2. *The Right to Informed Consent.* To give informed consent, potential subjects must be given sufficient information about the research, its purpose, and its method to be able to make knowledgeable decisions about whether to participate. This right is in large part formed by the need to inform potential subjects of the chances for any psychological or physical harm that they could experience as a result of their participation. Since the differences between individuals can be great, the researcher must address at least the most typical problems (or those most likely to occur). Here, again, a review committee may be needed to predict the likelihood of harmful effects and to assess the safeguards the researcher proposes. Finally, as part of the consent introduction, the researcher must offer a copy of any report of findings and conclusions at the completion of the research.

   In Patricia's research, there is no possibility of physical harm, but since much of her inquiry would be focused on family life, employment, use of medical care, and place in community life, answers to her questions could cause some psychological stresses. In order to anticipate this possibility, Patricia can give examples of questions that she will be asking and can inquire whether any degree of discomfort would result from answering them. She can also assure potential subjects that they can decide not to answer any question that would cause discomfort. The final offer she should make, one that all researchers dealing with human subjects need to make, *is the right of any subject to terminate participation at any point in the research process.*

3. *The Right to Confidentiality.* Potential subjects have the right to a guarantee that researchers will limit access to and ensure the security of information gained from subjects. We think again about a subject pool of employees of an organization; their consent to participate is qualified not only by the coercive factor in free consent discussed before but also by the legitimate worry that employers will be able directly to connect their answers to survey or interview questions to them personally. Researchers are bound to reveal the ways that they will limit access to and maintain security of data they collect: the names and positions of people who will have access to the data; the means of storage of data (secure computer files? bound and locked paper files?); and the saving or disposal of data, once it has been analyzed, and how that will be done.

   Since Patricia's interviews are to be tape-recorded, she will have to indicate who will be listening to the recordings, who will be transcribing them, and who will be analyzing them. She can keep the matter simple by declaring herself as the only person to do these chores. But most likely the transcription will be done by someone who is more expert than Patricia, and she may have to enlist the aid of a research assistant to do some of the analysis; in either case, Patricia is obliged to provide the names, positions, and addresses of those who will be doing the work.

In the case of storage of data, Patricia can guarantee that only she will have access to the tapes once they have been transcribed, analyzed, and stored. But she also has a particularly effective device that she can use to guarantee confidentiality: pseudonyms. By having subjects choose pseudonyms that they want to be known by in place of their own names, Patricia comes very close to a complete guarantee that subjects' identities will not be known and therefore that the information they give will be confidential.

4. *The Right to Privacy.* Sproull sees the right to privacy in very simple terms: "Research participants have the right to withhold information about which they feel uncomfortable. . . . Examples of information which is often considered private to some people include income, sexual activities or preferences and religious beliefs" (12). She suggests that researchers preview with potential participants the kind of information they are seeking and that they not ask questions in sensitive areas unless they are absolutely pertinent to the research at hand.

Patricia's moves to protect privacy are pretty much covered in two earlier categories: the rights of informed consent and confidentiality. She will let potential subjects know the kinds of questions she will be asking, telling them also that they don't have to answer questions they consider to be sensitive issues or an invasion of privacy. The use of pseudonyms could also be seen by some participants as a way of protecting their privacy, since their answers would not be connected to their real names. But, of course, others could find the questions invasive, especially given the face-to-face interview situation, regardless of pseudonyms. These people would have to resort to not answering questions in sensitive areas.

5. *Appropriate Methodology and Reporting.* In a discussion of research ethics, it has been perhaps implicit all along that research methods and the resulting analysis and reporting of findings need to meet the highest possible levels of responsibility. To Sproull, this means conducting "research systematically and objectively using as many controls as feasible (considering cost versus value) and acceptable research procedures appropriate for the specific project . . . [and includes] report[ing] research methodology, findings and conclusions in a complete and unbiased manner" (13–14).

Besides the need for researchers to present and follow procedures that are accepted by the community of researchers of the research field in question, they must also conduct their research in ways that do not allow personal biases to influence how they read and analyze the results of their work. Put in simple terms: *researchers should never try inordinately to find what they want to find.*

The same kind of admonition applies to researchers' reporting of results, analyses, and conclusions. Researchers must be willing to report all pertinent data, even when they are negative or are not in support of a hypothesis originally proposed. They must also report problems in the conduct of the research: errors in sampling, experimental procedure, or means of analysis and interpretation. This report would be accompanied by the usual complete review of actual methods of research used and the way they differ from the methods proposed, if in fact they are different.

Patricia's research situation is quite straightforward, especially with regard to appropriate methodology. Patricia has proposed to use participant observation and an open-ended interview schedule to produce tape-recorded responses on a variety of questions having to do with the economic survival strategies of the people of Indigo. Since Patricia has asked a research question about the nature of survival strategies and is not attempting to test a hypothesis, the question of appropriateness of methods is made easier to consider. The methods she proposes, especially participant observation, are time-honored by anthropologists. They expect the results to provide ethnographic information that will present a more complete picture of the survival strategies (specifically in this case, but other behaviors may also be of interest) of a particular culture or subculture. Instead of expecting data that will be statistically treated to determine a level of significance, anthropologists expect analysis of anecdotal evidence, both of the researcher's informants' activities and of her own. Reviewers of Patricia's methods would be looking at the ways she plans to become a participant observer and the specific questions on her interview schedule to judge whether her overall method will yield the kind of results expected to add information to what is already known about the poorer residents of Indigo.

Patricia's future reporting of the results, analyses, and conclusions of her work will also be fairly unconstrained by the demands of more highly structured research plans. Patricia may find that certain questions work less well in yielding useful information, which she would have to discuss, suggesting reasons for their limitations, but it is highly unlikely that these questions would yield completely useless information. And it is possible that questions can be refined during the course of interviewing in ways that would yield more useful information. As long as Patricia discusses these changes, she would be meeting her responsibilities for truthful, unbiased reporting.

CONCLUSIONS

It is entirely appropriate to end this long discussion on research with some final words about ethics. We don't mean to imply that we actually *have* the final words, but rather to suggest that we have come to understand the tremendous importance of responsible, ethical research in all sectors of the professional world. Without good research, no matter the field, no matter the magnitude of the effort, the value of the process of solving problems is seriously compromised. As professionals, we are certainly struck with the centrality of our collective, collaborative efforts to make research meaningful in order to solve problems, but at times we perhaps need reminders that these efforts are only as good as the individual convictions and endeavors upon which they are built. This goal requires "constant vigilance and a deep-rooted sense of ethical behavior" on the part of every professional (Sproull: 14). How else can the professional world be more than the sum of its parts?

WORKS CITED AND/OR CONSULTED

Beach, Dore. *The Responsible Conduct of Research.* New York: VCH Publishers, Inc., 1996.

Henninger, Maureen. *Don't Just Surf: Effective Research Strategies for the Net.* Second Edition. Sydney: University of South Wales Press, 1999.

Lang, Gerhard and George D. Heiss. *A Practical Guide to Research Methods.* Lanham, Md.: University Press of America, Inc., 1998.

Metter, Ellen. *Facts in a Flash: A Research Guide for Writers.* Cincinnati: Writer's Digest Books, 1999.

Sproull, Natalie L. *Handbook of Research Methods: A Guide for Practitioners and Students in the Social Sciences.* Second Edition. Metuchen, N. J., and London: The Scarecrow Press, Inc., 1995.

EXERCISES

1. At the beginning of any research project, researchers have to plan the methods they will use to collect data, both primary and secondary, and decide how to make those methods effective, reasonably comprehensive, and efficient. This preliminary work, which becomes part of what is called a preliminary prospectus, is important in making the proposal writing process that comes later easier and more effective (particularly in making the ultimate proposal more persuasive). In a memo to someone who would act as your research supervisor, write the portion of a preliminary prospectus that proposes the search methods you will use to find secondary literature on a research problem of your own choosing. Be specific about the kinds of search methods you will use (electronic methods for searching library holdings and Internet sources), why you will use them, and what you expect to find. This is particularly important in convincing your research supervisor that you are on a useful and productive track for collecting good secondary data.

2. Using the same research problem as you have used above, write the portion of a preliminary prospectus that proposes the methods you will use of collecting primary data. Be specific about the kinds of methods (experimental/empirical/quantifiable, nonexperimental/qualifiable) you will use, why you will use them, and what you expect to find. In this exercise, it is particularly important that you add an analysis of the ethical considerations of using the methods you propose.

Collaborative Writing and the Uses of Technology

PREVIEW OF THE CHAPTER

CASE STUDY

ADRIAN RIVERA'S OFFICE

HRP Systems, Inc. ▪ Southern California Operations ▪ San Diego, California

Adrian Rivera was on his way back to the office, and he had a lot on his mind. He had just attended a bidders' conference where he found out that he had three weeks to respond to a long, detailed request for proposal (RFP). His head was swimming as he tried to figure out how he was going to get the bid written in that short period of time, particularly when he still had other customers to attend to.

Adrian had worked for HRP Systems for over five years, in which time he had done quite well for the company. HRP Systems sold various software solutions to local and state law enforcement agencies. However, Adrian had never worked on a sale of this magnitude before. All of the law enforcement agencies in Southern California had agreed to implement a single "officer dispatch system," commonly known as a "911 system." Through the State Department of Administration, the cooperating law enforcement agencies had issued an RFP to a wide array of firms that could potentially provide a solution. The written RFP was over seventy pages long and included over 137 specific questions that must be answered. At the bidders' conference, which was a meeting between the state and the potential bidders held soon after the RFP was issued, bidders were allowed to ask questions to clarify various issues in the RFP. Between the written RFP and information provided at the bidders' conference, Adrian had a fairly good idea of what the law enforcement agencies wanted in a solution.

The system would have to track the whereabouts of thousands of law enforcement officers, allow a dispatcher to locate which officers would be able to respond most quickly to a call, record all phone or radio conversations between citizens, dispatchers, and responding officers, and ensure that all calls were resolved. HRP Systems had a product that met most of the customer's needs, but the size of this implementation was larger than any they had done before. It would take custom programming to meet all the needs, as well as a significant training effort to make the dozens of dispatchers capable of using the system. Clearly, Adrian was not going to write the entire bid by himself. He had neither enough expertise in some areas, nor enough time to get it done. Clearly, the need for collaboration helps define Adrian's problem in this professional situation.

BASICS OF COLLABORATION

The basics of collaboration assume that some problems can be solved only with the joint efforts of several people and that any resulting document or documents will have been produced for several reasons:

- Preparation of the document may require expertise that exceeds that of a single person.
- The document may be a formal report regarding a group's efforts in a particular area (e.g., a project report), and the stakes may be too high to leave responsibility for writing to only one person.

- The writing task may simply be too large for one person to complete in the time allotted.

These reasons for writing collaboratively occur quite often in organizational settings. A survey of seven hundred professionals in seven fields said that 87 percent sometimes wrote as part of a team or group (Lisa Ede and Andrea Lunsford, *Singular Texts/Plural Authors: Perspectives on Collaborative Writing,* Carbondale, IL: Southern Illinois Press, 1990: 60).

Once it has been decided that multiple people will be involved, someone must decide who needs to be involved (and who doesn't). Participants may be included for several reasons, including a strong interest in the topic, their expertise, or their position in the organization. However, the more people involved, the more complex the process of organizing the entire collaborative effort becomes. Although larger groups may be used during the technical problem solving phase, it is usually better to keep small the number of participants involved in the actual writing phase—perhaps in the range of two to five people.

Adrian decided to compose a bid response team consisting of him and three other HRP Systems employees:

- Cindy Samuelson, who knew the software comprising HRP Systems's 911 system better than anyone else
- Tony Pensiero, who managed the group of systems analysts and programmers who would have to develop any custom functions required by the customer
- Kristie Hanks, who has managed HRP Systems's training efforts

Both Cindy and Kristie worked in the same office as Adrian, but Tony was based in the company's headquarters. Although he knew it would be difficult to coordinate the efforts of the four employees, Adrian felt confident that this team could work together and provide sufficient expertise to respond to the bid accurately and professionally.

Although there are certainly advantages to working with others to complete a document, this method introduces a layer of complexity on top of the writing process. Group members need to know what is expected of them and should agree on such things as content, format, and deadlines. All these requirements take significant interpersonal communication among members—communication and effort that is not required when one is writing alone.

COMMUNICATE TO SOLVE PROBLEMS COLLABORATIVELY

As when writing alone, collaborative writers should keep in mind the two levels of problem solving discussed in previous chapters: technical and rhetorical. In a collaborative context, technical problem solving includes defining, researching, analyzing, resolving, and synthesizing solutions to the problem.

Collaborative technical problem solving requires extensive organizational efforts and communication to achieve these steps as part of a group. Collaborative rhetorical problem solving consists of coming to agreement among the group members about the way to communicate the results of the group's work effectively and about the actual process of organizing and writing the document. Also required is extensive communication among the group members to tie it all together.

Communicate about Content

When writing a document with others, it is extremely important that the decisions made in both the technical and the rhetorical problem solving stages be made clear to all writers involved. Everyone should have knowledge of, and ideally be in agreement with, the purpose, audience, context, and ethical stance of the document. It is very difficult to create a cohesive document if the various people involved with its creation are not in agreement on these matters that ultimately determine the content of the document.

Once Adrian got everyone's agreement to participate, he distributed copies of the RFP to each of them. Before any writing began, they had to agree on what products and services HRP Systems was going to bid on and how long they estimated it would take to customize and implement the solution. A thorough understanding of what the customer wanted was crucial to writing a successful bid, so Adrian suggested that each team member write down questions and comments regarding the RFP. He then called a meeting to discuss the issues. Tony was unable to travel to be with them, but the other three met in a room with a speakerphone so that Tony was able to hear and participate in the discussion.

During the meeting, the other team members relied on Adrian's familiarity with the customer to help guide their thinking. He made it clear that the customer had two overriding concerns: that the solution be able to accommodate growth and that the training be available on an ongoing basis as new people assumed dispatcher responsibilities. In addition, Adrian knew that the evaluation team would include a dispatcher who had long experience with a competitor's product. Whenever possible, the team should try to show in their document how their product improved upon the competitor's product, but without direct reference to the other product. No matter what else they did, Adrian knew that adequately addressing these issues was critical to winning this bid.

Once the team had developed a good understanding of what the customer wanted, they turned their attention to what products and services should be included in the bid. On the basis of this agreed-upon understanding, they then began to outline the structure of the bid document. Of course, some of this was governed by the RFP, which required certain sections and topics. But there was also quite a bit of creative freedom allowed, and the team spent over an hour creating an outline. After determining the content of the bid, the team needed to decide how to organize the process of getting the bid written.

Communicate about Tasks and Roles

In addition to agreeing on the content and purpose of the document, group members must be aware of what is expected of them. There are numerous tasks to be completed during both technical and rhetorical problem solving, and group members may not work on all of them. The tasks necessary to create a well-written, professional document include

- Researching the topic/problem (technical problem solving);
- Managing the project (coordinating activities among members);
- Writing, either a draft of the whole document or just a portion of it;
- Reviewing the drafts of the document and making suggestions; and
- Editing the document, including combining separately authored components and formatting the final document.

Group members must also organize their efforts by discussing such issues as

- Deciding who will do which tasks;
- Project management, including setting milestones and deadlines; and
- Document control, to ensure that editing changes are not lost and a cohesive document results from the group's efforts.

In general, collaborative writing requires more planning before the actual writing, and of course it requires significant communication throughout the planning, drafting, revising, and editing phases. In essence, the group must make as many decisions about what is to be written as possible before the actual writing process begins. Because in most collaborative writing efforts, group members will be writing in parallel (working on separate pieces of the document), they must be aware of what the others are writing so that they do not contradict each other or otherwise confuse the reader. When one is writing alone, it is much easier to change directions when writing—after all, the writer is the only one affected. If substantive changes to content or organization are made to a document that is being written by several people in parallel, these changes must be thoroughly communicated to others to avoid conflict and confusion when the various parts are assembled into the final document.

COLLABORATIVE WRITING STRATEGIES

Decisions about organizational issues result in a collaborative writing strategy for the group. Studies have shown that some common collaborative strategies used by professionals in a variety of fields include the following (Ede and Lunsford):

1. The team plans and outlines the task; then each writer prepares his or her part, and the group compiles the individual parts and revises the document as needed.

2. The team plans and writes as indicated, but only one group member revises.
3. The team plans and outlines the writing task, then one member prepares a draft, and the team revises and edits.
4. One member of the team plans and writes a draft, and then the group revises the draft.
5. One or more person(s) plan(s) and write(s) the draft, and then one or more revise(s) the draft without the involvement of the original author(s).
6. One member dictates, and another transcribes and edits.

Of course, several of these strategies can be used in combination. Which to use depends on many things, including the knowledge, writing ability, interests, and availability of potential group members. Use the skills and time of the participants wisely—it does not make sense to assign significant writing tasks to individuals poorly suited to complete them well. A group member who slaves over a portion of a document, only to have it completely rewritten by others in the group, will not be happy, and her or his time will have been largely wasted.

In keeping with the reasons for their involvement, the group members must also decide when each participant needs to be involved. In many cases, group members will participate in all phases of the effort, from technical problem solving, to planning the document, to writing, revising, and editing. In other cases, depending on his or her interests and expertise, the person may be heavily involved in only one of the phases, such as problem solving or editing. It is important that those who actually understand the topics written about in the document be involved in the final review. Collaborative efforts are often used in writing technical documents. A thorough understanding of the technical issues is necessary to provide competent review of the completed document.

After a short discussion, the group members were able to come to agreement on what roles each was to play. Because they had already agreed on a general outline of the bid, they decided to have each of the writers take a section of the outline that best fit her or his area of expertise and write a first draft of that section. Adrian was to write the introduction and serve as project manager and final editor. Although all group members would be able to read the various parts of the document and make suggestions, only the original author and Adrian would be allowed to change a portion of the document. This arrangement would make it somewhat easier to manage the writing and editing process. At the end of the two-hour meeting, Adrian felt that all members were "on board" and that they would be able to develop a professional bid in the time allocated. He wrapped up the meeting by getting all to agree they would have the first drafts of their parts to him within ten days, thereby allowing enough time for feedback from the others, revisions, and final editing and formatting. Putting an early but realistic deadline on completing the separate drafts was important because Adrian wanted to allow enough time to complete the whole bid even if the others missed their deadlines by a day or two.

Within a few days of their initial meeting, Adrian realized that the team had not discussed a few very important items. For example, they had not decided when they would next communicate with each other. As the team leader and project manager, Adrian felt he needed to know how the others were coming along on their parts in the time before their drafts were due. However, he had not formally asked for any interim status reports. In addition, each of the team members had other responsibilities and could go several days without seeing each other in the office or talking to each other over e-mail. To provide a mechanism for tracking progress and ensuring that everyone was moving ahead on the project, Adrian requested that everyone supply him with a draft outline of his or her portion of the document five days before their drafts were due. By creating at least one interim milestone, he had provided both motivation for the other team members to begin working on the project right away and had developed an "early warning system" of sorts to help him manage the project.

Additionally, Adrian had not specified in their first meeting in what form everyone would submit his or her drafts. Although everyone within HRP Systems was competent at using word processing software, not everyone used the same software. In fact, Adrian wasn't even sure whether they were going to give him an electronic "soft copy" of their drafts or simply going to print them out and hand them to him. There was no doubt that the team had to agree on how they would communicate with each other and share the outputs of their labors if they were going to avoid wasting valuable time. On the basis of his experience in working on other large projects, Adrian knew that the appropriate use of technology was vital to the success of their efforts.

TECHNOLOGY FOR COLLABORATION

Collaborative writing can be greatly facilitated by effectively using technology. Modern e-mail systems and project management software allow a group to work together without having to meet face-to-face. Telecommuting, flexible work schedules, and geographically dispersed corporations mean that people have to work together without actually being together.

Groups may work in any of several different ways, defined by their proximity in time and space. When working together at the same time, groups are said to be working synchronously. This is most often done through the standard face-to-face meeting. When collaborating by working independently and passing information to each other for use at different times, groups are said to be working asynchronously. This is often done by using e-mail or faxes to transmit information to one another. In addition to the time dimension, groups may be separated geographically. When this is the case, the inability to meet face-to-face may require that technology be used to facilitate collaboration. The following diagram shows how various technologies can be used to allow a group to work across time and place.

	SAME PLACE	DIFFERENT PLACE
Different time (asynchronous)	E-mail, notes written on printouts, etc.	Voicemail, e-mail, computer bulletin board systems
Same time (synchronous)	Face-to-face meeting	Telephone, computer chat, application sharing, video conferencing

There are numerous technologies that can be used to support various work arrangements. Each work arrangement and technology has its advantages and disadvantages. The following table describes some of these.

TECHNOLOGY	FUNCTIONS	SYNCHRONOUS OR ASYNCHRONOUS	ADVANTAGES	DISADVANTAGES
Face-to-face meeting.	"Traditional" meeting environment.	Synchronous.	Ability to see nonverbal communication (facial expression, body language, etc.). Immediacy of feedback.	Difficult to get all participants together. Unequal participation—some dominate, but quiet members may not participate at all. Time may be wasted because of discussions among only a few participants.
Video conferencing.	Allows distributed group members to see each other.	Synchronous.	Allows nonverbal communication (facial expression, body language, etc.) Immediacy of feedback.	Requires specialized equipment. Cheaper systems using PC-based systems attached to Internet are available, but video quality is poor.
Telephone.	One-to-one communication, telephone conference call.	Synchronous.	Immediacy of feedback. Supports geographically distributed groups.	Requires coordination of schedules. Conversations among more than two participants are difficult to support because of difficulty with turn-taking in telephone conferences. Participation by multiple people makes it difficult to identify who is speaking.
Computer chat.	Supports simultaneous communication among numerous geographically distributed participants (e.g., AOL Instant Messenger).	Synchronous.	Supports geographically distributed groups.	Sometimes difficult to follow the flow of conversation because of the absence of conversational turn-taking.
Application sharing.	Supports simultaneous communication among numerous geographically distributed participants, and allows them to view a shared copy of the document simultaneously. May use computer chat or audio to support verbal communications.	Synchronous.	Supports geographically distributed groups. Allows participants to jointly "navigate" the document by simultaneous scrolling and pointing with cursor seen by all participants.	Effective use of application sharing tools requires training and practice. All participants must have same software installed on their computers.

(*continued*)

TECHNOLOGY	FUNCTIONS	SYNCHRONOUS OR ASYNCHRONOUS	ADVANTAGES	DISADVANTAGES
E-mail.	Allows communication of a large variety of messages and ability to "attach" files (word-processed documents) to the messages.	Asynchronous.	Supports large and small groups that cannot physically meet together because of differences in location or schedule. Widely used and easily accessible by many people. Ability to attach documents to messages makes e-mail an excellent method for distributing co- authored documents to all participants.	Like all other written communication, e-mail is subject to misunderstanding. Once a message is sent, the sender has no control over it. It may be forwarded to others inappropriately. E-mail discussions may be difficult to follow, as messages may not always be read in the order in which they were sent.
Bulleting board systems, newsgroups.	Work similarly to e-mail, but require participants to go to a special electronic "location" to read the messages.	Asynchronous.	Same as e-mail, but also allow for easier discussions by presenting messages in the same order in which they were submitted (threading). Useful for long discussions with many participants.	Messages from a bulletin board system (BBS) or newsgroup are not presented with other e-mail messages, requiring participants to check an additional "location" for messages.

GENERAL GUIDELINES FOR USE OF TECHNOLOGY

There are obviously many choices when deciding how to communicate among a group of collaborative writers. The proper mix of synchronous and asynchronous communication, along with the technologies to support that communication, can greatly increase the effectiveness of a collaborative effort. To help you in that decision process, this section describes some of the differences between synchronous and asynchronous communication and gives some guidance about when each should be used.

Different forms of communication vary in the amount of information that can be delivered within a certain amount of time. Communication forms with more capacity for carrying information are considered rich (e.g., face-to-face interaction), whereas more restrictive forms of communication are considered lean (e.g., written letters). Generally, synchronous forms of communication (e.g., face-to-face, telephone, video conferencing) are considered richer than asynchronous forms (e.g., e-mail, written letters); and forms that include visual cues (e.g., face-to-face, video conferencing) are considered richer than those that don't include visual cues (e.g., telephone).

Face-to-face communication is generally considered the richest form of communication because it allows not only verbal communication but also a large amount of nonverbal communication. Nonverbal communication can include anything from voice inflection or tone of voice, to facial expressions and body language. It also has the advantage of immediate feedback, both verbal and nonverbal. For example, during a conversation, it is common for a speaker to watch the listener for nonverbal cues, such as a questioning look or a look of disagreement. On the other hand, listeners can often tell more from a speaker's tone of voice than from the actual spoken words—something parents reminded us all when they told us, "It's not what you said, it's how you said it!"

Thus, face-to-face communication is very effective whenever people are trying to develop a deep, shared understanding of a situation. When opinions are being expressed and judgments are being made, it is very useful to have as rich a form of communication as possible, as the lack of hard, objective "facts" makes it easier for misunderstandings to develop. Therefore, face-to-face communication is particularly useful at any point in the collaborative process where there may be a variety of opinions, and negotiation and compromise are required to come to a consensus. Such events often occur during the very early stages of a collaborative writing effort when group members are trying to reduce the uncertainty about what they are trying to do. It is during the early stages that the group must agree on the purpose, audience, context, and ethical stance of the document—points on which they all need to share the same understanding.

Although meeting face-to-face is an excellent form of communication for sharing opinions and coming to consensus, it is not always possible to do so, or it may not be the most appropriate method in some situations. Other, leaner forms of communication can serve a collaborative effort quite well. In those cases in which it is simply not possible to get everyone physically together, telephone or video conferences will often suffice, even if the group is trying to come to agreement on involved issues. And in situations in which the group is mostly exchanging facts (such as during early technical problem solving), lean, asynchronous forms of communication may actually be better than face-to-face meetings. Face-to-face meetings are notorious for their ability to waste people's time, since such meetings often lose focus or become discussions among a small subset of people attending. Asynchronous communication such as e-mail, though, has to be read only by those who need the information, without wasting the time of others. And in cases in which the group is geographically dispersed, collaborative efforts will proceed much more quickly through the liberal use of all forms of communication, rather than trying to arrange travel schedules to get everyone together in the same place and time.

After some thought, Adrian decided to implement a few things to help smooth the way on the project. Because Tony was located in another state and all the team members had busy schedules, Adrian decided that they would plan on meeting face-to-face only when necessary. For urgent communications, they would use the telephone and voice mail messages. For less-urgent messages and for exchanging drafts of their parts of their documents, they would use e-mail. Specifically, they scheduled a face-to-face meeting to be held three days before the final bid was due. By having everyone in the same room, they would be able to resolve any final issues and still have enough time to modify the document.

However, perhaps one of the more useful decisions they made was to ensure that everyone checked for messages frequently. It was agreed that phone messages should be returned within a few hours and that e-mail messages should be read and responded to within twenty-four hours. This schedule would minimize the amount of time that team members waited for information from each other, thus avoiding one of the biggest sources of delay in any team effort. The team decided not to use some of the other technologies, such as a bulletin board or computer chat systems, because the size of their group and length of their project did not warrant the extra effort to implement these technologies.

In summary, richer forms of communication are often most effective when a group is trying to reduce the uncertainty about what it is they are trying to accomplish. However, modern work environments often make it difficult to get all participants together. At the same time, we are learning that various technologies, such as telephones, video conferencing, e-mail, and application sharing can be very effective collaborative tools. Therefore, it is important that you become a knowledgeable, effective user of a variety of communication technologies if you want to succeed professionally.

TECHNICAL PROBLEM SOLVING AND PLANNING

The decision about how to communicate with each other was not taken lightly by the team. Adrian knew that there were times, such as when the team was trying to develop a common understanding of what they needed to accomplish, that it was important to ensure as rich communication as possible. For that reason, at the project kickoff meeting, everyone, with the exception of Tony, met face-to-face. Such meetings are often held during the technical problem solving and planning stage of a collaborative effort.

Using FTF Meetings Effectively

Face-to-face meetings are important in any collaborative effort. Although groups can work effectively without ever meeting together in the same place, it appears to take longer for the group to develop cohesion if some form of technology mediates all communication. Face-to-face meetings serve important functions, including both social and task-related functions:

- They help define the team. The ability to see everyone you are to work with in one place at one time allows the separate individuals to define a group identity and create a sense of working with a team.
- They allow for social interaction and team building. Informal, social discussions that almost always take place before and after (and in many cases, too often during) meetings allow group members to get to know each other, generating a desire to work hard for each other.
- Because of the richness (e.g., nonverbal communications, immediate feedback) of face-to-face communication, it is an excellent way for groups to share complex facts and situations that are subject to misunderstanding.

- Once again, because of the richness of communication, they are an excellent way to discuss critical decisions or contentious issues. Gaining consensus on difficult issues is often best done face-to-face. Seeing others while you're talking to them provides better discussion and more commitment to the decision.

Even though there are certainly good reasons for face-to-face meetings, most professionals find that much of their time in meetings is wasted. Too often, meetings are not well conducted, leading to little or no accomplishment. Group meetings are much more productive if someone takes responsibility for the success of the meeting. This meeting facilitator does not have to be the same person who typically leads the group, although it often is. An effective meeting facilitator makes sure that

- A clear objective is defined for the meeting and made known to all participants beforehand. This step allows the participants to prepare adequately and sets their expectations. If a meeting accomplishes a well-defined objective, those involved tend to be satisfied with the outcome and more willing to participate in the future.
- An agenda, laying out the specific topics to be covered and how long to spend on each topic, is distributed to all participants either well before the meeting or at least at the start of the meeting. The agenda should also include specific start and stop times for the meeting. This is an effective tool for the facilitator to help keep the meeting on track and moving ahead.
- Even if the agenda is not strictly adhered to (and often it may need to be deviated from), at least the meeting starts and stops on time. Waiting for late arrivals simply rewards their dysfunctional behavior, and they will tend to be late again in the future.
- The right people are there. Invite only those that need to be there either because they have some stake in the outcome of the meeting or because they have information useful to the other participants. And if the necessary people cannot attend, reschedule the meeting for when they can. If the right people are not there, it generally wastes the time of the other participants because they either cannot accomplish what is necessary or will simply have to rediscuss issues later when all the right people are present.
- No one is allowed to dominate the discussion, even if others seem reluctant to contribute. A good meeting facilitator knows when to cut off a verbose participant. Meeting participants find dominating talkers very frustrating, particularly if they feel that the dominant person is wasting their time.
- All group members participate in the discussion. If some people are not participating, the meeting facilitator should be aware that the discussion may have wandered into areas not relevant to all members or that some members are hostile towards the process. Of course, simple shyness may be why someone is reluctant to speak, but quiet members

should be drawn into the discussion so that their ideas are heard and they feel part of the group.

- At the end of the meeting, someone documents what decisions were made and who has responsibility for the various action items. Everyone should leave a meeting with the knowledge of what was accomplished and what they are to do next.

The team at HRP Systems followed these guidelines where appropriate. Adrian distributed an agenda beforehand, and made sure that everyone understood what they needed to do. Because they were meeting face-to-face, he could tell when Cindy and Kristie had a question or were unclear about something being discussed. However, he paid special attention to Tony because there were no visual cues to help facilitate the verbal communication. Adrian frequently asked him questions and gave him opportunities to ask questions. And when they finished the meeting right on schedule, Adrian quickly made notes about everything that had been decided and then sent the notes out to everyone via e-mail that same afternoon. This step reinforced the decisions that had been made and ensured that the other team members knew what they were responsible for.

Because the group was to have only one more face-to-face meeting before the bid was due, they would have to use e-mail and the telephone effectively. Although each person had a separate section to write, there were interdependencies among each of the sections. For example, Cindy and Tony would need to have numerous discussions about exactly what functionality would be met by the "standard" software that was being bid and what software would have to be written from scratch. Because of their respective locations in two different time zones and their otherwise busy schedules, it was likely that they would have to carry on much of this discussion through e-mail. Fortunately, the team members were all experienced e-mail users and had learned to use that medium to its full potential.

Using E-mail Effectively

For both personal and professional use, e-mail has become an extremely common means of communication. With regard to collaborative authoring, e-mail serves two purposes: as a way to communicate about the document and the task of writing, and as a way to distribute the actual document. This section describes the use of e-mail as a discussion and information-sharing tool, not for document distribution.

There are several points to remember about e-mail. First, business professionals tend to get lots of it—perhaps well over a hundred messages per day. Second, people tend to read e-mail "on-line" rather than printing it out, and reading on a computer screen is different from reading a printed page. Third, you never know where an e-mail message may get forwarded to. Fourth, e-mail is a sort of "interactive writing," where some rules of conversation and some rules of writing are combined. A set of conventions and guidelines regarding these and other e-mail issues has developed over the last twenty-five years. Following is a list of some of these guidelines.

- Because many people receive great volumes of e-mail, much of it may be deleted without even being read by the recipient. Make it easy for the recipient to tell what the message is about by using specific, descriptive subject lines.
- Different e-mail software may display a message differently. Stay away from complex formatting, such as tables or bulleted lists, in your e-mail message, unless you are sure the recipients are all using the same e-mail software as you are. If you have to use a significant amount of formatting, write your message in a word processor and then attach it as a file to your e-mail message.
- Notes should generally be short—keep them to just a few points. Because recipients may want to comment on your message, it becomes difficult if you have created a long, complex message. If you need to convey more than just a few points, consider breaking it up into numerous messages, which gives people a chance to reply more easily to the points they are most interested in. If the recipients are not expected to reply to the message, a long message is okay, but consider formatting the message as a word-processed document, and then send it as an attachment to the e-mail message. Proper formatting makes a longer document easy to read, but formatting is not preserved by many e-mail systems.
- E-mail is still written communication, and generally you should follow the rules of good writing, including proper punctuation, capitalization, and grammar. However, e-mail is also considered by most to be an informal form of written communication, perhaps something that falls between the "yellow sticky note" and a formal letter. If you are communicating with someone you know very well and you both understand the context of what you are writing about, the rules of proper English may be relaxed somewhat.
- Although grammatical rules may be relaxed among group members that are communicating frequently with each other, you can never be sure where a message will be forwarded to, so writing well, including proper spelling and grammar, reduces confusion and the potential for embarrassment. Those that do not know you well will form impressions of you based on your facility with language, so make as good an impression as possible.

  Frequently, you will not have time to craft finely worded e-mail messages. Many professionals may send dozens of e-mail messages in a day. Use spell checkers (usually available with most e-mail systems) to help you out whenever possible. However, remember that the main thing is to get your point across to the reader. If you have a choice between sending an important, but hurried message and no message at all, you should probably send the message.
- Consider breaking up long paragraphs whenever possible. When reading documents printed on paper, we are used to remembering where something is located by using our spatial memory (e.g., finding a word "about an inch from the bottom of the page"). However, reading an

e-mail message on a computer screen makes it more difficult to go back and find a particular word or sentence, as spatial references do not work as well when scrolling through a document on the computer screen.

- When carrying on an e-mail discussion, the context of the message is extremely important. All recipients need to know what you are referring to. It is usually a good idea to include the original e-mail, or if it is a long message, the relevant parts, in any reply, particularly if the reply is going to more than the original sender. Most e-mail software will allow you to do this automatically with any reply. At some point, however, this process can make the e-mail message very long and hard to read. Prudent cutting of old text from third or fourth generation replies during a discussion can improve the readability, but some important parts may be lost. This is simply one of the limitations of e-mail. Long e-mail discussions may be a sign that a face-to-face meeting is in order (if possible).
- Also related to conveying the context of your message, be careful with your use of pronouns. Make sure the reader knows who you are referring to before using such pronouns as "him," "her," "them," and so on.
- Emotions are difficult to convey in e-mail messages. However, there are numerous things you can do with the way you present your text to help make your point:
  - Use asterisks to emphasize a word or phrase.
  - Use all capital letters for added emphasis—this style has come to be interpreted as the e-mail equivalent of shouting.
  - Use white space (extra spaces and blank lines) to set a particularly important section apart.
- Although some think "emoticons" such as :-) help convey emotions, remember that it is difficult for the reader to understand the degree of your emotion. Sarcasm or other attempts at humor can be easily misunderstood, particularly among people who do not know each other well or who have distinctly different structural positions. If you are unsure of how an audience would take to smiley faces or any other emotions, don't use them.

TECHNOLOGY SUPPORT FOR DRAFTING, DISTRIBUTION, REVISION

In addition to using technology to communicate about the content of their document, the HRP Systems team made extensive use of technology to support the process of distributing drafts, commenting on them, and revising and editing them. When Adrian had suggested to the team that they use the telephone and e-mail as much as possible during the course of the writing process, they had also made some other important decisions. First, everyone on the team used the same word processing software to write their portions. Second, they all agreed to use the functions of the software to facilitate the collaborative process. Third, they all agreed on a standard way to name their files so that it would be easier to keep track of the various parts and versions of the document.

The decision that all team members use the same word processing software, even though seemingly straightforward, was an important one. Although most modern, full-function word processors can read files created with other word processing software, text formatting may be lost or seriously altered in the process. Most formatting can be saved when transferring documents between word processors by saving the file as a "Rich Text Format" (RTF) file, but it is likely that some formatting will be lost. Just as important, the advanced features to support collaboration (discussed later), although offered in various software packages, are not compatible across different word processors. To use these features, it was imperative that everyone on the team use the same software.

The features available to help them collaborate more effectively included

- Master document;
- Commenting feature; and
- Ability to track revisions and incorporate suggestions.

Their word processing software, like other word processors, had a "master document" feature. The master document approach allows a master outline to be created, and then portions of the outline can be parceled out to individual writers. This approach is particularly useful when there are many authors working on many portions of the same document. One advantage to this approach is that the general formatting decisions (heading styles, fonts, etc.) can be set in the master document and then this "style" will become the default for all the other subdocuments. This step saves time and effort in the final formatting process. However, the team decided not to use this approach because they did not create an outline detailed enough in their first meeting. In addition, they felt that with only four writers working on four different parts, it was not really necessary to use the master document approach.

The team decided to use the "commenting" feature available in their word processing software to share suggestions about each other's drafts. Most word processors now include this feature, which allows someone to insert a comment into the electronic version (file) of the document without actually changing the document itself. Accompanying is a view of a screen showing a comment inserted by Cindy into Kristie's training document. Cindy Samuelson's initials and comment do not appear in the actual document when it is printed—the "view comment" feature was chosen by Kristie so that she could read what Cindy had to say. Kristie could now address the comment and then delete it, or even completely ignore it.

The team decided that only the original writer and Adrian would be able to make changes to the actual document. Whereas Adrian was concerned about such things as style and consistency among sections, he deferred to the rest of the team when it came to the technical accuracy of their

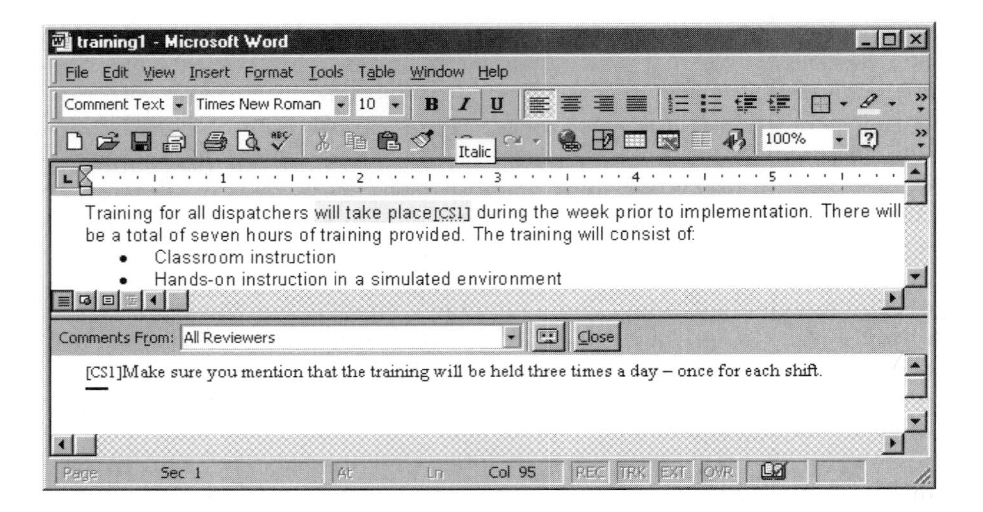

portions. Therefore, he wanted any changes that he made to be reviewed by the original authors. To help make it easier to manage this process, they used another feature common in word processing software—the ability to track revisions.

When Adrian made changes to a draft that he received from one of the others, he would add one change to the file name when he saved it. So, for example, when he made changes to *Training1.doc*, he saved the file with the name *Training2.doc*. Then he sent the file back to Kristie as an attachment to an e-mail message. She would open up the file and use the capability of the software to compare the changes between the two files. Any changes Adrian made would be shown as red text. Accompanying is an example of what the changes looked like when Kristie compared the two documents. Kristie was then allowed to either accept or reject the changes. If she rejected a change, software removed the additions from the latest copy of the document. If she accepted a change, it was no longer shown in red.

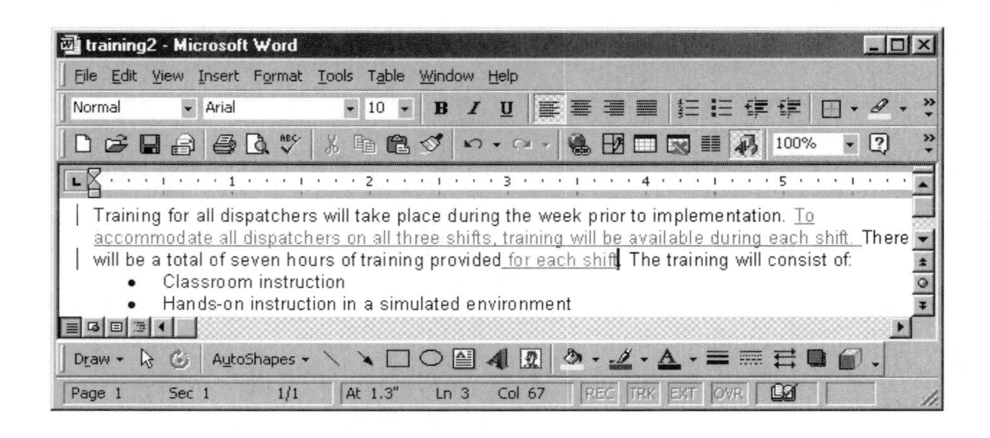

The two examples shown were created in Microsoft Word©. However, other word processing software, such as Corel Wordperfect©, provides similar functions. To learn more about the functions available to support collaboration, refer to the *Help* feature of your software.

Because the team distributed their documents to each other by attaching the documents to an e-mail message, they decided to name their files consistently to make it easier to manage them. Each writer simply created a separate document and named it after the major section he or she was writing. A digit was placed directly after the name to allow a sequential ordering of file names as revisions were made to the document. For example, Kristie, who was responsible for the section on training, named the first draft of her document *Training1.doc.* Then, as she or Adrian made major changes to the document, it was saved as *Training2.doc, Training3.doc,* etc. This method made it much easier to ensure that they were always working with the most recent copy of the file. Also, if they had kept every file named exactly the same, there would have been a chance that one of them would accidentally save an older copy of the file "over" a newer version, thereby possibly losing hours of work.

Adrian and the team made some important decisions that helped them use the technology to work together more effectively. These decisions were influenced by many factors, including the team members' familiarity with the technology and the size of the group. If the group had been larger or if group members other than the original writer and Adrian had been allowed to revise the document, the group would have had to implement additional mechanisms to manage those changes.

In large collaborative efforts, document control and routing procedures are typically somewhat more complex than in the case of HRP Systems. There are two basic problems associated with allowing multiple people to edit a document. The first is that as changes are made, there must be a way to ensure that others have access to the most recent copy of the document. For example, if a document is distributed to team members via e-mail and if one of the recipients makes changes to the document, the others have no way of being aware of the changes because they are viewing the original copy that was sent to them. Of course, the person making the revisions can send the document out to the others as soon as she or he finishes working on it. However, any delays in getting the new version of the document sent out to the others result in their seeing an old version of the document.

One solution to this problem is to provide a single place to store the document. File space that is shared among all group members (usually on a network file server) can be used as the single repository for all documents for the group. As group members edit the document, they save it to the store directory, where it is available to all other group members, thereby avoiding the problem of there being multiple copies of the document stored in different places. However, the procedure does not fix the second basic problem associated with collaborative efforts: how to ensure that two people do not edit a document simultaneously and so potentially "cancel out" the other's changes. This problem is best illustrated with an example.

1. Team member A creates a draft of a document (*Sample1.doc*) and stores it in a directory accessible to all group members.
2. Team member B accesses *Sample1.doc* and copies it to his own computer. He begins editing Sample1.doc.
3. Team member C accesses *Sample1.doc* and copies it to her computer. She also begins editing *Sample1.doc.*
4. Team member B finishes editing and then copies the changed *Sample1.doc* to the shared directory. This step overwrites the old version of *Sample1.doc.*
5. Team member C finishes editing and copies her changed version of *Sample1.doc* to the shared directory. Because this step overwrites the version of *Sample1.doc* that team member B placed there earlier, all of B's changes are lost.

Note that although changing the name of *Sample1.doc* to *Sample2.doc* does not really solve the problem, as team member C may not check to see if a *Sample2.doc* has been created, and she may just overwrite the changed *Sample2.doc* in the same way she could have overwritten *Sample1.doc.*

To avoid this type of problem, document management (or document control) systems are available that require documents to be "checked out" before they can be edited, then "checked in" before someone else can edit them. These systems help ensure that no one can be simultaneously edit a document and then overwrite each other's changes. Document management systems are an excellent way of ensuring that larger collaborative writing teams are able to coordinate their efforts.

In summary, technology provides numerous functions that can increase a collaborative writing team's productivity. However, just as the group must spend time planning the content of their writing efforts, it must also spend time coming to an agreement on how the technology will (or will not) be used, including

- Expectations for responding to phone and e-mail messages;
- Word processing software to be used so that compatibility problems can be avoided;
- How documents will be made available to group members;
- How comments and suggestions will be made and distributed regarding document drafts;
- Who will be allowed to edit documents; and
- Document control procedures so that changes to the documents are managed effectively.

Learning how to use the technology effectively will greatly enhance your ability to produce professional documents with others. Even widely available technologies, such as e-mail and word processors, are excellent tools for working collaboratively. You will be well served if you take the time to explore the collaborative functions available in these tools.

POINTERS FOR SUCCESSFUL COLLABORATION

Regardless of how technology is used to support the team's efforts, adequate attention to the following factors will help make the group successful:

- Manage the project.
  - Appoint or elect an individual to monitor the status of the project and to coordinate efforts among team members.
  - Ensure that everyone knows what he or she is supposed to do and when the assignment is due.
  - Create interim milestones for drafts and/or subportions of the document that are due before the "final drafts" are due. This step encourages people to begin working early and provides the project manager with a means of assessing progress.
  - Do not allow long periods to go by without communicating about the project. If project status meetings are not feasible, the project manager should frequently check in with team members to find out how their writing is progressing.
- Allow enough time for proper revision and editing.
  - Deadlines for drafts must be set early enough so that final revisions and formatting can take place. This procedure often takes longer than planned, since creating a cohesive document from parts generated by writers with different writing styles can be an arduous process.
  - Encourage everyone to get her or his part done by the deadlines, but realize that deadlines may be missed. If the final product has a deadline that simply cannot be missed, set deadlines for drafts to be completed earlier than absolutely needed. This step will provide some cushion if some parts are finished late.
- Have the right people involved, but not too many.
  - Too few people, or not having the people with the required expertise, will slow down the process and reduce the quality of the final product.
  - Realize that the more people there are involved, the more time is spent coordinating the efforts of the group.
  - Avoid adding people to the writing team late in the process, even if the project is running late. New members do not have a solid understanding of all the issues discussed by the group at the beginning of the process (e.g., the analysis of the audience and the purpose of the document), and if it is very late in the process, they will have missed other important discussions. Therefore, others in the group must either take the time to acquaint the new members with all previous discussions and decisions, or else risk not getting useful material from them.
- Use the team members' strengths.
  - Early in the process, find out what strengths everyone in the group has, and then assign roles and tasks on the basis of these strengths.

- Ensure that goals are articulated and shared.
  - Take enough time at the beginning of the process to ensure that everyone truly understands what the group is trying to accomplish.
  - Reinforce the purpose of the document during the collaborative process. For example, when distributing a draft of a portion of the document to the rest of the team, include a statement of the agreed-upon purpose of the document, and then ask the team members to assess the document using the statement of purpose as a guide.
- Create a work environment of open communication and mutual respect.
  - A cohesive group is likely to be a more effective group. Provide opportunities early in the project for team members to get to know each other.
  - Encourage frequent communication that includes all group members. If group members do not work in the same location, use e-mail and conference calls to keep everyone in contact.
  - Establish group norms that encourage critical consideration of each other's work, without the criticism's becoming a criticism of the other person.
- Allow writers to respond to others who modify the text.
  - Because writing takes tremendous effort, it is common for people to have a lot invested in their writing efforts. If someone else is allowed to alter a document, allow the original author to respond and/or disagree with the changes. Writers feel much better about collaborating with others if they continue to have input into the final form and content of their efforts.
- Have one person do a final edit to make the writing style consistent.
  - One team member should be tasked with ensuring that the document is consistent, in both content and style. This person should make sure that there are no contradicting statements. Although it may be difficult completely to eliminate differences in writing styles, the document should at least be internally consistent with regard to verb tense and pronoun use (i.e., first, second, or third person).
- Agree on a document control procedure.
  - Whether documents are distributed via e-mail, stored in a shared directory, or managed by a document control system, everyone must understand how the group will know which version of the document is the most recent, who is currently editing it, and so on.
- Use technology to increase communication and productivity.
  - Face-to-face meetings are excellent for helping the group come to a shared understanding of what is to be accomplished and for gaining agreement on difficult issues. However, in today's professional work environment, people's busy schedules or their geographic separation makes it difficult for everyone to get together on a frequent basis. Voice messaging and e-mail provide excellent means of communication that remove barriers of time and space.

- Modern word processors have available numerous features to improve collaboration, including the ability to embed comments and to track revisions. These features can greatly improve the effectiveness of the group, particularly when the documents can be easily shared with others through e-mail.

EPILOGUE

The team at HRP Systems completed the bid on time. The final product of their efforts was almost two hundred pages long. Each of the team members not only wrote his or her portion but also read and commented on the other portions as well. Because only Adrian and the original author actually edited each portion of the document, document management was simplified. By using e-mail and the commenting and revision tracking feature of their word processor, they were able to get feedback from the others quickly and to incorporate suggested changes.

Adrian spent over thirty hours editing and formatting the document, and he turned it in with only hours to spare. Unfortunately, the bid evaluation team took over two months to make a decision. In the end, though, HRP Systems won the bid, generating the biggest sale to date for the company.

REFERENCES

Beginner's Guide to E-mail: http://www.webfoot.com/advice/e-mail.top.html
Everything E-mail: http://everythinge-mail.net/

EXERCISES

Following are three socioeconomic problems of the workplace that call for a collaborative problem-solving effort. In groups of at least three members, consider the three cases, and then choose one to analyze and attempt to solve. Your analysis and response need to be done collaboratively, using the methods, techniques, and general technology discussed and explained in this chapter. Besides the response called for in each of the cases, prepare a memo report to your instructor that outlines specifically how your group collaborated on the solution to the case problem. Therefore, the ultimate products of this exercise will be a response to the case problem written in your capacity as a committee of the organization featured in the case and a memo report to your instructor written about your actions as a collaborative problem-solving group. The amount of time allotted for this exercise should be determined by the schedule as established by your instructor, but we suggest a week to complete the exercise to accommodate the process of collaboration.

Note: The cases may suggest some consideration of secondary research sources. Your group should decide on the extent of secondary source reading necessary to approach a solution to the problem.

The Case of Timothy & Thomas's Child Labor Guidelines*

The Situation

In the global marketplace, companies may have to work under different cultural and legal constraints than they work under "at home." In this case, the company must decide whether its guidelines for child labor are appropriate given the variables of child labor practices in this particular country, the effects of applying its current guidelines, and the ability of the company to stay competitive, to name a few of its concerns. The following section illustrates the complexities involved in this kind of issue.

The Problem

John Stein, vice president of operations for the garment manufacturer Timothy & Thomas, shifted restlessly in his airplane seat as he remembered the Pakistani girls who had looked no older than ten as they swept the floor between the rows of sewing machines that the women worked on. In that plant, the women and girls had been hard at work assembling the company's hottest item—T & T shorts. Like the rest of their products, the shorts had that wholesome American "feel good–look good" image.

But the image didn't fit Stein's memory of those girls at work, and the contradiction left him in a quandary. In keeping with Timothy & Thomas's reputation for social responsibility, the company's new Global Guidelines for Business Partners prohibited the use of child labor—with "child" defined as anyone under fourteen years old or the compulsory ending school age of the location in question. Stein had felt good about company policy until his trip to Asia, where he had found company policy no help at all.

Driving through Lahore, Yusuf Ahmed, resource manager, didn't mince words: "I should warn you. There's some confusion about the guidelines."

Stein stopped admiring the scenery. "What do you mean?"

"All the good contractors use kids. The little girls come to the plant with their mothers, and I know there are others on the machines who are younger than fourteen. That's just how it's done here."

"Haven't you told them they have to do it differently?"

"I'm not a cop," Ahmed said. "Besides, I'm not sure you'd really want me to do that. The situation is more complicated than you think."

"What's complicated? You say contractors aren't in compliance, so we threaten to cut off the contracts until they are."

Ahmed leaned forward impatiently. "We're lucky to have these guys, if you want the truth. They produce on time at competitive prices and

**Source:* This situation is adapted from Martha Nichols, "Third-World Families at Work: Child Labor or Child Care?" (*Harvard Business Review,* January–February 1993: 12–23).

with good quality, which is no small feat. Then here we come, the big company from the U.S., saying we won't buy what they produce unless it comes incredibly cheap, but of course they have to follow our company guidelines, even if it means that their costs go up—"

"Hold on," Stein broke in. "We have contractors in other countries who are complying with the guidelines, and they seem to be producing just fine."

Now Ahmed looked more tired than angry. "Sure. If we ask them to, the contractors here will fire any kids who are under fourteen. But that will affect at least sixty families, all of them very poor in the first place, do you realize that? And the contractors will still want assurances from us. These guys will want to know that we'll pay our share, even if the price goes up and we have to renegotiate the contracts."

"How much are we talking about?" Stein asked.

"The young kids who come with their mothers do more than you think, so labor will go up by at least a third to pay adults the minimum wage—or more for skilled work," Ahmed said. "Or if our contractors stick with the kids, the kids will say they're fourteen, because the families need them to work. What are their alternatives, really? They can hire themselves out as maids for almost nothing, or spend hours on embroidery at home, or they can go to a carpet factory, would you like that any better?"

Stein shook his head. "I don't feel good about any of this."

Ahmed turned away. "I can't help you there. I don't agree with the guidelines, you know, but since I work for the company, I'll do what I'm told." Ahmed stared out the taxi window. "Forgive me for being blunt, but sometimes I don't know what you managers in Boston are thinking, I really don't. As far as I'm concerned, imposing American values on the Third World just creates more problems."

Now, on the plane back home, Stein twisted around in his seat. If he didn't renew the Lahore contracts, the girls and their families would lose a major source of income. Or was that just a convenient rationale for looking the other way? After all, adhering to the company guidelines would send production costs for the T & T line, at least temporarily, through the roof, since they would have to pay more women. Timothy & Thomas's clothing empire was now scrambling to keep its edge in the North American and European markets.

The Pakistani girls posed the first test of the Global Guidelines for Business Partners (the company's policy on the prohibition of child labor). Yet being on the cutting edge of company policy just made his decision more difficult. John Stein realized that he was now responsible for the outcome. Unfortunately, all the talk of values and social responsibility in the world didn't erase the bottom line, the necessity for the company to make a profit in a competitive environment.

The Task

What should Stein do about the Lahore contractors? Your team of managers from Timothy & Thomas's International Division needs to write a response to Stein's concerns. Draft a memo to him that addresses his concerns and suggests a reasonable course of action.

Issues to Think About: What are the company's obligations to its employees (including the unofficial child employees) and to itself? Is the company in compliance with international agreements on child labor? What might be wrong about merely applying the guidelines as they exist now? How might the community in Lahore react to the implementation of the guidelines? Is "more of the same" an acceptable approach? How important is protecting the company's reputation? Is using children, thus economizing on wages, the only way for Timothy & Thomas to be competitive? Who should Stein consult about this problem?

THE CASE OF EMPLOYEE RIGHTS AND COMPANY OBLIGATION AT KIRKHAM MCDOWELL SECURITIES*

The Situation

Though only a fifty-person operation, Kirkham McDowell Securities, a St. Louis underwriting and financial advisory firm, is one of the region's leading corporate financial advisers. The firm's client roster includes established and successful regional companies as well as one of the country's largest defense contractors, a very conservative company for which the firm manages part of an impressive portfolio. As in most financial institutions, the corporate culture at Kirkham McDowell Securities is traditional.

The Problem

In this case, Kirkham McDowell's star employee is gay and wants to bring his partner to a company-client function, but his boss is concerned about how clients might react and what company policy should be on this issue. Following is a description of the situation.

George Campbell, assistant vice president in mergers and acquisitions at Kirkham McDowell Securities looked up as Adam Lawson, one of his most promising associates, entered his office. Adam, twenty-nine years old, had been with the firm for only two years but had already distinguished himself as having great potential. Recently, he helped to bring in an extremely lucrative deal, and in six weeks, he and several other associates would be honored for their efforts at the firm's silver anniversary dinner.

As Adam closed the door and sat down, he said, "George, I'd like to talk to you about the banquet. I've thought about this very carefully, and I want you to know that I plan to bring my partner, Robert Collins."

George was taken aback. "Well, Adam, " he said, "I don't quite know what to say. I have to be honest with you; I'm a little surprised. I had no

**Source:* This situation is adapted from Alistair D. Williamson, "Is This the Right Time to Come Out?" (*Harvard Business Review*, July–August 1993: 18–28).

idea you were gay." He looked at Adam for clues on how to proceed: his subordinate did seem nervous but not defiant or hostile.

Representatives of Kirkham McDowell's major clients and many of the area's most influential political and business leaders were expected to attend the banquet. This thought raced through George's mind as he asked Adam, "Why do you want to do this? Why do you want to mix your personal and professional lives?"

"For the same reason that you bring your wife to company social events," Adam replied.

A look of confusion flickered across George's face while Adam continued. "Think about it for a moment, George. Success in this business depends in great part on the relationships you develop with your clients and the people you work with. An important part of those relationships is letting people know about your life away from the office, and that includes the people who are important to you."

"But, Adam, a wife isn't the same thing as a—"

"It *is* the same thing, George. Robert and I have made a commitment to each other. We have been together for over five years now, and I would feel very uncomfortable telling him that I was going to a major social event alone—on a weekend, no less."

George thought for a moment. "Adam," he said slowly, "I'm just not sure you should try to make an issue of this at such an important time for the company. Why bring it up now? Think of our clients. We work with some very conservative companies. They could very well decide to give their business to a firm whose views seem to agree more with their own. You're not just making a personal statement here. You're saying something about Kirkham McDowell, something that some of our clients might fundamentally oppose. How are they going to react?"

Adam leaned forward. "This is an issue only if people make it an issue," he said. "I have resolved never to lie about myself or about anything that is important to me—and that includes my sexuality. It's not a decision that I've taken lightly. I've seen what has happened to some of my gay friends who have come out at work. Even at much less conservative companies, some are never invited to important social events with colleagues and customers, no matter how much business they bring in. They'll never know whether or not their bonuses have been affected by prejudice related to their sexuality. I know my career could be adversely influenced by this decision, but I believe that my work should stand on its own merits. George, I've been a top contributor at this firm. I hope I can rely on you to back me up in this."

"You've given me a lot to think about," George said. "And I don't want to say anything until I've had the chance to consider all the implications. I appreciate the confidence you've shown in me by being so open. I wish I had something conclusive to say at this point, but the fact of the matter is that I have never had to face this issue before. I'm concerned about how well this will play with our clients and, as a result, about how senior management will react. I personally don't have any problems with your being gay, but I'd hate to see you torpedo your career over this. It's possible that this could jeopardize some of our relationships with significant clients. Let me think about it for a few days."

After Adam left his office, George sat in silence for a few minutes, trying to make sense of the conversation. He was unsure of his next move. Adam clearly had *not* come into his office looking for permission to bring his partner to the banquet. George realized that he could do nothing and let events simply unfold. After all, Adam had not asked that Robert be included in his benefits coverage, nor had he requested a specific managerial decision. There was no company policy on paper to guide him through his dilemma. But Adam wouldn't have come to him if he hadn't wanted a response of some kind.

The Task

George Campbell has filled in your team of account managers about his concerns and has asked you to advise him on this matter. He is also concerned because the company has no policy that speaks to this issue. Write a memo to Karen Fieldman, Director of Personnel Relations, alerting her to the situation and making recommendations as to what the company policy should be.

Issues to think about: Should an employee's sexual orientation be an issue in dealing with clients? Should corporate etiquette require different behavior from employees on the basis of sexual orientation? Does the firm owe Adam the same opportunities given to his heterosexual peers? What will Kirkham McDowell be saying, both to Adam and to its clients, if it doesn't afford its gay employees the same opportunities?

The Case of SnowFun, Inc., Fargo, North Dakota

The Situation

SnowFun, Inc., of Fargo, North Dakota, is a medium-sized manufacturer of sleds, skis, snowshoes, and other winter recreational equipment. A major employer in Fargo, it has for the past ten years prided itself on the number of Native Americans it has hired—presently a remarkable 10 percent of its workforce. The firm has also prided itself on its careful consideration of working conditions and workers' rights. Although SnowFun is not unionized in any way, it has established an active Personnel Relations Committee that acts as liaison between employees and management on all manner of employment issues. You and four of your coworkers serve on the committee.

The Problem

Charlene Black Bear, as spokeswoman for the Native American workers of SnowFun, all of whom are members of the Sioux Nation, has presented a petition to allow the Sioux to a week off with pay in June so that they can take part in the religious Sun Dance festival, one of the rituals that bind the people together. The petition asks for this time off as equitable treatment, given that SnowFun closes for a week with pay between Christmas and New Year, in recognition of the importance of Christmas to Christians. The difficulty is that summer and early fall are the busiest times on the assembly lines of SnowFun as they try to fill orders for the fall and winter buying seasons. On the other hand, the week between Christmas and New Year has been historically slow, since many other businesses are shut down or are working at reduced levels, and since most of the orders for recreational winter equipment have already been filled. But the Sioux point out that if one group is given time off for traditional, religious observations, then so should another group be, notwithstanding the time it occurs. They also note in their petition that a complete shutdown of a week need not happen but that all Sioux should be allowed that time off if they want it.

The Task

As the Personnel Relations Committee, you are asked to present all such petitions to the management of SnowFun with analytical commentary: suggestions on the merit of the petition and on what decisions might be made. As the liaison between employees and management, you must take the concerns of both groups into account.

Write a memo to Gustaf Olafson, president of SnowFun, with a copy to Charlene Black Bear, in which you lay out your analysis and suggestions for action.

Issues to Think About: Acknowledging the legitimacy of religious observations for all workers; recognizing equal treatment of all workers; realizing the needs of productive, efficient management of SnowFun.

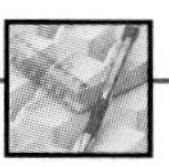

Solving Problems Through Proposals

CASE STUDY

PATRICIA WYNJENEK'S OFFICE

College Hall, 367 ▪ Department of Anthropology ▪ Hastings University ▪ Los Palmas, New Mexico

Patricia Wynjenek got to her office early. She was eager to get going on writing a proposal for more fieldwork. Although she had previously been to Indigo, Texas, to do some fieldwork in linguistics, she was excited to return to study the city and its poverty-level population more thoroughly, especially as to how people survive economically and still keep their families intact.

Indigo, Patricia believed, is unique among cities in the Southwest. Situated on the border between Mexico and the United States, it has a population of 100,000, the political and commercial infrastructure is predominantly Mexican-American, it has the lowest per capita income of any city of its size in the United States, and yet its family structure has not disintegrated as have so many in other American urban areas with similar problems of poverty. Patricia wanted to know how and why

PREVIEW OF THE CHAPTER

Indigo is different, and how Indigo might be used as a model for different kinds of social programs that would serve the needs of its people.

Patricia would need some money, of course, to do the necessary fieldwork, and she'd need some time off from Hastings; and the only way she could get some time off from Hastings would be to get some external funding. She thought of the sources that might fund her. Two came immediately to mind: the National Science Foundation (NSF), which had funded her Ph.D. dissertation fieldwork, and the Las Almas Buenas Foundation, which was privately endowed and which funded projects particularly concerned with the economic survival of Mexican Americans. Las Almas Buenas had a special emphasis on health care, regarding how that is taken advantage of and afforded by Mexican Americans—not only in the Southwest but also in other areas where migration has taken them.

Patricia knew she would have to write separate proposals for the two groups, with different emphases. But they both could be apprised by the information she already knew about Mexican Americans in the Southwest, particularly those of Indigo, and additional information that she could use from reviewing the work of others. She would have to talk specifically about poverty in Indigo—how it comes about, what form it takes, and how people deal with it. For Las Almas Buenas, she would have to direct her attention to health care—what access people have to established, modern medical facilities; and how they use traditional methods: remedies, cures, and people who bring them about apart from the establishment.

It would be a little tricky, but Patricia knew that if she defined carefully the problem and the objectives of her work; the background of the problem as she knew it and what others had done to research it, or problems like it; how she would go about performing the work and how long it would take; and how much it would cost, she could appeal to all parties: NSF, Las Almas Buenas Foundation, and Hastings University.

She certainly had some work to do, but she was buoyed by her conviction that her work was worthy of support and that in the end, she would get funding for the work that she believed she had to do.

FUNDAMENTALS OF A PROPOSAL

Hardly anything gets started or changed in the professional world without proposals. They may be written or oral, informal or formal, solicited or unsolicited, but whatever form they take, they become the basis for initiating practically any kind of work.

Purpose and Rationale

Proposal writing happens in all areas of the professional world. People in business, for example, may propose changes in the way something is made, entirely new products, new markets, and new ways of appealing to those markets. Successful proposals may alter office procedures; they may bring about changes in personnel or corporate moves to new locations. They are often written to improve financial arrangements or to change investment strategies. In fact, all areas of business and industry, if they are going to remain active and productive, at some time will require proposals for change. The opportunities for preparing proposals are almost limitless.

There are similar opportunities for writing proposals in other areas. Scientists regularly propose to research new avenues for cures for cancer and heart disease. Engineers propose new systems and modifications to equipment. Lawyers propose methods of handling their clients' legal problems. Teachers propose new courses of study, changes in administrative practice, and ways of expanding educational opportunities to neglected groups of the population. Government workers propose changes in bureaucratic functions, ways to make government services more efficient, appeals to foreign governments on future relations in trade, and cultural exchanges. And the list goes on. Every profession is involved at some time in writing proposals, because it's a professional responsibility to take the initiative.

Regardless of the subject, all proposals have the same basic purpose: persuading someone in a position of influence that something feasible needs to be done for a reasonable cost within a reasonable time.

This information explains why proposals offer a perfect example of the problem-solving activities of all professionals. Those who prepare proposals assume at the outset that some kind of problem exists—a procedure is inefficient, a market is untapped, or the course of a disease is not fully understood. They further assume that solutions to these problems are necessary—that with changes in procedure, or cultivation of new sources of revenue, or discovery of a turn in the way a disease develops, both the quality of professional work and the quality of life in its broadest sense will improve. Finally, they reason that they can do the work and come up with the appropriate solutions.

Kinds of Proposals

As we mentioned earlier, there are many kinds of proposals. Some are informal and mentioned only in passing, maybe over a cup of coffee. Most, however, are more formal than that, and require close attention to style, form, and content. The process of writing proposals takes up a large part of the time that professionals spend communicating—telling someone what needs to be done, how it can be done in the best way, and what the rewards are going to be for having done it. Proposals are categorized according to whether they are solicited or unsolicited and whether they are research or practical (see Figure 6.1).

Solicited Proposals. Proposals are *solicited* when they are requested by those who want certain work done or a certain problem solved. The impetus for solicited proposals comes, therefore, from the potential sponsors of the work rather than from the proposal writers themselves. The *felt need,* something we have mentioned several times before, lies with the potential sponsors, and the proposal writers, if they want the work, must respond to the specifications of the requested work as they are laid out in what is called a request for proposal (RFP), a request for quote (RFQ), or a request for bid (RFB). RFPs generally come from public and private foundations and from government agencies that usually, but not necessarily exclusively, are interested in having research done. RFQs or RFBs, on the other hand, generally come from business and industry sources that are looking for more specific or tangible work to be done.

Various professional areas request proposals. For example, government agencies like the National Science Foundation, the National Endowment for the Humanities, NASA, or private foundations like the Rockefeller Foundation all have programs that need to be carried on to fulfill their missions. To do this, they send out RFPs—extensive documents that outline the particulars of the agencies' programs and the areas in which they want work to be done. These RFPs are sent to research institutions, both public and private. Researchers then respond with proposals in their areas of expertise and are judged competitively on how well they understand the agencies' needs, how

Type of proposal	Solicited	Unsolicited
Research	■	■
Practical	■	■

FIGURE 6.1 **Kinds of proposals**

they propose to meet those needs—with what facilities and personnel, within what time span, and at what cost.

The same kind of interaction happens in the private sector. Visionometrics, Inc., in San Jose, California, for example, wants a new automated inventory control system to keep up with its growing business. The company sends out to private computer engineering firms, like Hewlett Packard, RFQs or RFBs that outline the extent and kind of inventory it has and is likely to have in five to ten years, and the kinds of problems that the present system has already caused or is likely to cause. Firms interested in getting the business then respond with proposals to solve the problems by means that lie within their capabilities. Contracts to do the work are awarded on a competitive basis to the proposers who, in Visionometrics's judgment, are best able to do the work at a reasonable cost and within a reasonable time.

Unsolicited Proposals. Unlike the solicited proposal, the impetus for *unsolicited proposals* comes from proposal writers, not from potential sponsors. In this case, the proposers, not the sponsors, perceive a problem or a felt need; they must convince potential sponsors that the problem exists and that they are capable of solving it, if they are to get the work. In this case, too, preparing proposals runs the gamut of the professional world, but the competitive edge, which we already know exists in responding to solicited proposals, is even keener. Government agencies like the National Institute of Education (NIE), for example, have a portion of their budgets set aside for work in which they have interest and for which they have set up broad guidelines, but for which they have issued no specific RFP. It is up to proposers, in competition with others, to convince NIE that a problem actually exists, that they have possible solutions to it, and that they can solve it at a reasonable cost and within a reasonable time.

The same is true in the private sector. Proposers have to convince potential sponsors, clients, or customers that problems or needs exist within their operations. Thus Visionometrics may be using an inventory control system that a proposer sees could be brought up-to-date and made more efficient but that Visionometrics itself has not yet thought to change. These situations, therefore, call for even keener persuasive techniques—again convincing sponsors that something is wrong or, if not exactly wrong, could be made better, when that need has not yet been seen or fully understood by the potential sponsors themselves.

Unsolicited proposals can be generated from both inside and outside an organization. They require that someone define a problem and devise a reasonable way of addressing or solving it—the basics of a proposal. Then the real work of writing a proposal, which will convince someone to support the work, begins.

Research Proposals. Research proposals address problems to which there is yet no tried or proven solution. This perspective does not mean that what is proposed is in some way unreasonable. It means, rather, that although the problem has been defined, the actual results of addressing the problem, or attempting to solve it, remain in an area of speculation. Researchers working on the AIDS epidemic, for instance, may take several different directions, not knowing exactly what they will find, in an attempt to find a cure for the disease.

Many of the projects sponsored by government agencies are generated from research proposals, and virtually all of them are solicited. Sponsoring agencies know that problems exist; in fact, these agencies have been established and mandated by Congress to solve them, and the agencies know that they must contract the best researchers possible to find the solutions. No one expects that a cure for cancer will be found tomorrow; nor is it known how a nuclear repository will ultimately affect an ecosystem, but everyone expects a careful search for cures and answers. Thus the National Science Foundation or the Environmental Protection Agency solicits research and later supports it, when, according to the best of its knowledge, that research is reasonable and promises progress toward cures and answers. In this way, research proposals address broad and prevailing concerns of our modern society: how to cure and prevent disease, how to educate significantly a large and heterogeneous population, how to sustain and improve the quality of life.

Practical Proposals. For our purposes, those proposals not based on the kind of research just described are called *practical proposals.* This is not to imply that research cannot be practical or cannot have practical results, because in fact, most of it is intended for ultimate practical application. But the focus of research proposals is on the study of what is not yet known or proven, whereas the focus of practical proposals is on a product or service within the known capabilities of the proposal writers. A computer engineering firm's proposal to Visionometrics to revamp its inventory control system, therefore, qualifies as a practical proposal. The proposers know what has to be done, know how to do it, and most often have already done it for other clients. Indeed, most of the work of business and industry is generated by practical proposals. They are also, in contrast to research proposals, generally shorter, less scholarly, and less fully documented, because the process that generates them operates within the private sector and is not subject to public, bureaucratic scrutiny.

ANALYZING THE PROCESS OF PROPOSAL WRITING

The process of writing proposals can be analyzed by asking several basic questions. The answers to them can translate into the various parts of a proposal. (See Figure 6.2.) These questions, considered as thoroughly as possible, not only will define the basic purpose for writing a proposal but also will provide an outline of the necessary content. We now need to consider the context of a proposal in detail.

Statement of the Problem

As we indicated in Chapter 1, the question that needs to start any problem-solving process is, What is the problem? Proposers need to define the problem clearly and precisely, because the definition establishes the basic reason

Questions That Proposals Answer

Statement of the Problem

- What problem or problems will the proposed project address or attempt to solve?
- In other words, what is the felt need?

Analysis of the Audiences

- Who will benefit from the proposed work?
- Who will review the proposal and either accept or reject it?
- Are the groups who will benefit from the proposed work and who will accept or reject the proposal the same?
- If the groups are different, what relationship do they have to each other?

Statement of the Objectives

- What are the objectives of the work proposed?
- What are the goals?
- What specifically will result if the work is done in the way it is proposed?

Review of the Literature and Previous Work

- What previous work has been done in the area of the problem?
- How does the problem fit into the larger context of work done in the area?

Methods for Doing the Work

- How will the work be done?
- What methods will be used to solve or address the problem and to achieve the objectives as defined?

Facilities

- What resources exist for getting the work done?
- Specifically, what facilities are available to bring about the objectives?

Personnel

- Who will do the work?
- How are they qualified to do the work?

Budget and Schedule

- How much will the work cost?
- How long will it take?

Expected Results and Evaluation Plan

- What results can be expected?
- How will the results be evaluated?

FIGURE 6.2 **Questions to ask and answer to define the contents of a proposal**

for proposing. Without a problem and an idea of who feels the need to have it solved, there is no need to propose.

Defining the Problem for a Solicited Proposal. Defining is easier in the case of solicited proposals, whether research or practical. The RFPs or RFQs are, by their very nature, defining documents. They set out the problem as it is perceived by the potential sponsors, and by virtue of the request also establish how it should be perceived by proposers. Proposers must then respond and show how their understanding of the problem coincides with that of the sponsors. So, if Visionometrics sees that it has an inventory control problem and calls for proposals to solve it, proposers must attempt to see the problem in the company's terms. If the proposers find that they need to redefine the terms of the problem, they should do so as long as they keep in mind Visionometrics's basic understanding of the problem. To do otherwise would be insulting. Organizations don't send out RFQs or RFPs on vaguely felt, ill-defined needs.

The same situation is true of a government agency like the National Endowment for the Humanities (NEH). Within its broad Congressional mandate, NEH has defined several areas of concern for furthering the study, knowledge, and appreciation of the humanities. One such area is secondary education and the need for high school teachers to maintain a high level of awareness of and education in the humanities. This, the Endowment thinks, can be done through collaborative efforts between colleges and universities and secondary school systems. So NEH calls for proposals on how such collaborative efforts could bring about a rejuvenated awareness and teaching of the humanities in the public schools. Universities and school districts, if they are willing and able to collaborate, would submit proposals that must demonstrate why they think collaboration would be beneficial and how it would be possible.

Defining the problem, or understanding the felt need, in the case of solicited proposals is not difficult as long as the areas of concern and expertise of both sponsors and proposers are evenly aligned. If proposers work actively in the areas in which sponsors want work done and if proposer and sponsor perceptions within those areas are similar, then this first step in the analysis of the problem of writing the proposal should be fairly easy to accomplish.

Defining the Problem for an Unsolicited Proposal. Defining the problem is harder when writing an unsolicited proposal, when a potential sponsor does not necessarily perceive a problem and must be convinced that one exists. In this case, the burden of proof lies completely with the proposers. They must have not only up-to-date knowledge of their own areas of expertise but also a full understanding of the areas of expertise of their potential sponsors. Let us return to our example of Visionometrics, Inc. A computer engineering firm that specializes in designing inventory control systems knows through contacts in the industry that Visionometrics uses a system that is outdated or inadequate, given the nature and growth of its business. The task for the computer firm, then, becomes one of convincing Visionometrics that the problem exists and that the proposers can solve it. The firm will have to show in a statement of the problem that it understands the peculiar needs that

Visionometrics has for inventory control; how its present system, though at one time adequate, is no longer so; and that the firm has designed and installed systems that would better meet Visionometrics's needs. Although the basic job of any proposal is to persuade, for the unsolicited proposal, persuasion becomes even more important. Proposers have to realize that ideas of what needs to be done may not be share initially by potential sponsors. They must, therefore, be sure that those ideas are clearly, thoroughly, and accurately delineated.

Analysis of the Audiences

Direct Beneficiaries of the Work to Be Done. Although the answer to the question of who will benefit from the proposed work may seem obvious if the problem has been defined clearly, it is still important to know who will be the *direct beneficiaries* of a solution or an address to a problem, and whether they are the same people who will review the proposal. In the case of Visionometrics, the groups are the same: Visionometrics will benefit from a new inventory control system, and it will also review the proposal for one. This circumstance makes the proposers' job somewhat easier, since there is only one audience that needs to be analyzed.

In the case of research proposals, however, especially those sponsored by public agencies, those who will benefit from the work to be done and those who review the proposal are most often separate audiences. If we talk about AIDS research, for instance, the direct beneficiaries are those who have the disease and those who are at risk of contracting it. The reviewing audience (in this case, perhaps a group from the National Institutes of Health) is made up of agency personnel and experts in medical research of various kinds, such as genetics, virology, and immunology. They will look at the proposal objectively and judge it on its general scientific, technical, and otherwise professional merit. The reviewers usually hold themselves apart from the direct beneficiaries of the research. It should be noted here, however, that a cure for AIDS will in the end benefit all of us, and to that extent, the beneficiary and reviewing audiences are the same. But for purposes of this discussion, it is useful to talk about the direct beneficiary and review groups as being separate audiences.

Analyzing carefully the needs of the group who will benefit directly from the work is important, then, for several reasons. A careful analysis helps proposers reflect on the problem and the felt need, to see that they are clearly defined. It also helps define specific objectives of the work to be done. And it may also suggest certain methods that need to be used to fulfill the objectives.

Reviewing Audiences. If the beneficiary and reviewing audiences are not the same, a close analysis of the reviewing audience of a proposal is fundamentally important. These reviewers will ultimately decide whether to offer a contract to do the work. Although defining the problem or felt need is the first step in writing a proposal, to do so without analyzing the reviewing audience carefully is akin to choosing the right vegetable to plant in the spring, given a particular climate and growing season, without preparing the soil. The choice of what will grow may be just right, but without a thorough preparation of the growing medium, the crop will yield only poorly or not at all. Good analysis of the reviewing audience is good preparation of the soil.

Besides the usual questions about audience that we have already considered in Chapter 2—What does an audience already know? What does it want to know? What does it need to know?—particular attention should be paid to understanding the reviewing audience's background. Proposers should know as thoroughly as possible the nature of potential sponsors' operations, what projects they are presently involved in, and what kinds of projects they are likely to support in the future.

To the fullest extent possible, proposers should understand what philosophical biases that sponsors are likely to have and how those might influence their review of proposals. Both public and private sponsors of research on viruses, for example, want to know how viruses operate so that vaccines may be produced, but may not want to support research that requires genetic engineering of some kind. Although information on biases may be hard to pin down, much can be learned from reading annual reports of an organization's activities or the policy statements produced in some form by all organizations, both public and private. Also, in the case of government agencies, reviewing past years' RFPs will be helpful in learning the history of interests and directions that agencies have taken. Personal contact with representatives of potential sponsors, if at all possible, will yield even more meaningful audience information.

The Review Process. Finally, and again to the greatest extent possible, proposal writers should determine how potential sponsors review proposals and award contracts. We do not claim that all proposals, written for either the public or the private sectors, would be reviewed in exactly the same way. Although individual differences exist, thc following discussion outlines general procedures and areas for consideration.

Public Organizations. Government agencies supported by public funds make the review process publicly known, and it usually happens on three tiers. After an initial screening process to see whether the proposals fit within the agency's general view and RFPs, they are sent to privately contracted reviewers, experts in the general field that the proposals address. These reviewers read the proposals according to guidelines supplied by the agency and supported by their own expertise, and then make recommendations about whether to pursue the proposals further. Proposals are then reviewed by a middle-level committee of agency staff that first considers the reviewers' recommendations and then, depending on the agency's mission, makes recommendations of its own. Finally, the decision to support the proposed projects is made at an executive level.

We take the time to mention this procedure because it so clearly illustrates the need for close audience analysis. In these cases there are three audiences reading proposals. All have similar interests and backgrounds, but each will use the information of the proposal differently. The outside reviewers are experts in the field in question and so will read the proposals with much greater concern for technical accuracy, the need for the proposed work, and its feasibility.

The middle-level staff committee, once satisfied from outside reviewers' comments that proposed projects meet certain technical criteria, will be concerned more with how the projects fit into the general mission of the agency. One way to do this is to see how closely proposers have responded to the letter of the RFP. It is at this level that all proposals live, die, or are given a second chance.

If the outside reviews are universally negative, however, there is virtually no chance a project will be supported. If the comments are universally positive, a project will be recommended for support, perhaps with slight alterations in procedure, budget, or timing. But if the evaluations are mixed and if the project otherwise fits into the agency's mission, the committee may turn down the proposal initially but may ask the proposer to revise and resubmit it for reconsideration.

A good percentage of revised proposals are funded, mainly because proposers get copies of the outside reviewers' comments (they are submitted anonymously) and can see where they have gone wrong. Numerous proposals may be rejected initially, but they are eventually funded because the projects had merit and the proposal writers were able to revise in areas that had caused problems.

But this is no invitation to sloppy preparation—a bad proposal, one that reflects no care in the consideration of its audience, will be flatly rejected. Those that show promise and otherwise good quality, however, are often reconsidered after changes have been made.

The last level of review—the executive committee—deals only with proposals that have been recommended for support at the first two stages of review. This committee, like its middle-level counterpart, is also concerned with the general mission and goals of the agency, and with the ideas of the outside reviewers, but ultimately it is concerned with costs. Although the executive committee is unlikely to turn down a proposal if it has reached the committee, it is more than likely to negotiate the budget—scrutinizing every item and its justification—that the proposer has supplied. It is rare indeed that a project is funded entirely without question, but the extent of the negotiations will be determined (and eased) in large part by how well and how reasonably the proposers have presented their requests for money.

Private Organizations. We have talked a good deal about how government agencies deal with proposals, but we have not considered how private firms and industry evaluate them. As we mentioned earlier, government agencies, by virtue of their dependence on public funds, generally have to make their actions much more public. The same is not true of the private sector, where proposals (either solicited or unsolicited) are reviewed on an individual basis, according to the firm's particular needs and wants.

Proposals may be evaluated by committees set up to evaluate a proposed project or by selected individuals. The main difference is that proposals in the private sector work through an internal procedure, with the company relying on its own technical and policy-making personnel to make recommendations and decisions.

Writers of solicited proposals must anticipate concerns of these audiences by paying close attention to the RFPs. RFPs are usually very specific, outlining in detail the work needed, within limits of cost and time. Proposers must address every specification within their capabilities and facilities to have even a chance at securing a contract. In the case of unsolicited proposals, however, the situations are less well defined and less formal. The chances of securing a contract are, therefore, more dependent on how well the proposers have researched the operations, have analyzed the needs of potential sponsors, and have established personal contacts with them.

It is clear that review procedures for both public and private sectors are complex and thorough, and the audiences multiple and demanding. Each audience wants a strong, global impression of the proposed project, but each also has special interests, which are determined by its purposes for reading, its background, and its biases. For proposal writers, there is no task more demanding than audience analysis. The audience must always be kept in mind; knowing the audience must thoroughly inform the entire proposal-writing process.

Statement of the Objectives of the Work Proposed

It is important for proposers to realize that defining a problem and defining what is to result from the proposed work are different. It is one thing to know and understand clearly what is wrong, lacking, or unknown; it is quite a different thing to know what realistic objectives, goals, and results can be achieved from doing the work proposed. These are separate issues, and they should be thought of separately.

Take the case of the Seattle-based educational consulting firm responding to an RFP from the U.S. Department of Education for drug education programs in public schools. The firm understands clearly the problem of drugs among America's youth. It knows the demographic and cultural backgrounds of those children who use drugs; it understands the effects that drug use has had on the conduct of family and public life. It has data on the effects that drug use has on children physically and psychologically. Armed with this body of information, the firm can write an impressively clear, thorough, and precise statement of the problem, and it will rely on this background to imply its objective: a generalized drug educational program.

The Department of Education wants to see, however, a specific list of realizable objectives. These might include workshops for parents to bring them up-to-date on the drug problem as it affects their children; in-service training for teachers for similar purposes, but stressing the physical, psychological, and sociological effects that drugs have on their students; and an integrated, graduated curriculum for grades K through 12 that would involve team-teaching specific units on various aspects of drug use in health, physical education, and social studies classes.

Proposers should remember a couple of points here. First, most proposed projects have a set of objectives that may make up one large objective. Reviewers want to know what that set of objectives is or how the larger objective is constituted, to see that proposers fully understand the problem and the task at hand. Thus in the preceding example, proposers have at least three subobjectives to meet before the larger objective—a generalized drug education program—can be achieved.

The second point to remember is that the objectives section of a proposal is reserved for telling reviewers *what* to be done, not *how* it is to be done. Proposers often confuse these ideas and try to answer both questions at this point in the proposal, whereas the answer to how something is to be done is reserved for the later section on methods. Again the drug education example is helpful. In thinking of the larger objective, proposers might mistake the three subobjectives as methods of achieving the larger objective and include information about how the workshops for parents would be taught, what the in-service training for teachers would require, and how the team teaching of the integrated curriculum would work. Although this information is necessary, it is not needed or expected at this stage of the proposal. Reviewers want and need a simple logical progression: What is the problem (definition)? What is to be done about it (objectives)? How that is to be done (methods) comes later.

Review of the Literature and Previous Work

The answers to the questions of what has been done in the area of the problem and of how the problem fits into the larger context of work done in the area suggest some important elements in the proposal-writing process: the proposers' experience, their awareness of what is going on in the field, and their notion of how the particular problem in question fits into a larger scheme of problems. The differences between research and practical proposals show up fairly clearly here and help illustrate how the answers to these questions add depth and credibility to the proposal.

Proposers of a research project want to show that they know what other work has been done in the area and how it helps define or influence the problem as they have proposed it. They therefore refer to and discuss in some detail the studies of other researchers—what they did, how they did it, and what their ultimate findings and analyses were. They also refer to their own past work, how it compares with or contrasts to the work of others, and how it has informed or influenced their present work. This discussion in research proposals is usually referred to as the review of literature, because indeed that is what happens: proposers review and analyze the published findings of their own and of others' research in order to present the problem in a larger context, and to demonstrate to reviewers that they understand that context. This is an important activity. It shows reviewers, depending on the literature reviewed, that the proposers are up-to-date on work in the field and that the proposers perhaps subscribe to a particular school of thought. It helps reviewers place the proposal in the appropriate context and therefore helps them decide whether to recommend it or turn it down.

A review of literature is not simply a list of sources or an annotated bibliography; it is a conclusive, evaluative discussion that must show the relationship of the proposed work to other work being done in the field. As such, it should follow all the conventions of documentation (citation) of secondary sources used in the field.

There is a similar task in writing the practical proposal. Proposers want to refer to previous work, but not in the way called for in a review of literature. They want to point specifically to projects that they have completed successfully for other clients or customers, projects that had similar problems as their central focus, similar objectives, and similar solutions. Proposers, of course, want to point to the success with which the problem was solved and the project completed. None of this process requires reference to published materials, though sometimes proposers will include brief testimonies to the success of a project from former clients or customers.

The previous work section talks only of the *proposers'* previous work or of their understanding of work that needs to be done on the basis of their experience. It does not usually discuss work that others have done; or if it does, it does so only obliquely, without specific references for obvious competitive reasons.

And thus in the case of the proposal to design and implement a drug education program for Seattle public schools, the group would speak of other special educational programs it had established. It would want to refer perhaps to the work it had done in sex education, another integrated and graduated program involving students, teachers, and parents at all levels, K through 12. Possibly it would point to the work it has done with school dropouts: identifying who they are and the kinds of families they come from, and the efforts made to get dropouts to return to school or to study for general education diplomas.

Proposers to Visionometrics, Inc., would want to demonstrate how they had set up new inventory control procedures for firms that were of a similar size, with similar growing pains.

A review of the proposers' experience is necessary in the practical proposal. The same idea applies as in the review of literature of a research proposal: to demonstrate that proposers are thoroughly up-to-date and have done the work necessary to solve the kinds of problems identified in the proposal at hand. The review, however, is rendered in a different way; instead of academic references, the review points to results that the proposers have already successfully realized, those well removed from theoretical speculation.

Methods for Doing the Work

Potential sponsors want to know not only how proposers conceptualize problems but also how they plan to solve them. The two processes clearly complement each other: a thorough definition and analysis of a problem will point to what should be done to address it; a clear and feasible plan to address a problem will demonstrate how well proposers have analyzed it. Proposers should be very careful, once again with audience in mind, to state *how* they want to achieve the objectives they have already set out. We are reminded at this point of the difference between the *whats* and the *hows* of a project: *what* needs to be done is the objective(s); *how* to achieve the objective(s) is the method.

As with all other sections of a proposal, the methods section should be written to persuade. Clear plans, with explanations and justifications for what methods will be used, need to be laid out. Reviewers will be aware of the best techniques for getting jobs done, and they will want to know whether proposers will use those techniques and how they will apply them. Even more specifically, they will perhaps evaluate the merits of one procedure over another, given the problem to be addressed or solved.

An example of this activity is the case of Patricia Wynjenek (which opened this chapter and which was discussed in Chapter 2). Patricia wanted to study the economic Sunnyvale strategies of low-income Mexican Americans living in Indigo, Texas. In her proposal to the National Science Foundation, she had to lay out specifically her methods of primary research, the ways she would collect her data in the field. First, she would become a participant observer, taking up residence in one of the poorer neighborhoods of the town, establishing contacts with key individuals in the neighborhood, and generally taking part in the life of the community.

But beyond participant observation, a time-honored method of anthropological research, she would have to collect specific information on how people, who were underemployed or unemployed and who had limited access to social economic programs, survived economically. To do this, she proposed to choose at random 102 households from the neighborhood in which she lived, to hire four residents from the neighborhood as interviewers, and then to have them interview the members of those selected households. Interviewers were to ask from a preestablished set of forty-five questions about such areas as employment (family income), education, medical care, and family life. The forty-five-minute interviews would be tape-recorded, and both interviewers and those interviewed would be paid $10 per interview. (The number of interviews would be more or less evenly divided among the four interviewers, thus allowing each to make about $250.)

The tape-recorded interviews would then be transcribed and analyzed to see what similarities, differences, patterns, or trends might exist among the answers. From this analysis, conclusions would be drawn about how this particular group of low-income families survives economically and what might be done to plan social programs for them and people like them in the future.

This example demonstrates the need for detail in statements of method. Reviewers have to be persuaded that proposers know how to address a problem and that the method for getting the necessary information will in fact accomplish that. In Patricia's case, for instance, reviewers would want to read and evaluate the forty-five questions planned for the interviews, in addition to the details already provided; they would then be able to tell more about her ability to analyze the problem in general and also about her ability to match the approach to that analysis.

We have spoken here mostly about research proposals, but the same generally holds true for practical proposals. Those who want to take on new approaches to problems, like that of Visionometrics, Inc., the firm that needs to update its inventory control system, have to know how proposers would bring about such change. They need also a detailed statement of method:

what new software will be necessary, why it will be necessary, to what degree it will replace or interface with existing software, what changes in hardware need to be made, and how the changes in either area or both areas will be coordinated while day-to-day business is still being conducted.

The approach that is needed to appeal to the audience in the methods section of a proposal is perhaps less theoretical than practical, but it is no less important, since practice always reflects theory. A forthright, detailed and logical statement of how to get the proposed work done is an essential part of the process of writing proposals.

Facilities

All work, of course, needs physical support; whether the proposal is based in research, practice, or both, proposers need to demonstrate that they have sufficient facilities to do the work they propose. Even in those proposals written to procure advanced equipment, proposers need to talk about existing facilities, if for no other reason than to show that the facilities are no longer adequate and thus to justify a proposal for new equipment.

Whatever the case, all physical facilities needed to perform or support the work of the proposed project need to be described. These would include libraries, their holdings and services; laboratories, including the essential equipment necessary for the project; and computing facilities, including the systems available to the proposed project.

Again, as with any section of a proposal, the intent here is to be persuasive. Proposers want to convince reviewers that the physical support is adequate— maybe even more than adequate—to perform the proposed work. If all preceding parts of the proposal, but especially the section on methods, are convincing and compelling, the use of facilities that may require slightly higher costs than those of competing proposers may be justified.

Some idea of how the facilities will favorably affect the duration of the project is also important to include here. Potential sponsors are especially keen to inefficiencies of time caused by use of outdated equipment or facilities with lesser capacities for doing prescribed work. But sponsors are also keen to demands put upon contemporary, up-to-date operations. Normally, workers on any individual project are not going to have unlimited access to an organization's facilities; therefore, a realistic projection of access and the way that influences duration should be reflected.

Personnel

Just as the physical resources that will support the work need to be described, so do the human resources. Even though earlier questions about previous work point to experience that proposers have had, reviewers have come to expect a full statement of the qualifications of those who will be doing the work.

Proposers should write biographical sketches that highlight the experience of each of the principals and the kind of work they will be doing for the proposed project. These sketches are then expanded into full résumés of all the principals that may be included in the personnel section or in an appendix at the end of the proposal.

Sponsors especially want to know about the personnel's experience in the field of the proposal: what previous research has been done; what has been published and where; and what other recognition has been gained professionally. The personnel résumés should emphasize such experience.

The same is not so true of the practical proposal, especially as prepared by large, well-established companies. Prospective clients or customers in business or industry will rely more on the reputation of the proposing organization than it will on those of the individuals actually doing the work. Although executive résumés are often called for, general employees' résumés are omitted. A section on personnel might even be replaced by a section on the history of the firm and its work in the field.

For smaller, younger companies, however, which depend heavily on personalized service to clients and customers, the personnel section remains centered on individuals. Résumés act as documentation of the qualifications of those who will be doing the work.

Budget and Schedule

Cost and time are obvious and very important considerations of any proposal. Although budget and schedule presentations are usually separate parts of a proposal, questions about each surface simultaneously, and with good reason. Potential sponsors want to know not only how much something will cost but also how long their money will be tied up paying for the proposed project.

The test of cost and time is reasonability. If proposal writers are realistic about what a project will cost and how long it will take, the chances for getting a contract, all other things being equal, are much greater. This consideration means that proposers must take care to list and justify each major expenditure and the time needed to complete each phase of the project.

Since it is the job of reviewers to know what things cost and how long they will take to accomplish, the proposal with the reasonable budget and schedule, given the problem and the methods of approach to it, will more likely be supported, even if it is not the least expensive. Reviewers are often suspicious of markedly low budgets or short time lines, because they can indicate several possibilities: the proposers are desperate for the work and are willing to cut corners; or proposers think they will be able to plead for additional funds and time after the contract is granted; or they have been simply naive or sloppy in making estimates. Whatever the case, it is very important that proposers not appear to be budgeting insufficiently, in terms of either money or time.

Having said this, we should note that as a practical matter, very few research proposals are supported at the full level requested, and most budgets are negotiated downward before contracts are granted. Practical proposals, however, may well be awarded to the lowest bidder, as long as the bid is reasonable; and, in fact, some public, governmental agencies are required to take the lowest bid, as long as the bidders meet the minimum requirements for doing business.

Expected Results and Evaluation Plan

All sponsors, no matter what the circumstances, want to know what the results of a project will be, but those organizations that solicit work, especially in the private sector, already have firmly in mind what they want to have at the end of a project. They expect to read a statement that summarizes the expected results, and of more importance, one that specifies how the results will be evaluated. These statements should lay out the means by which work will be evaluated and specify that the results will be presented in what are known as performance reports.

The same is true of the public sector, especially in the case of research proposals. Even when the work is solicited, potential sponsors expect a separate statement in which proposers detail what they expect to find, and of more importance, how they propose to disseminate that information to a larger community of researchers and scholars. Dissemination of information acts as a means of evaluation, since the community of researchers and scholars closely analyzes the proposed research and often will try to replicate it as it is done (or is proposed to be done) by those who are proposing the research. The process, whether done in the private or public sectors, should be evaluated.

Sponsors of publicly supported projects, in both research and practical areas, may expect a plan for third-party evaluation, whereby a team of outside evaluators gives impartial interpretations of the quality and impact of the work done. An evaluation plan needs to outline in detail how the evaluation will be made, how long it will take, and what the credentials of the evaluators are.

Although this kind of plan would not be required of practical proposals in the private sector, potential sponsors will want to know how well a project is progressing, and then once it is established, how well it works. Proposers must include, then, a plan for a series of periodic reports and performance reports. These plans would indicate how many reports would be written, at what intervals, and by whom. Though it is possible that outsiders would evaluate the work, most likely internal technical personnel would evaluate the project.

FORMAT OF A PROPOSAL

The preceding discussion of essential questions establishes the basic ingredients of most every kind of proposal—solicited or unsolicited, research or practical. But we should point out that not all proposals will answer each question in equal detail or necessarily in the order we have suggested. In fact, the formats will vary according to the emphases that the proposers want or need to make, and according to conventions of proposal writing within their fields of interest.

- Statement of the Problem
- Statement of the Objectives
- Methods for Doing the Work
- Budget and Schedule
- Expected Results and Evaluation Plan

FIGURE 6.3 **Elements included in the basic proposal format**

Basic Format

The elements included in the basic format (see Figure 6.3) are typical of most proposals. They may be arranged differently, however, to emphasize a particular element. With this basic format, most potential sponsors have what is necessary to make an informed and intelligent decision.

Expanded Format

In the case of more formal, solicited proposals, potential sponsors want additional information, usually specified in a section on format in the RFP, RFQ, or RFB. In the case of solicited practical proposals, in which the RFQ or RFB may not specify any format, potential sponsors nevertheless expect certain information beyond the basics to be included. Figure 6.4 lists the elements of this expanded format. Most of these elements have already been discussed;

Expanded Format → **Separate from the document**

- Letter of Transmittal
- Cover Page
- Abstract or Executive Summary
- Table of Contents
- Statement of the Problem
- Statement of the Objectives
- Review of the Literature and Previous Work
- Methods
- Facilities
- Personnel
- Budget and Schedule
- Expected Results and Evaluation Plan
- References Cited

Complete document

FIGURE 6.4 **Elements included in the expanded proposal format**

the following sections provide information on the additional elements included in the expanded format.

Letter of Transmittal. As its name indicates, the function of the letter of transmittal is to transmit, or send, the final proposal to the potential sponsor (reviewing audience). The essential ingredients of the letter should state that the proposer is sending a proposal (it needs a title), that the proposal addresses a certain problem and suggests what should be done about it (a summary of the problem and objectives), that the proposed work will take a certain time (total time commitment), and that it will cost a certain amount (total budget figure).

The opening of the letter should state the auspices under which the proposer is submitting the proposal. If the proposal is in response to an RFP, the proposer should say so and should specify the date of the request. If the proposal is unsolicited (no request for proposal), the proposer should say something to the effect that he or she is submitting the proposal on a problem that the reviewers would have interest in having solved. The closing of the letter should state (politely) the proposer's hopes for a favorable review. The sample letter shown in Figure 6.5 indicates the general format to follow.

Cover Page. Also called the title page, the cover page formally identifies the proposal. Usually in equal quarters of the page, it states the proposal's title; the name, position, and address of the recipient; the name, position, and address of the proposer; and the date. Figure 6.6 presents a sample cover page.

Abstract or Executive Summary. The abstract or executive summary is the first substantive statement, beyond the transmittal letter and the cover sheet, that reviewers read in an expanded proposal. It should always be conclusive, not merely descriptive. Since potential sponsors usually have numerous proposals to review at a given time, they want to know immediately, and in summary, all the substantive highlights of the project, from problem to objectives to cost to time. They are not interested in descriptive statements that set out only the general topics that the proposal will cover. These are not appropriate or useful, since preliminary decisions about continuing a review of a proposal are often made on the strength of the abstract or executive summary alone. Therefore, proposers need to speak of the best qualities of their proposed projects in a condensed form. Many RFPs call for limits of 150 to 300 words or will provide forms with designated spaces (not to be exceeded) that will not allow for more than 150 to 300 words. Thus, the abstract or summary must be truly conclusive and condensed.

Although the abstract or summary is among the first items of the proposal to be read, it is one of the last to be written. Proposal writers are usually not able to write abstracts or summaries until everything else in the proposal has been written. Although is among the last of the problems to be solved in writing proposals, it is often, because of the initial impression it makes, one of the most demanding.

Table of Contents. For long proposals, a table of contents should be included. It shows the contents and organization of the proposal, allowing the

THE AVERY MICROCOMPUTER LAB

Avery Hall Department of English Howell State University
Howell, Delaware 01164-5020

June 6, 2001

Donald L. Johnson. Vice Provost for Instruction
Driscoll Administration Building
Howell State University
Howell, Delaware 01164-4040

Dear Dr. Johnson:

Opening, stating what the proposal is and under what auspices it is being submitted (block paragraph)

Summary of problem and objectives (block paragraph)

Summary of time and costs (block paragraph)

Closing, asking politely for a favorable review (block paragraph)

Sincerely,

John P. McKenzie
Associate Professor

Encl.

FIGURE 6.5 **Sample transmittal letter for a proposal**

A Proposal to Research the Effects of Institutionalized
Computer-Assisted Instruction on the Writing
Program, Department of English,
Howell State University

Submitted to

Donald L. Johnson
Vice Provost for Instruction
Howell State University
Howell, Delaware 01164-4040

Submitted by

John P. McKenzie
Department of English
Howell State University
Howell, Delaware 01164-5020

June 6, 2001

FIGURE 6.6 **Sample cover page for a proposal**

reader to turn immediately to specific pages to read different parts of the proposal at will. It is sometimes written to the second level of specificity, section headers and major headers within sections, but often includes only the main sections of the proposal.

The table of contents includes lists of appendixes and may include lists of figures or illustrations, but sometimes the lists of figures and the like appear immediately after the table of contents. Note that the table of contents does not list material preceding it, such as the title page. The table of contents is paginated as is other front matter with lowercase Roman numerals. Like the executive summary, it is one of the last parts of the report to be written. Authors should wait until all changes have been made in the proposal before adding page numbers; otherwise, they may simply have to do it again.

Graphic Depictions of Schedule. Depending on the RFP requirements, proposers may find that they can organize material best by presenting schedule details graphically. For complicated schedules or schedules containing several subdeadlines, a graphic schedule is the only sort that will make much sense. Although schedule formats may be adapted for any audience, reviewers in industry are probably most accustomed to reading the three forms discussed here.

The first form is called a GANTT chart (named after Henry L. Gantt who developed it in 1917). The strength of the GANTT chart is that it can show the times of several simultaneous activities, such as those required to move a whole office full of people to another building (see Figure 6.7). The GANTT chart delineates time across the *x* axis and task down the *y* axis, and can indicate critical path activities with a *C*. The *C*'s mark activities that cannot be allowed to slip without jeopardizing the schedule.

A PERT chart is more appropriate for depicting sequential activities than simultaneous activities. It looks like a flowchart and is most useful when

TASKS	MONTHS FROM START					
	1 July	2 Aug	3 Sept	4 Oct	5 Nov	6 Dec
Administration						
Arrange Lease	C					
Arrange Phones, etc.				C		
Arrange Design Bid		C				
Design: Contracted						
Create/Submit Office Design			C			
Deliver Equipment						
Engineering						
Dismantle Conference Room						
Inspect Electrical System		C				
Assemble New Conference Room						C

FIGURE 6.7 **Example of a GANTT chart showing change of location**

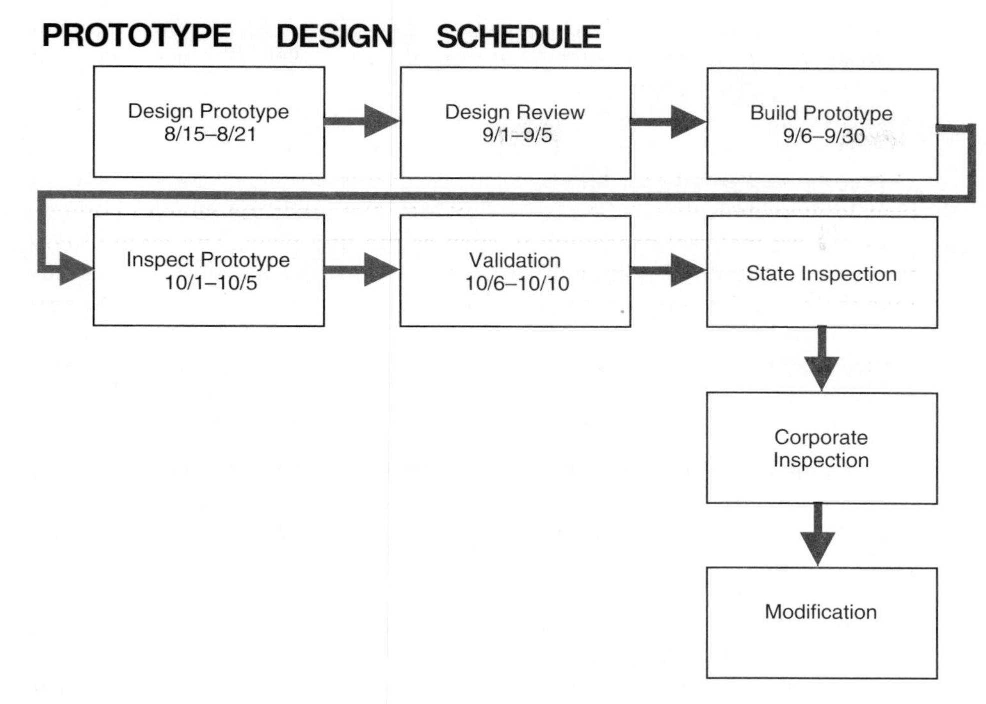

FIGURE 6.8 **Example of a PERT chart showing steps of a design procedure**

the process by which the schedule will proceed is complicated. For example, a schedule that depends on a number of approvals and validations would be effectively rendered by a PERT chart (see Figure 6.8). In order for this schedule to work, all involved will have to meet or beat their deadlines. Although they can try to alert people to impending action, they can't really carry out these actions simultaneously.

The third form is the milestone chart (see Figure 6.9). This chart is suitable for a long-range schedule composed of several phases. The milestone symbols indicate subcategories of actions (which may overlap) that feed into completion of the overall project. We can see from the chart in Figure 6.9 that these proposers have already gotten their work started, indicated by the filled-in milestone triangles. It doesn't make sense to use a milestone chart to indicate purely sequential actions; a PERT chart would be better for that kind of schedule.

References Cited. For proposals that require a formal review of literature or that cite other sources in other sections of the proposal, a list of references cited within the text should appear at the end of the proposal. Such a list provides reviewers with all the pertinent information about the sources at a glance. Since there are several bibliographical formats in use, proposers should use the one considered standard in their field. These formats can be found easily in standard reference books or in the lists of references cited at

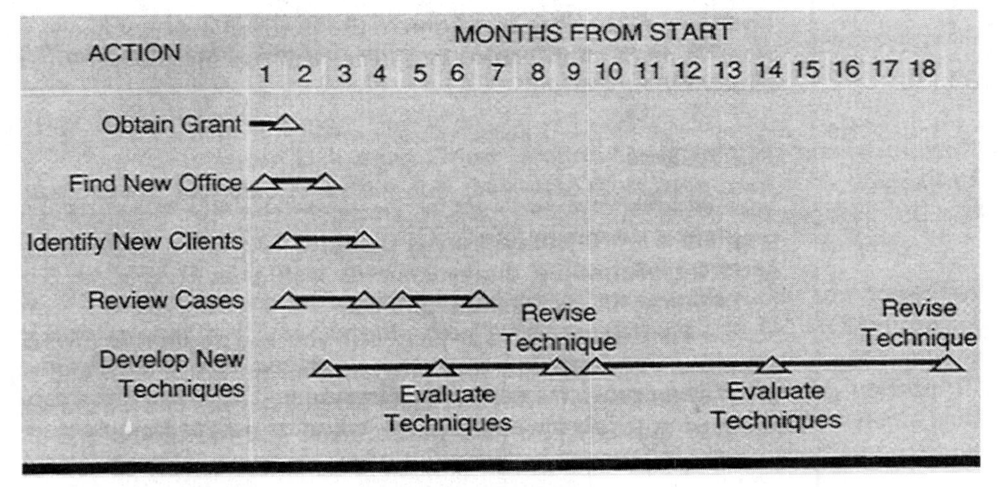

FIGURE 6.9 **Milestone chart showing the establishment of a new drug program**

the end of articles published in professional journals. There are also now computer programs that properly format such lists.

A list of references cited differs from the annotated bibliography discussed in Chapter 2 in that it is just a list, without annotations, of only the sources actually used and cited in the text of the proposal. Most lists of references cited are made up of only printed or published sources, but for those who have used interviews with experts to write their proposals, the names of the persons interviewed, their positions, addresses, and dates of the interviews should also be listed.

CONCLUSIONS

The process of writing proposals is varied and complex, and since virtually all the professional world's work is initiated or changed by proposals, there is really no way to avoid writing them. There may be some consolation, however, in knowing that proposals are often written with others, or collaboratively. We see evidence of this tendency in the profiles of professionals in Chapter 1, and we have taken up the topic of collaborative writing in detail in Chapter 5.

But as demanding as the work is, writing a successful proposal—convincing a potential sponsor that the proposer can recognize a problem and suggest reasonable, effective ways of addressing or solving it—can be one of the more rewarding tasks of professional life.

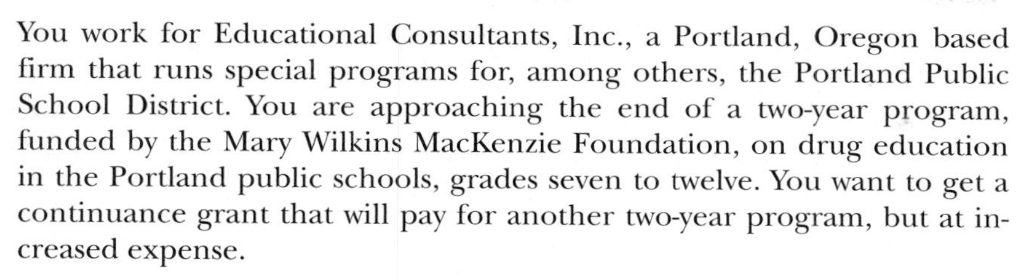

EXERCISE

1. Case Study

The Situation

You work for Educational Consultants, Inc., a Portland, Oregon based firm that runs special programs for, among others, the Portland Public School District. You are approaching the end of a two-year program, funded by the Mary Wilkins MacKenzie Foundation, on drug education in the Portland public schools, grades seven to twelve. You want to get a continuance grant that will pay for another two-year program, but at increased expense.

The Task

You and your team need to persuade the MacKenzie Foundation that your program is worthy of additional funding. Therefore, you will have to demonstrate the need for continuing the program as well as at least some degree of successful performance in the existing program.

To start planning your proposal, you had support staff collect some data on what has taken place in your program over the last two years and what has happened in other programs during the same time. They have listed the data that follow. You will need to evaluate the data for accuracy and usefulness and then decide where certain pieces of information would go in the format of your proposal. Although you will have to keep all sections of the proposal in mind, you may find that some sections are not covered by the data. You will then have to instruct the support staff to collect that information.

The Data

A. John C. Babes in *Drugs and Youth Today* (New York: Harper & Row, 1992) says that alcohol abuse among teenagers is more serious than other drug abuse. There has been a 35 percent increase in the number of adolescents who drink to excess in the last three years. (Our program has put little or no emphasis on drinking as a problem.)

B. Government findings (1998) show that marijuana use is on the decline. Our program has found the same: from a high of 45 percent of schoolchildren, ages thirteen to seventeen, using marijuana once a week in 1997, the percentage has decreased to 27 percent by June 1999. (Use of marijuana has been one of our focal points.)

C. Although use of cocaine has decreased among Portland students, use of crack has increased. Three years ago, crack was hardly known among school-aged children, but since then it has grown dramatically in popularity. (We need to realign our emphases in the cocaine area.)

D. Our use of team teaching involving teachers of biology, social topics, history, and physical education has worked well for special workshop presentations: students' attention is centered on the drug problem at least four times a semester. But reports from the Bay Area and Los Angeles school districts indicate that they are having success integrating drug topics directly into the curricula of the various subject areas.

E. Tricia C. Blount reports in her article "Drugs Are Here to Stay" (*Newsweek*, October 30, 1998) that most education programs still have to fight the notion adolescents have that drugs are a way of life—either because they themselves take them or because their friends do.

F. Surveys of parents of children in grades seven to twelve need to be continued to determine attitudes towards drugs, knowledge about drugs, and firsthand experience with children using drugs. This information is valuable for continued updating of our program.

G. Our "Celebrities Talk about Drugs" feature has been quite successful, with majorities of students reporting that they were most impressed with that part of the program. But preliminary investigation indicates that speakers' fees will rise about 15 percent in the coming year.

H. The Department of Health and Human Services reports that amphetamines are being used by 15 percent of schoolchildren in grades seven to twelve. This represents an increase of 3 percentage points over two years. A combination of amphetamines and barbiturates is being used by 12 percent of the same age group, up from 10 percent over the same period. (Our records indicate that Portland students rank with the national average, even though our program has placed good emphasis on this area of the problem.)

I. In-service training for teachers needs to be continued. They need to be updated on all aspects of the drug problem, how well our program has been received, and how well it has worked.

J. The overall program will see an average increase in costs of 15 percent over the next two years. This increase is compared with an average increase of 11.5 percent for other programs like it in the Northwest.

K. Since the program began, those whose grades had dropped 5 or more percentage points since they began using drugs now report that their grades have resumed to former levels but have gone no higher. They also report that they still use drugs at least once a week.

L. Drug rehabilitation centers in the northwest report that they have seen an increase in enrollment of 22.5 percent among schoolchildren ages fifteen to seventeen over the past two years.

The Final Product

At the end of the exercise, you and your team will have produced a proposal outline, specifying which data belong in which sections of the proposal, as they have been outlined and discussed in this chapter, and which additional data need to be collected. You may also decide that some data are useless and that some are potentially useful—if they were to be developed further. Specify both cases and your reasons for the determination, and instruct your staff accordingly.

As we have pointed out earlier, proposals can be written individually or in groups, collaboratively. Choose a problem, either one you want to work on and have already defined, perhaps in your work in Chapter 2, or one that you and others want to work on. Plan and write a rough draft of the four main parts of a proposal: statement of the problem, objectives, review of literature or previous work, and methods. Each individual or group will submit the draft to a review team for evaluation. Use the peer review worksheet provided in Figure 6.10 to organize comments on the drafts.

PEER REVIEW WORKSHEET FOR PROPOSALS

Reviewer's Name: ______________________________

Author's Name: ______________________________

Answer the following questions as thoroughly as possible.

1. Statement of the Problem
   a. Does the statement of the problem actually define a problem? Can you identify that something is wrong, lacking, or unknown? In a sentence or two, restate what you think the problem is.
   b. Does the author provide sufficient background to the problem? Do you know how the problem developed, when, where, to what extent? Suggest improvements, if necessary.
   c. Is it clear who will benefit from a solution to the problem? Specify who will benefit and how.

2. Statement of the Objectives
   a. Do the objectives actually state what will be done through the research? Or does the author somehow confuse what will be done with how it will be done (method)? Explain.
   b. Do the objectives follow logically from the statement of the problem? Do they seem reasonable goals to achieve? Explain.
   c. If the objectives are listed or bulleted, are the entries in parallel construction? If the objectives are written in narrative, or paragraph form, is it still clear what will be done? Suggest improvements where necessary.

3. Review of the Literature and Previous Work
   a. Is the review a discussion, that is, is it integrated, and does it cohere, or come together, as one piece?
   b. Does the review expand on the statement of the problem, that is, does it show clear connections to the problem? If not, suggest improvements.
   c. In general, does the review discuss the pertinent highlights of the problem the author is pursuing? If so, summarize those highlights here. If not, suggest those that need to be addressed.
   d. Is the review conclusive or purely descriptive? If it is only descriptive, suggest ways of making it conclusive.

4. Methods for Doing the Work
   a. Do the methods grow naturally and logically out of the objectives of the author's project? That is, would the objectives be met if the methods were performed?
   b. Are the methods presented in logical order? That is, is there a sense that once certain tasks are done, others should follow? Explain.
   c. Given the reviewing audience for the proposal, are the methods fully explained and detailed? Is there any doubt about how something will be done? Explain.

FIGURE 6.10 **Example of peer review worksheet for proposals**

ASSIGNMENTS

1. Following this "Assignments" section are examples of proposals written by students (see Figures 6.11 and 6.12). They are not perfect examples and shouldn't be used as models to be copied, but they are representative of what some students can do when assigned to write their first formal proposal. Consider them carefully, and decide how closely they adhere to the principles discussed in this chapter. What are their strengths and weaknesses? Please use the Peer Review Worksheet provided in this chapter to evaluate the proposals.

TERM PROJECT OPTION

Write a proposal on the research problem that you have decided to work on this semester. Depending on the time allotted by your instructor and the purpose of the proposal, you may follow thc basic or expanded format provided and discussed in this chapter. According to your instructor's wishes, you may write the proposal to your instructor (who may not be an expert in the field in which you are proposing the work) or to another audience, one that would likely be in a position to support the proposal work if you were writing the proposal in the world of work.

SAMPLE STUDENT PROPOSALS

Proposal to Investigate the Effect of AIDS Knowledge and Concern on the Sexual Behavior of ISU Students.

Sheryl Soderstrom

English 307
Spring 1991
Dr. Tracy Montgomery

40 Oakwood Dr. #114
Pocatello, ID 83204
233-5854

FIGURE 6.11 **Student proposal 1**

Abstract

AIDS is a growing threat to the heterosexual young adults of Idaho due to a feeling of immunity and continuing risky sexual behaviors. Studies and documentation are needed to establish that risky behaviors are engaged in here and that students at Idaho State University are susceptible to contracting the HIV virus in spite of a high level of knowledge of transmission and protection. In addition, information is needed to isolate what discourages people from engaging in safe sex, what affects changes in sexual behavior, and how to best proceed with education efforts. It is expected that future education must center on self-esteem, decision-making skills, negotiation skills, and behavior modification to be effective in preventing the further spread of HIV.

This proposal requests $1,486.00 to undertake a survey that addresses the issues of AIDS knowledge, perception, concern, opinion, and behavior at ISU. A journal article will be prepared from the survey results and will be submitted to the instructor on April 22, 1991. This article will subsequently be submitted to Sociology and Social Research. Results will also be released to the ISU Bengal and the ISU AIDS Education Committee.

FIGURE 6.11 ***(Continued)***

PROPOSAL TO INVESTIGATE THE EFFECT OF AIDS KNOWLEDGE AND CONCERN ON THE SEXUAL BEHAVIOR OF ISU STUDENTS.

1.0 Statement of the Problem

1.1 Today, having unprotected sex is like playing Russian roulette because of a small, fragile, but deadly virus known as HIV. (See 1.2 Glossary) The HIV virus causes a breakdown of the human immune system. When this happens, opportunistic diseases and infections invade the body. The presence of these specific diseases and infections leads to a diagnosis of AIDS. AIDS is a terminal condition. There is no cure.

The cause of AIDS is a virus, not a lifestyle or a particular activity. But HIV can be transmitted by the body fluids exchanged in unprotected sexual activities (semen, vaginal fluid, blood) (9.9, 9.22). Since information about how to protect oneself has been available since 1983 (9.3, 9.9), there is no reason for anyone to contract HIV (and subsequently receive a death sentence) through sexual activities.

The presence of HIV in our society and its fatal consequences should cause responsible people to change their sexual practices to protect themselves and others. However, articles in Mademoiselle (9.14), U.S. News & World Report (9.2), and Newsweek (9.20) indicate that AIDS is spreading into the heterosexual population, and misinformation about AIDS is rampant. Results of a Centers for Disease Control (CDC) study of student blood samples from universities across the country show that 1 in 300 students are HIV positive (9.3). It appears that the threat of AIDS has done little to curb sexual activity among heterosexuals and especially among college students (9.26).

In conservative southeastern Idaho, one might question the validity of these observations. We've somehow gotten the idea that AIDS only happens somewhere else, to certain other groups of people. We think we are immune because of the heavy religious emphasis in this area, our conservative nature, our isolation from urban areas known to harbor "at risk" populations, and because "we don't engage in any risky behaviors."

However, a 1989 survey of ISU students indicated that risky sexual behaviors are prevalent (9.27). Additionally, there have been 81 people diagnosed with AIDS and there are 280 +HIV infected people (that we know of) in Idaho (9.9). We are also aware that there are HIV+ people and PWA's going to school at ISU (9.3, 9.9). In fact, if we extrapolate from the CDC study (9.2, 9.3, 9.20) and another blood sample study (9.12), ISU with a student enrollment in excess of 10,000, probably has 20–30 HIV infected students. The scary side of this is that most of those people don't know they are infected and are probably continuing to engage in risky sexual behaviors.

As long as people in Idaho and people at ISU continue to believe they are not at risk of contracting HIV, and that they as a whole don't engage in any risky behaviors, they will continue to take the information they are given about AIDS and toss it away as not applying to them. Until it is brought home to individuals that it can happen to them, we will not see a change in behavior and AIDS will spread exponentially in Idaho as it has in other states.

FIGURE 6.11 ***(Continued)***

As someone who has friends who are HIV+ and who have AIDS, and as someone who has lost friends to AIDS, I am concerned that no more people be sentenced to an early death because of sex. An investigation is needed to study the effect of the threat of AIDS on populations of sexually active, mobile, young adults who still naively believe it will never happen to them (9.17, 9.23, 9.26:52). A local study will (hopefully) be taken more seriously by Idahoans and may convince them that they are at risk, just like anyone else in the U. S. engaging in the same risky activities. Information gathered in this survey may better equip us to deal with AIDS in Idaho, will tell us what people think about it and are doing to protect themselves, and how we as educators can best reach Idahoans and get them to protect themselves.

While the HIV virus is transmitted in ways other than just sex (blood, shared needles, mother's milk), the proposed survey only looks at sexual transmission. This survey is intended to show ISU students that they are not so different from students elsewhere. It is intended to serve as a baseline to determine how successful AIDS education has been, how concerned people are, and whether their sexual behavior has changed as a result of the threat of AIDS. It also looks at people's opinions on a couple of AIDS-related issues and PWA's.

The final product, an article reporting survey results from ISU, may alert not only ISU, but other schools/students in isolated areas to the relevancy and need for AIDS education that concentrates not only on providing information about transmission but also speaks to behavior modification.

1.2 Glossary

1.2.1 AIDS—Acquired Immune Deficiency Syndrome. A condition characterized by a compromised immune system (caused by HIV) and the presence of specific opportunistic diseases/infections, including, but not limited to, Kaposi's sarcoma and Pneumocystis carini pneumonia (PCP).

1.2.2 HIV—Human Immunodeficiency Virus. The virus identified as the cause of AIDS.

1.2.3 PWA—Person With AIDS.

1.2.4 PLWA—Person Living With AIDS.

1.2.5 Knowledge—For this survey, knowledge refers to answers to items testing knowledge of protection, transmission, and outcome.

1.2.6 Perception—For this survey, perception refers to answers to items testing level of concern about personally contracting AIDS/HIV, level of concern about children contracting AIDS/HIV, and perception of AIDS as a health problem both nationally and in Idaho.

1.2.7 Behavior—For this survey, behavior refers to answers to items testing willingness to discuss the problem and solution with one's children, willingness to participate in identified safer sex practices, and avoidance of risky behaviors.

1.2.8 Opinion—For this survey, opinion refers to answers to items testing personal view of three controversial issues: condom machines on campus, HIV infected children in public schools, and AIDS as a punishment from God.

FIGURE 6.11 ***(Continued)***

2.0 Objectives

2.1 To write an article for Sociology and Social Research.

2.2 To report the results of my survey to the students, faculty, and staff of ISU, via the ISU Bengal, and to the ISU AIDS Education Committee.

2.3 To determine if increased knowledge about AIDS is correlated to changes in sexual behavior.

2.4 To determine if increased concern about AIDS is correlated to changes in sexual behavior.

2.5 To discover how ISU compares to other universities regarding knowledge. concern/perception, opinions, and sexual behavior.

2.6 To establish a baseline of proof (if any) for the need for continued education about AIDS.

2.7 To determine the direction future AIDS education at ISU should take.

2.8 To determine if students at ISU do engage in risky sexual behaviors and are therefore susceptible to contracting HIV.

2.9 Expected results:
- 2.9.1 Basic knowledge will be high.
- 2.9.2 Personal concern will be low.
- 2.9.3 Level of knowledge will have no effect on behavior.
- 2.9.4 Level of concern will have some effect on behavior.
- 2.9.5 People will support condom machines on campus and HIV infected children in schools.

3.0 Review of Literature

My research for this project centered on four categories of material, which provided me with background information and comparison figures focusing on AIDS and the college student.

The first category, and by far the most extensive, is previous surveys. Most of the surveys I found were done at other universities in the U.S. (9.1. 9.7, 9.8, 9.11, 9.17, 9.18, 9.25, 9.27, 9.28, 9.29, 9.31). Of these, two surveys were specifically aimed at measuring knowledge (9.1, 9.31). and two dealt with sexual behavior (9.7, 9.18. 9.29). One paper reviewed a number of surveys (9.11) some of which I have included in my research, one looked at peer pressure and homophobia (9.17), and one article reported on two ISU surveys concerning condom machines on campus (9.25). The final source in this section is a paper on a small survey (n-84) done at ISU that looked at perception, concern, and some safer sex practices (9.27). A few surveys were done at universities outside the U.S. (9.13, 9.15, 9.24). and some were done with specific populations (other than college students) (9.4, 9.6, 9.21). In general, these surveys report that basic knowledge is good. but concern is low, and sexual behaviors are not changing.

The second category is information about AIDS. This includes the need for education (9.16, 9.23), reports on misconceptions, precautions, fears, hysteria, prejudices, and risks (9.10, 9.14, 9.30), facts about AIDS (9.9, 9.22), and activities on some campuses to fight AIDS (9.5).

The third category features statistics about AIDS. Two articles report on the Centers for Disease Control (CDC) study of the incidence

FIGURE 6.11 ***(Continued)***

of HIV on campuses in the U.S. (9.2. 9.20), and another reported on a different blood sample survey also documenting incidence of HIV infection in college student populations (9.12). One source is my interviewee, Dr. Larry Farrell, who provided incidence statistics both nationally and locally (9.9). I also include as a source in this category a local epidemiologist; Jack Bennett, who contributed Idaho AIDS/HIV Statistics (9.3).

The final category focuses on sex and the college student and includes just two resources. The first reports on sexuality and sexual behaviors of college students in 1980, before anyone knew about AIDS (9.19). The other article reports on the spread of AIDS among heterosexuals, especially on campus (9.26). It was published in 1987 and shows relatively little change (if any) in sexual behavior since the threat of AIDS.

4.0 Method

4.1 Research previous surveys done at ISU, other universities, and the general public.

4.2 Research information on AIDS, sexual behavior of college students.

4.3 Test a preliminary survey for understandability, needed changes, etc.

4.4 Design and administer survey dealing with knowledge, perception, opinion, and behavior.*

4.5 Analyze returned surveys for frequency.

4.6 Cross-tabulate data.

4.7 Analyze cross-tabulated data.

4.8 Interview Dr. Larry Farrell, microbiologist and AIDS lecturer.

4.9 Write up findings for journal article.

4.10 Report findings to ISU Bengal, ISU AIDS Education Committee.

4.11 Submit journal article to Dr. Tracy Montgomery for evaluation.

4.12 Present oral report of survey/article to class.

4.13 Submit article to Sociology and Social Research.

5.0 Facilities/Equipment

Facilities:
Eli M. Oboler Library,
Idaho State University

Equipment:
ABsurv-survey analysis program
First Choice-word processing program
Computer
Printer
Paper
Reference Books
Copy Machine

*See copy of survey attached at end of section 4.0.

FIGURE 6.11 (*Continued*)

CONFIDENTIAL AIDS SURVEY

1) Did you attend a lecture by Richard Carper during the week of October 22–26, 1990?

___YES ___+ When?

___NO ___> Go to #9.

___Wed. night public lecture

___In a class

___Friday luncheon

2) Why did you attend?

___extra credit

___information about AIDS

___interested in subject

___curiosity

___normal class period

___other

3) Did you learn anything new?

___YES ___NO

___About AIDS

___About risky behavior

___About how to prevent AIDS

___About how it feels to have AIDS

___Other

4) Prior to the lecture by Richard Carper how did you feel about people with AIDS?____________________

____

____

____

5) Do you think Richard Carper was effective in his presentation of the subject?

__YES ___NO ___SOMEWHAT

6) Has your opinion of people with AIDS changed since hearing Richard Carper?

__YES ___NO ___SOMEWHAT

7) Have you made any changes in your lifestyle because of Richard Carper's presentation?

___YES ___NO ___SOMEWHAT

8) Additional comments about Richard Carper and/or his presentation(s):

____

____

____

____

____

____

____

____

____

____

____

____

9) The only safe sex is_________.

10) The next safest sex is_______.

11) AIDS is transmitted by:

___coughing

___giving blood

___kissing

___sharing needles

___tears (drugs)

___TV

___mother's milk

___hugging

___semen

___vaginal secretions

___breathing

___blood

___sharing eating utensils

___blood transfusions

___toilet seats

FIGURE 6.11 ***(Continued)***

12) How have you learned about AIDS?

___class ___friends

___special lectures ___magazine articles

___parents ___children

___books ___other

13) Do you personally know anyone who is HIV+ or who has AIDS?

___YES ___NO

14) There is a cure for AIDS.

___YES ___NO ___UNDECIDED

15) AIDS is something most people should be concerned about.

___YES ___NO ___UNDECIDED

16) People can catch the AIDS virus from public toilet seats.

___YES ___NO ___UNDECIDED

17) AIDS can be transmitted from females as well as from males.

___YES ___NO ___UNDECIDED

18) Most people who become infected with the AIDS virus will not get fully developed, fatal AIDS.

___YES ___NO ___UNDECIDED

19) Would you stop being friends with a person if you discovered that he or she had AIDS?

___YES ___NO ___UNDECIDED

	STRONGLY AGREE	AGREE	DISAGREE	STRONGLY DISAGREE	NOT APPLICABLE
20) AIDS is the most urgent health problem facing the U.S.	SA	A	D	SD	NA
21) AIDS is the most urgent health problem facing Idaho.	SA	A	D	SD	NA
22) AIDS affects only limited, specific groups of people.	SA	A	D	SD	NA
23) I am personally concerned about contracting AIDS.	SA	A	D	SD	NA
24) I am concerned about my teenage children getting AIDS.	SA	A	D	SD	NA
25) I have or will discuss AIDS with my children.	SA	A	D	SD	NA
26) I have or will discuss safer sex practices with my children.	SA	A	D	SD	NA
27) Because of AIDS I have increased the length of time I date someone before engaging in sexual intercourse.	SA	A	D	SD	NA

FIGURE 6.11 (*Continued*)

	STRONGLY AGREE	AGREE	DISAGREE	STRONGLY DISAGREE	NOT APPLICABLE
28) Because of AIDS I am more likely to ask a potential sex partner about his/her sexual history.	SA	A	D	SD	NA
29) Because of AIDS I have made the decision to remain celibate prior to marriage.	SA	A	D	SD	NA
30) Because of AIDS I am more likely to use condoms when having intercourse.	SA	A	D	SD	NA
31) Because of AIDS I am less likely to engage in oral sex.	SA	A	D	SD	NA
32) Because of AIDS I am less likely to engage in anal sex.	SA	A	D	SD	NA
33) Condoms should be available in the bathrooms on campus including in the dorms.	SA	A	D	SD	NA
34) HIV infected children have a right to go to public school.	SA	A	D	SD	NA
35) There is little that can be done to prevent the spread of AIDS.	SA	A	D	SD	NA
36) AIDS is a punishment sent by God because of people's immoral behavior.	SA	A	D	SD	NA

37) AGE:

38) SEX: ___MALE ___ FEMALE

39) MAJOR:

40) HOME STATE:

41) CLASS: ___FRESHMAN ___SOPHOMORE ___JUNIOR ___SENIOR
___GRAD STUDENT

42) MARITAL STATUS: ___MARRIED ___HOW LONG? ___YEARS
___DIVORCED ___SEPARATED ___SINGLE ___WIDOWED
___LIVING WITH SOMEONE (SEXUAL RELATIONSHIP)
___HOW LONG?

43) SEXUALLY ACTIVE? ___YES___+ # OF PARTNERS IN LAST YEAR
___NO

FIGURE 6.11 ***(Continued)***

44) RELIGION:______________________ ___ACTIVE ___INACTIVE

45) To what extent have your religious beliefs affected your sexual behavior?

___NOT AT ALL ___SOMEWHAT ___VERY MUCH

6.0 Personnel

6.1 Sheryl Soderstrom, Principal Investigator

I began this investigation out of my own curiosity about whether the AIDS education we have access to has affected people's behaviors, perceptions, concerns, knowledge, and opinions. My active membership in both the ISU AIDS Education Committee and The Southeast Idaho AIDS Coalition further spurred my desire to obtain some measurement of where students on our campus were at on these issues. I have had classes in Social Research, Social Statistics, and AIDS, and I have participated in a team research project involving a small survey on behavior changes because of AIDS here at ISU. I have also completed successful independent research projects. I am qualified to handle this project and my resume will support my suitability.

6.1.1 My resume is attached.

6.2 Dr. Larry Farrell, Interviewee.

Dr. Farrell is a professor with the Department of Biological Sciences at ISU, specializing in microbiology. He teaches a class on AIDS at ISU and has been hired by the State Board of Education to do in-services on AIDS for teachers in the public school system. Dr. Farrell has also done a number of community presentations on AIDS. He has been at ISU since 1972 and so is familiar with the school and the community. He is an active member of the ISU AIDS Education Committee, the Southeast Idaho AIDS Coalition, the Idaho AIDS Foundation. and the State Board of Education's Drug, Alcohol, and AIDS on Campus Committee. He is generally considered one of the area's leading authorities on AIDS. Dr. Farrell's background and involvement clearly establish his expertise in the field of AIDS and his value and authority in speaking about ISU students.

6.2.1 Dr. Farrell's Curriculum Vitae is attached.

FIGURE 6.11 ***(Continued)***

SHERYL SODERSTROM

40 Oakwood Dr. #114
Pocatello, ID 83204
(208) 233-5854

EDUCATION Current Status: Senior 4.0 GPA

Idaho State University—Pocatello, ID 1989–Present
Bachelor of University Studies; Major—Thanatology, Sociology: Minor—Psychology, Philosophy. Dean's List every semester. Trained as AIDS/Safer Sex Peer Educator.
> ASISU Scholarship—Fall 1990
> Independent Research (not for class): Confidential AIDS Survey (n-245)

Hospice Volunteer Training—Pocatello, ID 1991

University of Alaska—Fairbanks, AK 1987–1989
Major-Psychology: Dean's List every semester.

Life, Death 8 Transitions Workshop—Wasilla, AK 1988
Elizabeth Kubler-Ross 5-day Retreat and Training Session

Headquarters Beauty Academy—Fairbanks. AK 1981–1982
Licensed in State of Alaska 1983–1986.

Galileo—Adams Community College—S.F., CA 1980–1981
Took courses in Licensed Vocational Nursing, Certified as Nurse's Assistant 1981.

Flathead High School—-Kalispell, MT
Graduated with Honors 1974; State Speech Champion 1974.

EXPERIENCE-VOLUNTEER

Muscular Dystrophy Association Telethon

Phone Supervisor—Idaho Falls, ID 1990
Recruited, trained, scheduled, and managed over 80 volunteers. Coordinated with local radio and TV personalities. various businesses, and organizations. Assisted Telethon Coordinator.
Worked independently.

Mail Room Supervisor—Fairbanks, AK 1984–85. 87–88
Located, trained, and managed 5–10 volunteers.
Handled and readied pledges for mail out.
Worked independently.

Phone Volunteer. Mail Room Volunteer 1980, 1983

The Compassionate Friends

Co-leader—Pocatello, ID 1989–Present
Organize, plan and present the monthly topic.
Facilitate and conduct support group meetings.
Assist with public relations and presentations.
> Wrote Dedication speech for the Children's Memorial Garden at BRMC-1990.
> Designed birth-death announcements.
> Met with the Mayor and other city officials about cemetery conditions/maintenance.

Member—Fairbanks. AK; Pocatello, ID 1985–1989

FIGURE 6.11 (*Continued*)

EXPERIENCE-PROFESSIONAL

LDS Church: full-time Missionary, PA — July 1986–July 1987
Enhanced public relations. Wrote weekly reports. Wrote, updated and conducted surveys. Designed pamphlets. Kept records. Team taught a series of six presentations.

The Prime Cut. Fbks, AK: Hairdresser — Nov. 1985–July 1986

Scheduled appointments, sold products, did end-of-day bookkeeping. Helped plan advertisements, specials, and promotions.

Happy Faces Day Care. Fbks, AK — Aug. 1985–Nov. 1985

Licensed Operator. Self-employed. Cared for 5 children daily. Kept the books. Wrote the policy statement and parent contract. Kept required state records. Wrote reports.

Gooden Services. Fbks, AK: Bus Aide — Jan. 1985–July 1985

Assisted driver in providing safe transportation to and from school for mentally and physically handicapped people, ages 5–20.

Wedgewood Hairstyling Salon and The Split End.
Fbks, AK: — Aug. 1982–Jan. 1985

Hairdresser

PAPERS & PRESENTATIONS

"Death and Dying" Guest Lecture. — 1991
Dr. Larry Farrell's AIDS class—ISU, Pocatello. ID.

"Belief in Life After Death as a Measure of Death Attitudes as Established through Epitaphs" Original Research. — 1990
> First Place—2nd Annual Idaho Conference for Students in the Social Sciences and Public Affairs—Boise State University, Boise, ID.
> Honorable Mention—The National Undergraduate Social Science Conference—Weber State College. Ogden, UT.
> 17th Annual Western Anthropology/Sociology Undergraduate Research Conference—Santa Clara University. Santa Clara, CA.

MEMBERSHIPS

Southeast Idaho AIDS Coalition

Served on Mission Statement/Goals Committee. Designed and marketed "Condomgrams."

ISU AIDS Education Committee

Student Representative to the State Board of Education's Committee on Drugs, Alcohol, and AIDS on Campus—1991. AIDS/Safer Sex presentations.

Foundation of Thanatology

Association for Death Education and Counseling

Elizabeth Kubler-Ross Center

REFERENCES Furnished upon Request.

FIGURE 6.11 ***(Continued)***

LARRY DON FARRELL
CURRICULUM VITAE

Personal

Born-November 5, 1942. Woodward, Oklahoma.
Married-August 8, 1965. Two Children.
Present address: Residence-843 North 10th
Pocatello, Idaho 83201
Telephone: (208) 234–0334
Business-Dept. of Biological Sciences
Idaho State University
Poncatello, Idaho 83209
Telephone: (208) 236–3171

Education

B. S. May, 1964, University of Oklahoma, Microbiology.
M. S. August, 1966. University of Oklahoma. Microbiology.
Major Professors, Drs. John Lancaster. Donald Cox. Robert Collier.
Research topic, Phage replication and sporulation In Bacillus subtilis.
Ph.D. August, 1970, University of California at Los Angeles, Bacteriology.
Major Professors: Drs. Frederic A. Eiserling and William R. Romig.
Research Topic, Development of the genetic system of Bacillus subtilis bacteriophage SP02.

Academic Experience

1989–present	Professor, Dept. of Biological Sciences, Idaho State University. Pocatello, Idaho 83209.
1988–present	Assistant Chairman, Dept. of Biological Sciences, Idaho State University. Pocatello, Idaho 83209.
1984–1988	Associate Professor. Dept. of Biological Sciences, Idaho State University, Pocatello. Idaho 83209.
1978–1984	Associate Professor. Dept. of Microbiology and Biochemistry. Idaho State University. Pocatello. Idaho 83209.
1977–1984	Department Chairman. Dept. of Microbiology and Biochemistry, Idaho State University, Pocatello, Idaho, 83209.
1972–1978	Assistant Professor, Dept. of Microbiology and Biochemistry, Idaho State University. Pocatello, Idaho 83209.
1970–1972	Instructor, Dept. of Microbiology, University of Illinois, College of Medicine, Chicago. IL 60680.
1966–1967	Teaching Assistant, Dept. of Bacteriology, University of California, Los Angeles. California 90024
1964–1966	Teaching Assistant, Dept. of Botany and Microbiology, University of Oklahoma. Norman, Oklahoma 73019.

FIGURE 6.11 (*Continued*)

Courses Taught:

Introductory Microbiology
Acquired Immune Deficiency Syndrome (AIDS)

General Virology
Bacterial Virology Laboratory
Survey of Electron Microscopy
Microbial Genetics
Medical Microbiology
Advanced Bacterial Virology
Practical Electron Microscopy
Principles of Molecular Biology

Research Experience

1967–1970 Graduate Student, Dept. of Bacteriology, University of California, Los Angeles, California 90024.

1970–1972 Postdoctoral Fellow, Dept. of Microbiology, University of Illinois College of Medicine, Chicago, IL 60680. Mentor, Dr. Harvard Reiter.

Since 1972, I have directed my own laboratory at Idaho State University. A total of eleven students have completed MS degrees under my direction, seven have gone on to Ph. D. programs and one to a D.O. program. I have been awarded a total of five research grants by the Faculty Research Committee of Idaho State University and one grant by the Academic Dean's Council of Idaho State University.

My primary research interests are in the area of microbial genetics, as specifically applied to bacterial viruses. I am currently working on the viruses of Sphaerotilus natans, a bacterium of commercial importance because of the problems it causes in sewage treatment plants.

Professional Organizations

International Society for AIDS Education
Society of the Sigma XI
American Society for Microbiology (both national and Intermountain Branch)
American Association for the Advancement of Science
Involvement in Professional Organizations:
Member-Volunteers Committee for 78th annual meeting of the American Society for Microbiology, Las Vegas, NV, 1978.
Papers presented at several meetings, Intermountain Branch, ASM.
President, Intermountain Branch, ASM, 1980–81.
Chairman–Volunteers Committee for 85th annual meeting of the American Society for Microbiology, Las Vegas, NV, March, 1985.

Professional Activities

Organizer and Liaison for the Idaho Committee of Correspondence. National Center for Science Education.
Idaho AIDS Foundation. Member of the Executive Board, Co-Director of Education Task Force.
Pocatello AIDS Task Force. Interim Director (1988)
Invited reviewer for chapter on Microbiology in Campbell's Biology, Second Edition, Benjamin/Cummings Publishers.

FIGURE 6.11 (*Continued*)

Special Recognition

Listed in 14th edition of "American Men and Women of Science"
Listed in 22nd edition of "Who's Who in the West"
Listed in 6th edition of "Who's Who in Technology Today"
Listed in 1st edition of "Who's Who in Pocatello"
Listed in 2nd edition of "Who's Who of Emerging Leaders In America"
Listed in 2nd edition of "Five Thousand Personalities of the World"

Publications

"Observations of single phage-infected cells of Bacillus subtitles" Journal of Ultrastructure Research _25:501–506 (1968).

"Phleomycin-stimulated degradation of deoxyribonucleic acid in Escherichia coli" (With H. Reiter) Antimicrobial Agents and Chemotherapy 4:320–326 (1973).

"Phleomycin-stimulated degradation of deoxyribonucleic acid in Escherichia coli. II. Inhibition of solubilization by bacteriophage T4" (With H. Reiter) Canadian Journal of Microbiology _22:645–53 (1976).

"The effects of bleomycin on Ehrlich ascites carcinoma in suspension culture" (With J. T. Eells and L. J. Fontenelle) Proceedings of the Western Pharmacological Society _20:199–203 (1977).

"Handwashing: Ring Wearing and Number of Microorganisms" (with G. Jacobson, J. Thiele and J. McCune) Nursing Research _34:186–188 (1985).

"Use of Serological Techniques for the Detection of Bacteriophages in Dairy Products" (With M. Oswald) J. Idaho Academy of Science _22:15, 1986. ABSTRACT.

"Electron Microscopic Studies of Phage-Infected Sphaerotilus natans " (with R. Pithawallá) J. Idaho Academy of Science _22:16, 1986. ABSTRACT.

"Isolation and Characterization of Bacteriophages for Different Strains of Sphaerotilus natans" (With D. Roberts) J. Idaho Academy of Science 24:10, 1988. ABSTRACT.

"AIDS Education—An Opportunity for Interdisciplinary Education" (With J. Girvan) Accepted for publication in Education.

"Developing an Effective Half-Day Secondary School Inservice on Acquired Immune Deficiency Syndrome (AIDS)" (With J. Girvan) The Clearing House 62:381–383 (1989).

Non-Refereed Publications

"AIDS Update" (With J. Girvan). A series of articles appearing in the Idaho State Journal during 1987–88.

"AIDS in Idaho: The Facts, the Fears. and the Information Resources" Idaho Librarian 40:32–34. 1988.

FIGURE 6.11 ***(Continued)***

7.0 Budget

7.1 Personnel

7.1.1 I am the only investigator working on this project. My research and analysis fee is $15.00/hour. I foresee a total of 80 hours of labor to complete the project.

80 hours $15.00/hour—	$1200.00
7.1.2 Dr. Larry Farrell, consultation fee: 1 hour X $50.00/hour—	$ 50.00
	$1250.00

7.2 Supplies and Equipment

ABsurv program—	$ 160.00
Survey copies (275 copies of 2 page survey) 550 pages X $0.05 per page—	$ 27.50
Printer ribbon (1)—	$ 10.45
Computer paper 600 sheets X $0.0125—	$ 7.50
3-hole punch (1)—	$ 12.95
Hi liter (1)—	$ 1.00
Pens, pencil—	$ 2.50
Binder (1)—	$ 5.35
Columnar pad (1)—	$ 2.50
	$ 229.75

7.3 Research Material

Copies of articles 125 pages X $0.05—	$ 6.25
7.4 Total Budget—	$1486.00

8.0 Schedule

10/90–11/90. 2/91–3/91 Library Research, reading articles
10/27/90 Preliminary survey
11/1/90–11/15/90 Administer Survey
12/15/90–1/15/91 Analyze for frequency
2/1/91–3/25/91 Analyze cross-tabulated data
3/8/91 Interview—Dr. Larry Farrell
3/14/91 Proposal due
3/15/91 Personal Conference—defend proposal
3/25/91 (week of) Interview report due
4/15/91 (week of) Progress report due
4/22/91 Major project due
4/29/91 (week of) Oral presentation due

9.0 References

9.1 AIDS Survey, 1990. AIDS Task Force. Lewis-Clark State College, Lewiston. Idaho.

9.2 A Scary Little Survey of AIDS on Campus. 1988. U.S. News & World Report. 14 November: 12.

9.3 Bennett, Jack. 1991. Lecture notes. Guest speaker in Dr. Larry Farrell's Biology 469 Microtopics: AIDS class. Idaho State University. Pocatello, ID. 29 January.

FIGURE 6.11 ***(Continued)***

9.4 Bezilla, Robert, ed. 1988. The Gallup Study on America's Youth 1977–1988. The Gallup Organization, Inc. Princeton, NJ.

9.5 Bruno, Mary, et al. 1985. Campus Sex: New Fears—AIDS is the latest STD to hit the nation's colleges. Newsweek. 28 October: 81–82.

9.6 Bryan, Clifford. 1989. Study sponsored by NAACP and CDC and Idaho Department of Health and Welfare. Reported in the Idaho State Journal. Found on Sociology Department bulletin board, Idaho State University. Fall.

9.7 Carroll, Leo. 1988. Concern with AIDS and the sexual behavior of college students. Journal of Marriage and the Family. 50:405–411.

9.8 College Students only 'Paying Lip Service' to Safe Sex Message. 1990. Pediatric News. Orlando, Florida. May.

9.9 Farrell, Larry D. 1991. Personal Interview. Idaho State University. Pocatello, ID. 8 March.

9.10 Fear of AIDS virus spreads to colleges nationally, 1985. CPS, carried in The ISU Bengal. 1 November.

9.11 Fennell, Reginald. 1990. Knowledge, Attitudes and Beliefs of Students regarding AIDS: A review. Health Education. July/Aug. 21:20–25.

9.12 Gayle, Helene D., et al. 1990. Prevalence of the human immunodeficiency virus among university students. New England Journal of Medicine. 29 November. 323:1538–41.

9.13 Greatorex, Ian Frederick & John Michael Valentine Packer. 1989. Sexual Behavior in University Students: Report of a Postal Survey. Public Health. 103:199–203.

9.14 Hacinli, Cynthia. 1988. AIDS, straight: A Heterosexual-risk update. Mademoiselle. August: 138.

9.15 Hanna, Jack. 1989. Sexual Abandon: The Condom is unpopular on the campus. Maclean's. 25 September:48.

9.16 Hepworth, Jeri and Michael Shernoff. 1989. Strategies for AIDS Education and Prevention. Marriage & Family Review. 13:39–80.

9.17 Hirschorn, Michael W. 1987. AIDS is Not Seen as a Major Threat by Many Heterosexuals on Campus. The Chronicle Of Higher Education. 29 April. 33:1, 32–34.

9.18 Ishii-Kuntz, Masako. 1988. Acquired Immune Deficiency Syndrome and Sexual Behavior Changes in a College Student Sample. Sociology and Social Research. October. 73:13–18.

FIGURE 6.11 (*Continued*)

Diane Maree Desselle
421 Campus Vista
Pullman, WA 99166

Dr. P. R. Gavin
Department Head
Department of Veterinary
Clinical Medicine and Surgery
College of Veterinary Medicine
Washington State University
Pullman, WA 99164-7012

August 1, 1999

Dear Dr. Gavin:

The attached document is a proposal to conduct a study on alternatives to live animals in student surgery laboratory. The study will consist of interviews with professionals and other concerned parties and research into various types of alternative models. The results of the study will determine the best alternative methods available, depending upon usefulness and price.

I hope you will realize the need for this problem to be addressed. I feel confident that this study will prove beneficial to the veterinary community.

Sincerely,

Diane Desselle

Diane Desselle

FIGURE 6.12 **Student Proposal 2**

PROPOSAL TO STUDY THE METHODS AND EFFECTS
OF ALTERNATIVE TRAINING IN VETERINARY
SURGICAL LABORATORY

Submitted to the Washington State University
College of Veterinary Medicine

FIGURE 6.12 ***(Continued)***

Submitted by Diane Desselle on August 1, 1999

FIGURE 6.12 (*Continued*)

Table of Contents

FIGURE 6.12 ***(Continued)***

Abstract

This proposal will address the need for finding the best alternative models for live animal use in student surgery laboratory. The demand for decreasing the use of live animals for training purposes is steadily growing, and the best training possible is necessary to produce quality veterinarians.

Several alternative models are in use today, but this study proposes to find the best model both in quality and in cost. I propose to conduct interviews with various people in the field including veterinarians, veterinary students and veterinary professors. I will ask students and veterinarians how they feel about alternative methods, why they feel that way, and whether to continue in this manner. I will ask professors what kind of effects those methods have on the quality of education.

The proposed study will last 13 weeks and will cost approximately $1,560.

FIGURE 6.12 (*Continued*)

Statement of the Problem

In veterinary schools around the country there is a growing concern among students that doing live animal surgical practice for the sake of learning does not mesh with the ideals of medicine. As students of veterinary medicine their goal is to promote and maintain the good health of the animal, not to cause pain.

Performing surgery on live animals which don't require the surgical procedures at best makes some students uncomfortable and at worst makes some students abandon veterinary medicine.

In order to remedy the situation a number of substitutes to live animals for surgical training have been introduced, in practice and in theory, ranging from the use of cadavers to highly technical holographic techniques. Substitutions using these techniques can be costly and/or time consuming, both to the veterinary school and to the students.

These substitutions could also have the effect of not adequately preparing the veterinary students for actual surgery once outside of the academic world. Surgery in actual practice has to be accurate, yet quick, because mistakes and/or wasted time could mean the death of the animal. However, in a surgery in which no live animals are used the surgeon may have the tendency to relax and make more mistakes.

Finding the best substitution for live animal surgery available will prove useful to veterinary schools as a way to better the quality of the education available. It will also prove useful for veterinary students who will be able to learn surgical techniques without having to harm live animals. The veterinary community as a whole will benefit because students who learn surgery using these techniques will most likely become better veterinarians. Animal deaths because of training will be lessened and animal rights groups may be placated.

Good training is important to the well-being of the animals and to the livelihood of the veterinarian. This good training can only be achieved if the best methods available are utilized.

Objectives

One of the major objectives of this research would be to determine which alternative training method or methods would be the best academically and financially. This determination would be based upon information obtained on each type of method currently used or proposed. There would be a cost analysis, and a summary of the pros and cons of each method. Also, interviews with those in the field would provide a background of which method they feel is the best among those they use.

A second objective of this research is to determine the effects of alternative training on the performance of the veterinary student both pre-graduation and post-graduate. If the effects are negative, different alternative approaches need to be found.

The final product would be an article proposing the best alternative technique and would give the veterinary colleges the chance to compare their own surgery laboratory practices with those proposed in the article.

FIGURE 6.12 (*Continued*)

Review of Literature

The ideas for using surgical substitutes gained recognition in the mid 1980s. Veterinarians, veterinary colleges and the non-veterinary community all have opinions on live animal use. There is a common theme among all: veterinary surgical training needs to decrease the use of live animals through alternative methods, but complete eradication of live animal use is not an option.

Dr. P. B. Jennings, Jr. (1986) listed several alternatives to the use of live animals in student surgery laboratory. These techniques ranged from the use of cadavers and synthetic models to the training of veterinary students in the same way medical students are trained, with lots of internship and residency which is most costly and time consuming. ("Alternative to the Use of Living Animals in the Student Surgery Laboratory." *Journal of Veterinary Medical Education* 13: 14–15.)

Dr. Eberhard Rosin (1986) finds that although using live animals in surgery laboratory raises ethical questions, alternatives fail to teach the necessary skills required of professional veterinarians, and should not be the only training available. He feels strongly that live animal use is necessary in order for the skills to be taught adequately. ("The Importance of Using Living Animals in the Small Animal Student Surgery Laboratory." *Journal of Veterinary Medical Education* 13: 11–13.)

DeYoung and Richardson (1987) find that the use of bone models for orthopedic teaching is superior as a primary teaching device for students. They claim that the students' self-confidence is bolstered and control and comprehension are superior to those who do not use models. ("Teaching the Principles of Internal Fixation of Fractures with Plastic Bone Models." *Journal of Veterinary Medical Education* 14: 30–31.)

Dr. Franklin M. Loew (1989) found that Tufts University has implemented a program in which live animals are not used for unnecessary medical practices. This was an action requested by students. To make up for not using live animals the students who take the course will be required to spend extra time in surgical and clinical rotation. ("Tufts Develops Alternative Program for Teaching Surgery." *Journal of the Veterinary Medical Association* 195: 155.)

Bernard Z. Rollin (1990) advocates the rights of the students who find live animal surgery morally wrong. He feels that the Veterinary profession will benefit from these students because they show the kind of concern necessary for a caring practice. ("Changing Social Ethics on Animals and Veterinary Medical Education." *Journal of Veterinary Medical Education* 17: 2–5.)

Dr. James Lincoln, in an interview (1991), discussed various techniques used at Washington State University for surgery laboratory. He stressed that live animal use should not be completely taken out of the program because they are necessary in some cases. He believes that the quality of education when using alternative methods is dependent on the students' attitudes.

Finally, Nils Soren Peterson, a computer programmer and representative, (1991) discusses the use of computers in the place of live animals. He points out that the computer is far superior in such exercises as the study of Mendelian genetics. He also states that many new

FIGURE 6.12 ***(Continued)***

innovations are just prototypes but advances are being made rapidly in order to improve education while reducing animal use. ("Education, Computer Software and Animal Welfare." *Animal Welfare Information News Letter 2: 3+.)*

Methods

This study will gather most of its data from interviews with veterinary professors, veterinary students and veterinarians in the community. The interviews will be one on one. The questions will be designed to discover how the interviewee feels about live animal surgery laboratory and whether he/she feels that substitute methods give accurate training. In addition, professors will be asked what kinds of alternative methods are used and which ones they feel are most useful.

Answers will be evaluated by myself based on a compilation of all interviews. These interviews will provide useful information not only about how the general veterinary community feels about alternatives to live animals, but also what alternatives professors feel are the most useful.

The rest of the data will be gathered at the Washington State University College of Veterinary Medicine's library or at the Owen Science and Engineering Library as secondary information.

Facilities

The various facilities of Washington State University will be utilized. Research will be conducted primarily in the Veterinary Medical/Pharmacy Library and in the Owen Science and Engineering Library, with the other three library facilities employed as needed. These libraries allow access to books, journals, magazines. newspapers, government documents and other materials, and have over three million items on hand. The various computer services that the 1ibraries give access to, such as Coug-a-log and the night search, will also be utilized. The Veterinary Medical/Pharmacy Library has many useful journals and monographs on hand that may be employed to conduct research. Washington State's Computer Science Department will also be used to complete the finished article. The CMS system will be employed to format the text and it will be printed using the laser printers available for student use.

In addition, interviews will take place in the Washington State University College of Veterinary medicine and at various veterinary clinics around Pullman. Dr. James D. Lincoln, an associate professor in clinical medicine and surgery at the College of Veterinary Medicine will be one of the interviewees. Dr. Lincoln has been with Washington State University since 1977. He has been on the faculty of the College of Veterinary Medicine since 1980 and teaches small animal surgery laboratory with an emphasis on orthopedics. Dr. Lincoln's laboratories use alternatives to live animals for most surgery laboratory periods. Other professors and students will be interviewed. An interview may also take place at the Washington State University College of Engineering and Architecture.

FIGURE 6.11 ***(Continued)***

Personnel Report

Diane Desselle: Investigator. Zoology major at Washington State University. Bachelor of Science anticipated in May, 1993. Applied for admission to the College of Veterinary Medicine, October, 1991. Worked in a veterinary clinic as an assistant from 1988 to 1991. Will do research, evaluation, compilation and final draft of report.

Budget

Salary

My fee for design, research and evaluation of the information is $12/hour. I foresee 130 hours of work for a total of $1,560.

Overhead Costs

Overhead costs include computer use at $250 and the preparation of the final report (including labor and printing fee) at $70.

Total Cost

The total cost of the project is $1,830.

Time Line

The schedule for this article will be as follows:

- Research of previous work ....2 weeks
- Interviews ....4 weeks
- Compilation of results ....3 weeks
- Examination and interpretation of results ....4 weeks

The project will begin in early September and will end in mid December. Some parts of the schedule, such as the research and the interviews may run concurrently.

Expected Results

Several methods have come forth in previous years as alternatives to using live animals in surgery labs. It is important to do a thorough review of these methods to find the best alternative available and how alternative surgical training affects student performance. The importance of proper training cannot be understated. I expect the results of my work to highlight the most workable alternative method available in order to increase training productivity.

Evaluation

Once a new program based on the results of this article is implemented, it will be necessary to evaluate its usefulness in an actual situation. The most important evaluation would be student performance. Student performance could be evaluated during the fourth year of classes. During this year, students actually work on clinical cases, where their performance is already evaluated. The performance of students trained on the alternative method described in the article could be compared with the performance of students trained in other alternative methods.

FIGURE 6.11 ***(Continued)***

Another method of evaluating student performance is through employers after graduation. Employers are only going to hire the veterinarians they feel are most qualified to work for them. If employers are more pleased with the veterinarians trained using the new alternative methods than those trained on methods already in use. then the new program is useful. Employer responses could be gained through surveys over a three year period, comparing the performances of new graduates trained on the alternative method described here to those trained in other ways.

Curriculum vita

DIANE M. DESSELLE

Campus Address
SW 421 Campus Vista
Pullman, WA 99163
(509) 332-2301

Home Address
3011 Boundary St.
Olympia, WA 98501
(206) 943-6312

Education 1991– Washington State University
Bachelor of Science in Zoology
Expected in May 1993

1988–1991 Centralia Community College
Associate of Science in Pre-veterinary medicine
Received in June 1991
G.P.A. 3.04

Experience 1988–1991 Veterinary assistant, Kitty Klinic
Lacey, WA 98503
Assisted in reception area, examinations, and surgery.
Filled prescriptions. Did laboratory work. Cleaned cages.

1990–1991 Stall cleaner, James Gang Ranch
Lacey, WA 98503
Cleaned horse stalls. Assisted in training.
Assisted in veterinary work. Fed, watered and blanketed horses.

Activities

1991– Partnership for Equine Therapy and Training.

1991– Organization for Future Veterinarians.

1991– People-pet partnership.

FIGURE 6.11 ***(Continued)***

Solving Problems Through Periodic (Progress) Reports and Completion Reports

CASE STUDY

LEAH FELDSTEIN'S OFFICE

Educare, Inc. ■ 1200 Stone Way, Suite 500 ■ Seattle, Washington

Leah Feldstein was nearing the end of the third quarter of the second year on a $300,000 grant from the John T. Elliston Foundation, a grant that she had had a big part in writing the proposal for. Although she had already written six quarterly reports on EDUCARE's Program Drugfree, she was particularly concerned about this one, because Elliston was embarking on a close review of all funded projects, with an eye to possible cuts. She knew that if she weren't persuasive about the progress EDUCARE had made in drug rehabilitation, the firm could lose part or all of its funding on the renewal grant that had to be proposed next quarter.

She also knew that EDUCARE had what Elliston saw as problems with Program Drugfree. Her readers in Washington would certainly have a keen interest in what EDUCARE had done to solve the problems and what the future looked like for continued improvement.

PREVIEW OF THE CHAPTER

(continued)

Leah quickly reviewed the problems. First there was the relatively static client population. In the almost two years since this funding period began, the number of clients had stayed right around 150, even though she had hired three new counselors for a total of eight. The ratio of 18 to 20 clients per counselor she believed was favorable—counselors could spend more time with each client—but unfortunately the people in Washington were often more concerned with funding formulas and wanted to see higher ratios; more clients per counselor meant to them greater impact for total expenditures.

This reality led directly to the second problem. Leah did not believe in aggressive recruitment. She thought that enlisting clients should be done by existing means: referrals by teachers, parents, and other social service workers in the Seattle area. There was also, of course, the occasional self-referral. Through these means the program had enlisted a total of 198 new clients, replacing those who had passed out of the program, either successfully or unsuccessfully, over six quarters. Although she knew Elliston would like to see higher enlistment rates, she wanted her staff to spend their energies working with the existing clients, not on beating bushes looking for new ones. Leah thought that good work done with the existing pool was more important than expanding the pool, and thus perhaps risking a dilution of their primary effort: good and effective counseling, which meant getting clients off drugs for good.

This problem of recruitment was related, as was the first, to the last problem, which in many ways was the most difficult to confront. Leah wondered whether she was going to be able to convince Elliston that Program Drugfree indeed was effective in getting clients off drugs and then in keeping them off drugs. The amount of time clients spent in counseling both in and out of school, the number who had gotten off drugs, and the number who had stayed off drugs were all comparable to figures of other programs that she knew about. In some cases, EDUCARE's success rate was even better. But she was not sure, given the problems of counselor-client ratios and of recruitment, that the results would be viewed as favorable by her readers. That perception, she concluded, she would have to keep foremost in mind while she wrote her report.

There was a lot riding on this report, but having defined the problems at least, Leah decided she was ready to start consulting with her staff and then to begin drafting the report.

FUNDAMENTALS OF A PERIODIC REPORT

Purpose and Rationale

Periodic reports are also referred to as *progress reports.* This term can be misleading because many people think of a progress report in only one way: a document reporting positive change in the condition of ongoing work over time. The difficulty hinges here on the notion of "positive change." Not all progress reports can document positive change over time. Sometimes they must report negative change; for example, instead of being able to report on how well a new marketing strategy has worked for a new product, some manufacturers may have to admit that the product had less appeal than they originally thought it would or that the strategy to market it was poorly conceived or timed. But those who are responsible for documenting the progress of the marketing campaign (progress thought of here in the most general of terms—how the campaign is getting along over time) must still write reports.

These reports may be written weekly, bimonthly, quarterly, or on any other schedule that seems to fit the needs of the situation. For that reason, many have come to call progress reports *periodic reports,* a more useful term that describes the fact that they are written at distinct intervals and that removes the idea that they must report only positive changes.

Periodic reports, then, are written to document changes in conditions, either positive or negative, of any ongoing work of an organization that is crucial to its operation. They are not usually written to document what might be called the everyday work of an organization—work that does not essentially change over time or does not affect major policy decisions.

Another important element of periodic reports is that they are written to those in managerial positions so that these people can make decisions; therefore, the reports must be informative and persuasive in a way that makes options for action clear. Management relies on these reports to decide whether a project is on the intended track, whether stated goals are being met, whether the budget is sufficient, and finally, whether the project or plan of action is on schedule.

One additional and important factor having to do with purposes for writing and reading periodic reports needs to be considered, especially in Leah Feldstein's case. Most writers of periodic reports write only for the managerial audience of the organization they work for, who for Leah was John

Fiske, Director of EDUCARE. But because EDUCARE is funded by external means, it is obliged, often by contract, to report periodically on the progress of projects to agencies that fund it. This was Leah's situation. John Fiske always expected Leah to write the periodic reports to Elliston, even though he was the signatory of the report. So Leah had to write for not only John Fiske but also the program officers at Elliston.

She found, however, that she could take good advantage of necessity. John Fiske became, in a significant way, a reviewer and an editor of her periodic reports to Elliston. He could consider not only problems that were strictly internal, such as those having to do with scheduling and personnel, but also those problems that Elliston would be concerned about, like client-counselor ratios, recruitment, and the overall success rate. Leah found that writing for John Fiske, which she had to do anyway, was a good built-in editing device that made for better reports for Elliston.

In general, periodic reports, whether they are written strictly for internal audiences or also for external audiences, play an integral part in decision making. They become, therefore, one of the more important kinds of writing that professionals are required to do.

QUESTIONS THAT PERIODIC REPORTS ANSWER

Different audiences will require different formats, but essentially, periodic reports should cover the following topics:

1. *Orienting information:* project addressed, time reported on, background, changes in report format.
2. *Evaluative overview of work done for current reporting period:* the report's thesis and a conclusive statement of what happened.
3. *Current status:* a review of the major actions and their significance in moving the project along, problems encountered, and recommendations.
4. *Conclusions:* a summary of the significance of what has happened, including a preview of future work and objectives.

The answers to several questions will help specify the purposes for writing and reading periodic reports, and therefore help determine the basics of their contents (see Figure 7.1).

What Project or Work Is Being Reviewed in the Report?

Most organizations have multiple projects going on at the same time, so it is necessary to identify the one the report discusses. This step can be done simply by using a familiar title or name, sometimes in combination with a code number. In writing for Elliston, Leah has to identify not only EDUCARE, the group responsible for the work, but also the project, Drug Rehabilitation (as distinguished from Drug Education, which is a different project), as well as

Questions That Periodic Reports Answer

- What project or work is being reviewed in the report?
- What period of time does the report cover?
- In summary, what has happened during this period?
- What work has been accomplished?
- What problems arose while doing the work?
- Is the work on schedule and within budget?
- What work has to be done in the future?

FIGURE 7.1 **Questions to ask and answer to define the contents of a periodic report**

the contract number Elliston has assigned to it. Since Elliston has hundreds of contracts with many organizations, it would have trouble finding the documentation in its database without the contract number.

Leah usually identifies her reports in this way:

EDUCARE, Inc., Seattle, WA, Drug Rehab, Ell DRA x15-023

What Period of Time Does the Report Cover?

Since periodic reports are written after distinct periods of time, it is important to identify the particular reporting period in order to keep the report separate from previous and future reports. Leah is about to report on the activities of the third period, but it is the third period of the fiscal year that begins in July, the customary working calendar of govemment and many industries. She will write the following, then, very simply:

Period Covered
Third Quarter, January–March, 2001

In Summary, What Has Happened during This Period?

Readers of all reports need to read a general summary of the activities and work of a project and an evaluation of their results. As in the case of proposals, discussed in Chapter 6, an abstract or *executive summary* lays out the conclusive results—what actually happened and what it means—for readers. Readers, but especially those in upper management, can, after reading the summary, decide what part they have in continuing the work or furthering the goals of the project; they can evaluate better who should know about the activities and what kinds of questions about completing the project successfully need to be asked and answered and by whom.

Some readers, depending on their positions, will make decisions about what has to happen immediately, on the basis of what they learn from the summary. Others will pass the report along to others for them to make decisions.

In either case, the executive summary is important at initial stages in considering what has to happen and who is responsible for making it happen.

In this section, Leah has to cover the whole gamut of concerns: the specific activities of her staff, the changes she and her staff have proposed in completing those activities, the results of their activities to date, and the likely results of future activities, both as they have been carried out in the past and as they may change, in view of new activities proposed. But Leah has to remember that *as conclusive as this section is supposed to be, it is still intended to be a summary,* not a detailed account of the activities, which is left to the next section, Work Accomplished.

What Work Has Been Accomplished?

Clearly, readers of periodic reports need to know what has happened during the reporting period. They look to see whether the activities projected for the period actually conform to what was done and whether anything was done differently. But of most importance, readers want to know whether overall goals of the project are being met by the work of the period. Therefore, writers must not only describe clearly and thoroughly what has happened but also explain how that has furthered the purposes and objectives of the project.

In her report, Leah has to review the usual activities of her counseling staff and show how they have been promoting the goals of drug rehabilitation as defined by both EDUCARE and Elliston. But for this report in particular, since so much is riding on it, she has to stress also the initiatives she and her staff have taken to move clients successfully out of the program. Her report, then, has to include discussion of the changes in group counseling: the formation of smaller groups (no more than seven per group, rather than the usual ten to twelve) that met more frequently (on the average of twice a week, rather than once a week). Although she cannot point to a significant increase in the number of clients moving successfully out of the program, reports from her staff on the progress of their individual counseling activities indicate that their clients enjoy their new group sessions because they are more intimate. The clients are also glad to meet twice a week, because they think the sessions are becoming more helpful. Her staff predicts that with these generally more positive attitudes, the success rate will be higher during the next quarter.

What Problems Arose While Doing the Work?

In many ways, the problems section of a periodic report is more important than the section on work accomplished, because problems, if they exist, require management to make immediate decisions about how to solve them and get the project back on track, and then, perhaps, to make future decisions about how to avoid similar problems with similar projects in the future.

The problems section requires analysis and decision making not only from the readers but also from the writers. In fact, it is the job of the writers to help their readers make decisions about what to include in this section. If a relatively minor problem has cropped up, say some procedural snare in the routing of information that was solved quickly and without implications for the future, writers should know not to include discussion of it. But if a procedure that has been followed for some time has become increasingly more burdensome as more work is accomplished or has become more complex—for example, the ways that payments are made to vendors or consultants—then writers should analyze and describe the problem in detail: what has gone wrong, what the source of the problem is, and how progress specifically and generally has been affected.

But the analysis should not stop here. Good reporters of problems will also analyze and suggest possible solutions. This step is very important for two reasons: it shows that the writers are doing their jobs responsibly and not expecting upper management to do their thinking for them, and it helps readers decide more efficiently on a solution or on a plan of action. Even if the ultimate solution is not one that writers suggest, readers at least have some options, so that they can better decide that something else needs to be done.

When planning what she would include in the problems section of her report to Elliston, Leah had a slightly different situation to analyze from the one she had in writing for EDUCARE's internal review. Since she did not see the static population or lack of aggressive recruitment of clients as real problems, even though she thought her readers would, she decided that she could not reasonably discuss them as problems. If she did, she would have to suggest some solutions, like changes in methods of observation and referral, that she didn't believe in. The only legitimate problem, in her view, was the rate at which clients successfully left the program. She knew that the rate could be higher, even though EDUCARE's success rate was comparable to several other programs of the same size. And further, she believed that the initiatives she and her staff had taken in group counseling would soon show more positive results.

She decided, then, to discuss the success rate as the main problem of this reporting period and then to refer to the initiatives she had discussed in the work accomplished section as a solution already being tested. She could also project an increase in the success rate for the next quarter on the basis of reports from her staff. In this way, she would demonstrate that she had assessed the problem intelligently and was doing something reasonable about solving it. This solution would also leave room for the program officers at Elliston to suggest some fine-tuning of the initiatives, if they saw fit.

In summary, there are four things to remember about writing the problems section of a periodic report:

1. If a problem has already been solved and is not likely to become a problem again, it shouldn't be mentioned. Mentioning it just wastes time for both writer and reader.
2. If a problem exists that can be solved easily without other managerial intervention, it shouldn't be mentioned either. To do so would also waste the reader's time, and worse, would show lack of initiative on the writer's part.

3. If there is likely to be a significant difference between what the writer(s) and readers perceive as problems, it is the writer's responsibility to analyze the difference carefully and try to address it with some kind of reconciliation or resolution in mind, but without seriously compromising either side's position. Once again, Leah's case serves as a good example. Leah knew that Elliston wanted to see the population in the program grow, but as we have already pointed out, Leah didn't want to take part in aggressive recruitment. She decided to emphasize the allied issue of an increase in the success rate as a reasonable rhetorical compromise.
4. If legitimate problems exist that impede the crucial work of the organization, they should be analyzed and described in detail: what needs to be done, how it should be done, who needs to do it, and when. If the solutions will cost additional money or take more time, a general statement outlining the additional costs and time should also be included. If no problems exist, they shouldn't be created.

Is the Work on Schedule and within Budget?

If the work is on schedule and within budget, then a simple statement to that effect is all that is necessary for this section. But if work has fallen behind schedule or is likely to do so in the near future, or if costs are exceeding projections or are likely to do so in the near future, these realities should be part of the discussion of problems. Management appreciates as much honest troubleshooting as possible about cost and time so that disastrous consequences can be averted.

This section, then, should be reserved for how far the work has fallen behind or will likely be exceeded. Writers need to supply detailed statements, including timelines or budgets that compare the original projections for time and costs with those that are now in place or soon to occur. Refer to Chapter 6 for a discussion of schedule graphics.

Although suggestions on how to minimize time and cost overruns should be part of the discussion of solutions in the problems section, amplifying them here in detail is a good idea. This step not only will reinforce the discussion of problems but also will allow readers who are most interested in time and costs to skim the problems section and get to the details of the overruns in this section.

Leah's situation with regard to costs and time is slightly different from that of others. Since EDUCARE is wholly dependent on external funding to carry out its programs, it is also bound by negotiated contracts that stipulate exactly how much will be spent in what areas and how long the project will last. If Elliston agrees to pay $300,000 over a two-year period, that is all it will spend in that time. Although it may agree to a new contract, that will have to be negotiated, and once it is agreed to, it allows for little or no flexibility.

Leah, then, does not have to worry much about this section of the periodic report, but writers in the private sector, who generate their own funds and can alter spending and schedules to meet contingencies they hadn't planned for, often find this section very important. They want to suggest reasonable alternatives to the original plan, to get the project back on track with as little disruption as possible. Thoughtful, careful analysis and appropriately addressing readers' concerns can accomplish that goal.

What Work Has to Be Done in the Future?

Not only does management need to know what has gone on in the recent past and what is going on in the present, it also wants projections about what will likely happen in the future. Since planning is of crucial importance to all organizations, the need to know about future activities and how they fit in with what has already happened is compelling. Writers should talk about the specifics of what will happen—what work has to be completed, who will do it, how it will be done—all with the general objectives of the entire project in mind.

For Leah, the task is fairly simple. She will talk again about the initiatives taken in group counseling and suggest, on the basis of further information she gets from her counselors and from clients, how the success rate can be improved. This discussion will include plans for changes in individual counseling sessions and staff development.

In these ways, Leah thought she could bolster the success rate of her clients in the future. Her plan certainly makes good sense philosophically and is probably practical and sensible, but it will have to be tried, an undertaking that is within her abilities and resources, and that she needs to lay out for both John Fiske's and Elliston's considerations in this section.

FORMAT AND SUBSTANCE OF A PERIODIC REPORT

Leah wrote her report in a format that she and Elliston had agreed to in the past, but certainly it is not the only format that can be used. Often periodic reports are less formal than Leah's, especially to in-house recipients. In roughing out the draft, Leah was grateful that she didn't have to deal with the endless forms that government agencies, like Health and Human Services, required of their funded programs. There is generally no informal agreement on format when it comes to reporting for the government: format is described in minute detail, not allowing for the kind of narrative discussion that Leah felt more comfortable with.

SAMPLE PROFESSIONAL PERIODIC REPORT

Figure 7.2 is the rough draft of Leah Feldstein's periodic report to the Elliston Foundation on the activities of EDUCARE's drug rehabilitation program for the third quarter of 2001. Leah will send the report with a letter of transmittal (for more specific information on letters of transmittal, see Chapter 6) signed by John Fiske, even though Leah is the author of the report. She has also identified the report as a draft and has dated it. This step will allow Fiske time to make changes before the report is due at Elliston, and more practically, it will help ensure that the report will not somehow be sent out in its present form, that is, without editorial review.

This first material is the orienting information. Because John Fiske, the Executive Director, is the signer of the report, Leah shows his office as the report's origin. Other information designates the period of reporting, the recipient, and the project name and contract number. All of the identifying information shown is necessary, because organizations like Elliston fund numerous projects and need to be able to draw up information from their files quickly and accurately.

The summary of project activities lays out all the significant activities of the period. As is true of all conclusive summaries, this one will allow readers to make decisions about how the project should proceed, whether new directions need to be taken, whether existing directions need to be bolstered, whether everything should remain basically as it is established. In solving the rhetorical problem of writing any report, but perhaps especially of a periodic report, a conclusive summary anticipates questions readers will have and allows them to decide whether the project is meeting the goals as set out in a proposal or through other less formal means.

EDUCARE, INC.

1200 STONE WAY, SUITE 500, SEATTLE, WA 987116 206-457-7789
FAX: 206-457-9900

OFFICE OF THE EXECUTIVE DIRECTOR

DRAFT—MARCH 15, 2001
THIRD QUARTERLY REPORT—JANUARY–MARCH, 2001
THE JOHN T. ELLISTON FOUNDATION ON
PROGRAM DRUGFREE, EDUCARE, INC.

Project

EDUCARE, Inc., Seattle, WA, Drug Rehab. El 1 DRA x15-023

Period Covered

Third Quarter, January–March, 2001

Summary of Project Activities for the Period

Program Drugfree is making the kind of progress projected for this quarter. And in an attempt to boost our success rate, we have begun two changes that affect both clients and staff. Group size has been reduced and frequency of meetings has increased. Since we retain the same number of counselors, we have rearranged their coverage so that each counselor now meets on the average with two groups, twice a week, for an hour each session. Counselors have found these smaller group meetings to be positive. In general, the sessions are more client-directed than counselor-directed.

The cumulative success rate has remained at 27% for the third quarter, but the positive responses to the new initiatives in group counseling encourage us to think that the rate will increase to 38% next quarter.

The staff is considering two other modifications to the program: additional client-directed individual counseling sessions and increased staff development. An additional individual session per month would be directed mainly by clients and would focus on role playing. In staff development, the staff plans to meet weekly instead of every 2 weeks, devoting the additional time to reviewing new findings in the field and reporting on professional meetings. We hope in this way to improve our service to EDUCARE's clientele.

FIGURE 7.2 **Leah Feldstein's periodic report**

Work Accomplished

Individual and Group Counseling

The drug rehabilitation counseling staff of EDUCARE, Inc., has continued to meet regularly in both individual and group sessions with 152 registered clients. The individual sessions have been conducted, as usual, both in the schools and in the four outreach centers in Seattle and larger King County. They continue to stress the four primary areas of concern: family relations, relations with friends, relations with persons of authority, and self-esteem building. Clients still meet counselors individually at least four times a month for an hour each time.

The conduct of group sessions has changed, however. Instead of meeting once a week in groups of 10–12, the groups have been reduced in size to no more than seven clients, and they meet usually twice a week. This means that we have had to rearrange counselor coverage so that each counselor now meets, on the average, two groups, twice a week, for an hour each session. Accounting for preparation and debriefing time, counselors are now spending an average of 7 hours a week on group counseling, roughly twice what they were spending under the old plan.

In devising the new plan for group counseling, we took special care not to break up the old groups arbitrarily. We took into account friendships or good relations that seemed to have been formed among clients, and no individual has been reassigned to a smaller group made up totally of relative strangers. We have also tried to keep the same counselors for the new groups: in every case at least half the group has worked with the same counselor in the past. These measures reassure us that clients' useful contributions to group discussions will not be unduly restrained.

In general, we have retained the format and content of group sessions. Each session is based on a theme derived from one of the four counseling areas mentioned above. But the additional meeting per week allows for more immediate reinforcement or continuation of discussion that may have been artificially abbreviated under the old plan.

Counselors have found that clients are in general more forthcoming in the smaller groups. They tend to talk more, more specifically, and to the point. The greatest, most positive change, all counselors agree, is that clients are directing their comments more often to each other, and not strictly to the counselor. Discussions have become friendlier, more expansive, and more spontaneous; the general atmosphere has become more client-directed, and less counselor-directed. Instead of saying what they think the counselor wants to hear, clients are now saying what they want to hear from themselves and from others, and are comparing and contrasting their experiences, failures, and successes. One counselor reported recently that in a discussion of relations with parents, her clients seemed to have forgotten her, and that there was so much talk back and forth among the members of the group that she could barely get a word in edgewise. She said, "I felt left out, rather useless, but I was thrilled. My clients got

FIGURE 7.2 (*Continued*)

The first paragraph reviews the highlights of the period, establishing (or reestablishing/some of the basics of their work: how many clients they see, how often, where, and for how long. She also is sure to reiterate that the primary areas of concern in their individual counseling approach remain the same. Thus, she continues to orient the readers to a context that might be difficult to remember otherwise.

The remaining paragraphs of this section lay out the new approach to group counseling. Keeping her audience in mind, and the problem they may see with client-counselor ratios, Leah is quick to point out that the new arrangement requires about twice the time of each counselor for group counseling.

In paragraph three, Leah details the configuration of the new groups, which on the surface might seem unnecessary. Since this is a new arrangement, she wants to show that the configuration was based on good theory; familiarity and friendship are effective means of getting people to open up, to say what they have to say without embarrassment or intimidation.

Leah talks in paragraphs four and five about how meeting more often and in smaller groups has encouraged more open responses from clients, making their experiences more client-directed than counselor-directed: a distinct and sound goal of the program, in both EDUCARE's and Elliston's eyes.

Her concluding paragraph for this section does what all good conclusions do. It looks ahead, inviting the readers to think of possibilities that lie beyond the scope of the report.

The success rate is important for Leah to discuss. It is something the project must look out for and something the readers of her report at Elliston will want to know about.

She talks about the "slips" in clients' performance outside the program, specifying what happened to them and their decisions to return to the program. She even tries to inject a little humor (although understated and dry), recounting one client's substitution of marijuana for crack and the results: he reported that "he was feeling better," but he "was also invited to return to the program."

Leah sums up the current client enrollment and the success rate in percentages for at least ninety days. She also mentions, however, and this is important, that she sees the new initiatives as promising improvements in the success rate. She wants to put the best face forward on the work of the period and show that there is promise for the future.

more said, and in more particular and articulate ways, than they had said in all sessions in the last 2 months."

Counselors have also reported from individual sessions that clients like the intimacy of the new group sessions, that they feel less shy and more willing to take part, and that they actually look forward to meeting twice a week with their groups. Though many admit that their individual battles with drug abuse have not yet been won, the same number agree that they are more able and willing to discuss their battles, and that in itself has given them greater confidence that they will eventually win.

Success Rate

We have moved clients out of the program successfully at about the same rate as last quarter. In the first third of the quarter, twenty-eight clients left the program with staff consent and approval. Follow-up interviews, conducted 1 month and 2 months later with clients, teachers, parents, and other social service workers indicated that twenty-three clients had remained free of drugs. Five had slipped "once or twice," but showed no recurring habitual pattern. And three who had slipped often and were returning to habitual patterns were invited to come back to the program. Two have returned within the last 10 days.

In the second third of the quarter, twelve clients left the program with staff consent and approval. Follow-up interviews with all concerned indicated that nine have remained drug-free; two have slipped once or twice, and one reported that he had substituted marijuana for crack and was feeling much better. He was also invited to return to the program.

The percentage who have left the program unsuccessfully, that is, without staff consent and approval, and who, again with information from follow-up interviews, we believe to remain on drugs, stays at about 9%.

With returnees and new clients, current enrollment is 152, and the success rate remains, as it had for the second quarter, at about 27% for those who have been in the program for at least 90 days. But we believe that with the initiatives we have taken in group counseling, the success rate will increase. The change in attitudes toward counseling in general has been outstanding. Clients are particularly pleased that they are able to feed back information from their group sessions to their individual sessions, not only because they feel a new camaraderie and openness with their fellows in the groups, but because they are also usually dealing with the same counselors; they don't have to start from the beginning—most of what they think about and understand about their problems with drugs has already been brought up and discussed in some way.

On the basis of this, counselors now predict an increase in our success rate. They have identified nineteen clients beyond those who would normally leave successfully who will be ready to leave the program successfully in the next quarter. That, with the cumulative 21-month success rate factored in, would raise our success rate to almost 38%, an 11% increase in the rate we have been achieving, and substantially more than the rate for other programs of our size.

FIGURE 7.2 (*Continued*)

Problems Encountered

Our modest success rate remains our biggest problem. Though the rate has increased from 21% in the first quarter of 2000 to 27% in the first quarter of 2001, it has remained at 27% for the last two quarters. But as we have indicated, new group counseling procedures have already had a positive impact, based to date on observation and anecdotal evidence. We look forward with confidence to the next quarter when we think we can expect the rate to increase to as much as 38%. Though 38% is still not ideal, it is a respectable and encouraging figure, and if it can be gradually improved, the program will have met the long-term objectives that were set at the beginning of the funding period, and will have exceeded the quarterly objectives by 11%.

The staff is currently considering two other modifications to procedures that are likely to have a positive effect on the success rate. The first has to do with individual counseling sessions. We are thinking of adding one session per month, either at the beginning or at the end of the month, depending on scheduling needs, that will be almost entirely client-directed. Clients will choose from the previous month's discussions a topic that they want to pursue further. The topic can have anything to do with the four major areas of concern: family, friends, authority figures, or self-esteem, but clients are supposed to ask of counselors the same kinds of questions normally asked of them. Counselors will respond in ways that their clients would: sometimes reticently or incompletely, sometimes volubly or expansively. The point is to allow clients to ask their own questions and to receive the kinds of responses that they themselves are likely to give.

This kind of exchange has been suggested as a way for clients to investigate areas of insecurity or indecision in their own thinking and to see how adults, who have no familial or other extra-therapeutic ties to them, respond to problems they feel and have confronted in their own lives. It is also a means by which clients are able to exercise some measure of control, something, we have always known, clients lack and sorely feel the need of.

The other change has to do with staff development. Though we conduct staff meetings weekly, the time is usually spent discussing administrative and procedural matters: revising schedules, reporting on meetings with school officials or other social service agencies, and documenting clients' progress. Little time is spent discussing the theoretical or practical approaches to drug counseling, or the results of professional meetings staff members have attended. The result is a certain stagnation and lethargy in our thinking.

To correct this situation, we have decided to devote at least one meeting a month to seminar discussions, reviewing new findings and happenings in the field and suggesting how they might be applied to EDUCARE's clientele. This will add no additional hours to already full schedules, and the staff has agreed that the active exchange of ideas, in an arranged setting, will be beneficial.

The project remains within the contractually set, limits of schedule and budget.

FIGURE 7.2 *(Continued)*

In the problems section, Leah identifies the problem that she is willing to admit is a problem. She does not mention client counselor ratios or recruitment, issues that Elliston may want to consider problems but that she does not. She is specific about dates and figures, but she is quick to predict how the problem may be solved with the next group counseling procedures. She suggests that the predicted success rate of 38 percent is one that Elliston would be pleased with.

In the second paragraph, she begins to discuss other measures she thinks will help improve the success rate.

The other measure responds to another problem, counselor stagnation, certainly related to the success rate. but perhaps more indirectly. The staff's decision to meet more often and to devote the additional time to discussing developments in the field is both a good idea for the obvious benefits that can accrue, but also as a demonstration of the staff's commitment to the program. This change in priorities Leah hopes will positively impress her readers and perhaps effectively reduce criticism of client-counselor ratios and unaggressive recruitment.

The schedule and budget information speaks for itself, and the plans for the future are itemized, specific, and clearly connected to what has already been discussed. There can be no questions about what is supposed to happen, including EDUCARE's intentions to

seek additional funding, clearly something Leah and EDUCARE want Elliston to look upon favorably.

Plans for the Next Quarter

While plans for the next quarter have been detailed in other sections of this report, the following summary lists them.

1. We will continue the initiatives taken in group counseling to boost our success rate to 38%.
2. We will begin additional individual counseling sessions that will be more client-directed than counselor-directed.
3. We will increase the number of staff meetings from twice a month to four times a month, with at least one of the meetings devoted strictly to reviewing findings and developments in drug counseling.
4. In addition to a discussion of how these program changes have worked, we will produce a final report on the activities of EDUCARE's drug rehabilitation program (HHS DRA x15-0023) for the funding period, 1990–91.
5. We will propose a renewal of the grant for the funding period 2002–2003.

FIGURE 7.2 (*Continued*)

EXERCISES

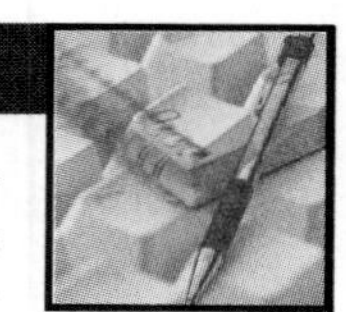

1. You are a counselor on Leah Feldstein's staff at EDUCARE. She has asked you to help her plan the final quarterly report to Harper St. John at Elliston on the activities of EDUCARE's drug rehabilitation program. You are at the end of the second month of the quarter, so you are certainly able to comment on your own activities for the period and for the year, but you need to talk to your colleagues about their activities. You decide to get together with two or three others to discuss what the report should contain.

   Given your knowledge of what was said in the last quarterly report, decide, with the others, what the final report should discuss. Be sure to think about the basic elements of a periodic report: work accomplished, problems encountered, and the future, given what you think the future might hold. An outline of the contents of the report should be the result of your discussions.

2. Put yourself in Harper St. John's position at the Elliston Foundation. You receive periodic reports from many programs reporting on their activities and have to evaluate and make judgments on them all. You want to do the best job as easily and efficiently as possible in the best interests not only of Elliston but also of the programs. You therefore

look for and admire thorough but efficient reporting, because it makes your job easier and your work more effective.

Given what you know of the activities in drug rehabilitation that EDUCARE has taken on in the past, especially in the last quarter, what would you expect to read in the last report of the funding period? What would you expect to be able to do, given what you have read? An outline of the expectations should be the result of your thinking.

ASSIGNMENTS

1. Your discussion group at EDUCARE has decided to write a memo to Leah Feldstein on its thoughts about what should be included in the final periodic report to Harper St. John at the Elliston Foundation. You want to make sure that you have represented your activities well but also that you have taken into account the difficulties that normally arise in the course of your work. In addition, you want to make some intelligent suggestions for the future. Write the memo to Leah, and send a copy to John Fiske, Executive Director of EDUCARE.
2. As Chapter 2 discusses, people often write memos for purposes of documentation only; they are called memos "to file." Among other purposes, they serve as reminders of what has been done about a particular problem, how successful the action has been, what problems have arisen (and how they might be solved), and what needs to be done in the future. Since EDUCARE is on the questionable list for renewed funding, Harper St. John wants particularly to document his reactions to what EDUCARE has been doing to keep to the goals that both EDUCARE and Elliston have agreed to for Program Drugfree. In that way, St. John can call up the information on EDUCARE quickly and determine more efficiently whether continued funding is in order.

   Assuming Harper St. John's position, write his memo to file.

TERM PROJECT OPTION

Write a periodic report on the project you are currently working on. It should be directed to the same audience who received your proposal. Remember to make your report conclusive and appropriately explanatory about the significance of the actions you have taken. After you have drafted the report, edit it once for passives and nominals, which tend to abound in periodic reports. Refer to the Appendix on selected problems of usage and style if you aren't clear about these constructions.

CASE STUDY REVISITED

PATRICIA WYNJENEK'S OFFICE

Department of Anthropology ▪ Hastings University
Las Palmas, New Mexico

Patricia Wynjenek felt good. She had recently returned from a highly successful sabbatical leave to the Mexican-American border, specifically to Indigo, Texas, where she had studied economic survival strategies of low-income Mexican Americans. Through her research protocol, which included participant observation and interviews with about one hundred residents of one of the poorer barrios of Indigo, she was able to collect much useful information on the ways that the residents of Indigo survived economically and kept their families intact in the face of the kind of adversities that have fractured similar kinds of poverty-stricken communities in many other urban areas. Patricia thought rightly that Indigo needed to be studied, both to determine the factors that make it unique and to suggest social policies and programs that state and federal governments might adopt to serve more efficiently their poorest citizens.

Patricia's sabbatical was partially funded by outside sources—government agencies and private foundations, who have interests in the kind of work she does. One such group, Las Almas Buenas Foundation, which is particularly concerned with medical problems and access to medical facilities among Mexican Americans, had funded 25 percent of her expenses while she was in the field. She needed now to report finally on her findings of how low-income Indigoans use and pay for medical faculties to treat ailments and stay healthy.

Patricia felt pretty confident that she could report intelligently. She had collected a lot of good information on health-care practices in Indigo, had analyzed it, and thought she was able to draw some meaningful conclusions. She actually felt not only confident but also pleased about the report she would write. Her contact at Las Almas Buenas, Dolores Sanchez, had written her to suggest the form that her completion report might take. Even though Dolores had received periodic reports from Pa-

tricia on her work in Indigo, she wanted a final report that would appeal to the Board of Directors of Las Almas Buenas, a report that would lay out specific information in terms that they could understand. That undertaking meant she would have to repeat information that Dolores and other readers of her periodic reports had already read. But in so doing, Patricia could write the report for a broader professional audience, who would need the background information.

Patricia saw the possibility of meeting the needs of two situations with one report, an efficiency she was rarely able to practice, but one that she was eager to try. She sat down at her computer to start writing the report with more enthusiasm than she could remember having for such a job in a long time.

FUNDAMENTALS OF A COMPLETION REPORT

Purpose and Rationale

There are basically two kinds of completion reports: those written for an audience in the private sector (sometimes called final reports) and those written for a more academic, research-oriented audience (sometimes called research reports). They both fulfill the same purposes:

- To define and/or address professional problems
- To explain what work was done to solve or address the problems
- To conclude what the work ultimately means and what its significance is
- To make recommendations for future work, if necessary

Although completion reports can take different forms, the rhetorical basis for them is the same. Someone who has been principally responsible for work on a specific project needs to report on what has gone on, why it has gone on, what the results were, and what the results mean; and someone closely interested in the work, usually a managerial audience, needs to know the same. Just as they do with periodic reports, the rhetorical concerns on both sides of completion reports—writer and reader—readily coincide.

The important difference to remember between completion reports and periodic reports is that the work for better or for worse has come to an end by the time a completion report is written. The goals of the project may or may not have been met, and there may be more that can be done in the future, but the immediate need is to report on the end, or completion, of the project. Both writers and readers need to tie things up to see the end of

serve as academic reports on research, readers want to know how the work was done—specifically, the methods, materials, and conditions under which the work was performed. This information permits readers to analyze and evaluate the results in light of how they were achieved, thus allowing a closer scrutiny of the reporter's interpretation of them.

In the case of OUTERWARE's installation of ENCONTRO for the capitol building in Hartford (discussed in Chapter 3), upper management wants to know how the design of the program was adapted to fit the capitol building's needs—what assessments of present systems for environmental control and physical inspections took place and how existing computer facilities were adapted to accommodate the new system. It wants to know what training and orientation were necessary for the capitol building personnel to get the new system up and running. Finally, it wants to know what problems arose in doing the work, whether and how they were solved, and what predictions for the success of future work of a similar kind can be made.

The case of Patricia Wynjenek is similar. She needs to report on the ways she determined that segments of populations in urban border cities were observed and interviewed, how the samples of population were made up, what questions of the interview schedule were asked, who did the interviewing and why, and what the results were. These considerations require further explanations of the various parameters: the kinds and frequency of participant observation (meetings, social gatherings, informal contacts), the number of respondents to the survey (including how they were chosen—at random or by stratification as to age, gender, and other demographic characteristics), the same kind of demographic description of the interviewers, the questions that were included in the survey, and the raw data by numbers and percentages that resulted from the answers to the survey questions.

For both cases, the answers to these questions map out the work that was done and show how it was done, so that readers can begin to determine whether the conclusions of the report seem reasonable, given the methods of collecting the data. The ultimate validity of interpretation is a separate issue.

What Conclusions Can Be Drawn from the Work That Was Done? How Can the Data be Interpreted? What Decisions, If Any Are Necessary, Can Be Made, Given What Is Now Known?

The answers to these questions form the third element of the rhetorical basis for the completion report. It is the most important because it explains the significance of the work. Without having an educated evaluation of what the work accomplished, organizations can't know whether the directions they are taking make sense, whether they are effective, or whether they can be made effective. Upper management makes important decisions on the basis of information provided in this area—decisions that affect not only the conduct of business but also the future of research. These decisions need to be made if solving problems through good analysis is going to continue to be effective.

OUTERWARE, therefore, uses Mary Anne Cox's interpretation of how ENCONTRO has worked at the capitol building in Hartford to make decisions about similar projects. Understanding what went on, what problems arose, how they were solved—many of the same problems that periodic reports discuss—and finally, what the whole project means for the present and the future, helps question and perhaps affirm or change basic directions of the company. Such activity also helps establish initiatives and new directions for the future.

The same is true of Patricia Wynjenek's work. Her understanding and discussion of the dynamics of economic survival along the Mexican-American border is important in two ways: it helps fill out the knowledge of Mexican-American culture, the fastest-growing minority group in the United States today, especially as it exists in border areas; and it suggests ways that social programs might be made more responsive to the needs of those whom they are supposed to serve.

FORMAT AND SUBSTANCE OF A COMPLETION REPORT

As we have indicated, completion reports are written for special purposes—purposes that go beyond the conduct of everyday business—and they therefore require some special treatment. The quality of the production is dependent on several factors: time, costs, and, of course, purposes for which the reports are written. They are often specially printed, complete with four-color graphics and illustrations, and then bound. The binding can range from three-hole ring binding to sewn book binding; again, the method of binding is dependent on resources and on the purpose of the report.

Although the four points in rhetorical problem solving—purpose, audience, context, and ethical stance—are the most important elements to consider in any technical and professional writing, different formats are helpful to define and organize those rhetorical elements. We suggest two generic formats for completion reports; these elements can be used in the absence of formats supplied or required by those who have called for the reports. The first format would be appropriate for a report written for private agencies, government, or industry—a nonacademic audience; the second, for an academic, research-oriented audience. (See Figure 7.4.)

Elements of the Format for a Nonacademic Audience

Letter of Transmittal. The letter of transmittal records the production and delivery of the entire finished report. (For a full discussion of letters of transmittal, refer to Chapter 6). If the report is intended for an internal audience, one within the organization that the author works for, the letter may be written as a memo; but if the report is being sent to an external source (in Patricia Wynjenek's case, for instance, Las Almas Buenas Foundation), it is written in conventional business letter format. The letter formally identifies the report, reiterates why the report was written (including reference to any contractual obligations), and summarizes the highlights of the work that was

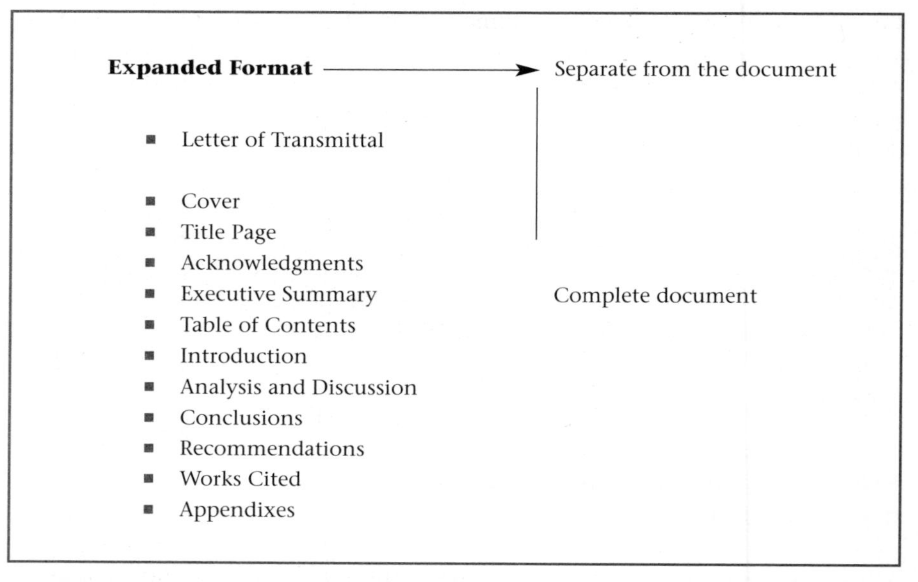

FIGURE 7.4 **Elements included in a completion report for a nonacademic audience**

done, the main conclusions that were drawn, and any recommendations that may have been made. If the report's organization or structure departs from what would normally be expected, the letter identifies the departures. It also identifies any discrepancies between work proposed and what was actually produced.

Writers usually close letters of transmittal with an invitation to the reader to comment on the report and with some indication that they look forward to continued interaction and service in the future.

As we have also indicated in Chapter 6, the letter of transmittal includes a list of those to whom copies of the report have been sent. The list may be short—only those who have a vital interest in the report—or it may be long, including those who have a vital interest and those, who by virtue of their position or as courtesy dictates, should get a copy. The author of the report has to decide, in keeping with the purpose of the report and with protocol, who will receive copies.

One structural element in the production of the report is useful to remember. As noted before, completion reports are often specially printed and bound. Annual reports to stockholders are a good example, because they go out to a wide variety of readers. The letter of transmittal is not normally bound with the rest of the report; it is attached to (or covers) the report; thus, another term for a letter of transmittal is the *cover letter.*

Cover. The cover of the report, no matter the range of appropriate production styles available, must contain some basic information: the title of the report, sufficiently descriptive of the contents; the name and corporate address of the author or the group that produced the report; and the date. Sometimes the name of the primary recipient of the report is also included. Cover graphics, if appropriate, should also suggest the basic purpose of the report and reflect the desired corporate image of those who have produced it.

Title Page. The first page of the report is the title page. It contains the title of the report (exactly as it appears on the cover), the name and address of the recipient, the author's or the originating organization's name and corporate address, and the date. No graphics are expected or necessary, but convention nevertheless calls for a title page. It reflects the formality of the report and helps identify what the report is, if for some reason parts of it get separated.

The title page shows no pagination, but it is often counted as the first of the total number of pages of front matter, that is, material that precedes the main substance of the report.

Acknowledgments. If writers of completion reports have received extraordinary help from others to write the report, they should acknowledge that help formally. But it is important to understand what extraordinary help is. If people go out of their way to supply information or are consulted on matters that lie within their knowledge and expertise, but also go beyond their usual structural and professional responsibility to supply that information, they should be acknowledged.

Most completion reports written in the corporate sphere involve the help of those who are expected to supply it. It is a part of their professional responsibilities, and they supply information routinely. They need not be acknowledged formally; in fact, to do so would look silly. Acknowledgments should be reserved for those special cases of extraordinary help.

Under these circumstances, readers expect to know in what areas the help came. Therefore, it is customary to specify not only the person(s) who supplied the help but also what kind of help it was: "For her help in suggesting appropriate investment strategies for the coming year, I must thank Jane Drussel of Smithson and Company, Toledo, OH."

If an acknowledgment page is included, it is counted in the total number of pages of front matter and paginated with a lowercase Roman numeral, either (i) or (ii), depending on whether the title page is also counted in the total number of pages of front matter. The numeral is placed either in the bottom center of the page or in the upper right-hand corner.

Executive Summary. The executive summary is an abridged version of the whole report. Because decisions, even if they are preliminary, are often made on the basis of summaries, they need to be conclusive rather than simply descriptive. Conclusive summaries tell the whole story in brief: what the problem was, how it was approached or solved, what conclusions were drawn, and what recommendations were made. Relative to the length of the whole discussion, it is fairly short, usually running no more than five hundred words (about two double-spaced pages), but it can run as long as 10 percent of the length of the body of the report.

Since the executive summary is a condensation of the entire report, it should be one of the last parts written. It is also considered front matter and is paginated in lowercase Roman numerals in the count that begins with the title page.

Table of Contents. For longer reports, a table of contents is absolutely necessary. It shows the contents and organization of the report, allowing the reader to turn immediately to specific pages to read different parts of the report at will. It is usually written to the second level of specificity: section headers and major headers within sections.

The table of contents includes lists of appendixes and sometimes lists of figures or illustrations. If the lists of figures and tables don't appear in the table of contents, they should appear separately with their own headers. Lists of figures and the like appear immediately after the table of contents. The table of contents does not list material preceding it, such as the title page.

The table of contents is paginated as is other front matter, with lowercase Roman numerals. Like the executive summary, it is one of the last parts of the report to be written. Authors should wait until all changes have been made in the report before adding page numbers; otherwise, they may simply have to do it again.

Introduction. The introduction educates and prepares readers for what is to come in the report. It outlines the purpose and scope of the report, and provides background material so that readers can read the discussion intelligently. The introduction is clearly very important. Beyond the executive summary, it is the first substantive piece in the report, and as such, it needs to lay out the reasons for the report, the way that it should be used, and any particular information necessary to place the subject and purpose of the report into a larger context.

The section can be headed "INTRODUCTION," in all-capital letters, and centered as a major heading, or it may be called something else, as long as the heading clearly identifies the material that follows as introductory. The heading, whatever it is, should follow the title of the report (this is the third time, if covers are used, that the title appears).

Pagination of the main body of the report, as separate from the front matter, begins with the first page of the introduction, but it is customary not to show a numeral on the first page. Therefore, Arabic numerals begin to appear in the upper right-hand corner of the second page: 2. Pagination continues throughout the body of the report, through any appendixes and the index (if one is necessary).

Analysis and Discussion. The analysis and discussion form the main narrative of the report and provide all the data, evidence, and interpretation needed to show the following:

- What the authors were trying to do
- Why they were trying to do it
- What they actually did
- What they found out
- What it all means
- What, if anything, the work implies for the future

The section is normally broken down into smaller units or topics for discussion, each given a descriptive header (or topical heading) to identify it. It is important that the breakdown be logical and consistent, showing a hierarchy of development. Since not all topics require equal billing or are themselves thought of as distinct subtopics of other larger topics, a series of major and minor headers should be established and then consistently used.

The choice of centering major headers in uppercase letters, as in

GENERAL USE OF MEDICAL FACILITIES ALONG THE BORDER

and of justifying (aligning) minor headers to the left-hand margin in uppercase and lowercase letters, as in

Use of Conventional Facilities

is really up to the author. You will notice that the documents in this book use different systems of heading levels. Whatever system an author decides on should be clear to readers, and it must be used consistently throughout the report. To do so is rhetorically persuasive. It helps organize reading and thinking, and it allows, as is so important in technical and professional writing, retrieval of information in a nonlinear way.

Conclusions. Once again, in order to provide easy access to important parts of the report, conclusions, even though they may have been implied in the preceding analysis and discussion section, should be summarized in a separate section. Many busy readers will read the executive summary, skim the introduction, and then turn directly to the conclusions to see what the authors have decided about the problem that is discussed in the report. Readers can then decide whether they need to fill themselves in on how the conclusions came about and read parts of the analysis and discussion at will.

Separating the conclusions from the analysis and discussion is a good idea not only for readers but also for authors. Authors, in summarizing conclusions in a separate section, are forced to look at the report as a whole and decide whether what they have implied as conclusions is reasonable. Writing a section on conclusions helps authors specify their thinking about the problem and the implications that have arisen from it.

Recommendations. If the analysis and discussion and the conclusions suggest that specific actions need to be taken, the actions should be presented separately as recommendations. They can be listed and enumerated or bulleted, but they need to be specific as to what should be done in the future. Vague notions of what should be done are of no use to most readers; they want and need the particulars, so that they can make decisions more easily. The finer details of how a plan will be carried out can be worked out later, but the basic outline of what needs to be done and why should be stated clearly and concisely.

Works Cited. Most completion reports require the use of secondary sources of some kind. The material can be obtained from printed sources or from the personal commentary of people considered experts in the area that the writer is investigating. In either case, courtesy and professional integrity require acknowledgment (or as it is often called, citation) of those sources. (A similar section, called "References Cited," is discussed in Chapter 6.

The acknowledgment of printed sources is more formally prescribed. Not only must borrowed material (either quoted or paraphrased) be cited parenthetically in the text, but in addition, full information on the source of the material must be included in this section in a list, usually headed "Works Cited." The basics required are the bibliographic information: author's name (listed alphabetically by last name), title of the work, and publishing information (place of publication, publisher, and date). In the case of articles published in professional journals or in edited collections, inclusive page numbers in which the article appears are also provided. Citation styles, which vary from field to field, are often published in individual style sheets; writers should use the conventional citation style of the field in which they are reporting.

Citing information gleaned from personal interviews is less formally prescribed, but no less necessary. Writers often name the source in the text and parenthetically cite the place and date of the interview. Then, in this section, sometimes in a list separate from the printed sources, writers include the name, position, and professional affiliation of the source; sometimes they repeat the date and place of the interview. This separate list is often headed "Other Sources."

Appendixes. An appendix (singular form), or appendixes or appendices (plural forms), are added-on pieces of information that enhance the presentation of the report but are not absolutely necessary to readers' understanding of it. They come at the end of the report and take the form of charts, tables, photographs, specifications, test results, or computer program printouts, sometimes with explanatory narratives. They are not included in the text because they are not immediately pertinent and would be disruptive to the arrangement and flow of the discussion.

Appendixes are meant to bolster the main discussion of the report and should not be thought of as a dumping ground for irrelevant data, impressive only as padding. If there is more than one appendix, each is individually lettered A, B, C, and so on, and titled as follows:

> APPENDIX A: CENSUS DATA ON USE OF MEDICAL FACILITIES
> FOR AMERICAN-MEXICAN BORDER MUNICIPALITIES, 2000

or

> APPENDIX B: CORRELATIONS BETWEEN NUMBER OF YEARS OF
> FORMAL EDUCATION AND USE OF MEDICAL FACILITIES AMONG
> AMERICAN-MEXICAN BORDER POPULATIONS

If there is only one appendix, it is simply labeled as such, with no A or B designations:

> APPENDIX: CENSUS DATA ON USE OF MEDICAL FACILITIES
> FOR AMERICAN-MEXICAN BORDER MUNICIPALITIES, 2000

Appendixes are paginated only if they are longer than one page. If, as in some cases, they are to be paginated separately from the rest of the text, the visible pagination begins with the second page and continues for the length of the appendix. The upper right-hand corner of the second page would show Appendix A-2, the third page, Appendix A-3, and so on.

Thus, the basic format is complete; it organizes and explains the whole process of problem solving for any particular project from beginning to end. Although some organizations may design and require authors to use their own formats for writing completion reports, the format we have just discussed serves all the rhetorical needs of completion reports and may be used in the absence of company formats.

Elements of the Format for an Academic, Research-Oriented Audience

As suggested at the beginning of this discussion, completion reports are also often written for more academic, research-oriented audiences. Although the rhetorical needs are essentially the same, reports on research place more emphasis on two areas: the theoretical underpinnings and environment of the work, and the methods and means used to carry out the work. Readers need this additional emphasis so that they can more completely assess and evaluate the reporter's interpretation and claims of significance for the work completed. (See Figure 7.5.)

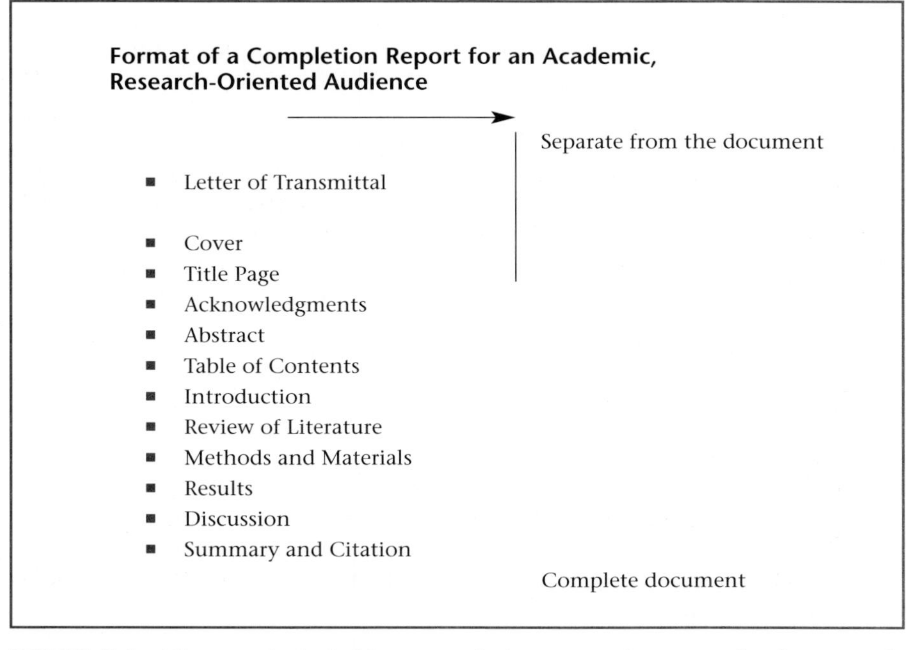

Format of a Completion Report for an Academic, Research-Oriented Audience
Separate from the document
Letter of Transmittal
Cover
Title Page
Acknowledgments
Abstract
Table of Contents
Introduction
Review of Literature
Methods and Materials
Results
Discussion
Summary and Citation
Complete document

FIGURE 7.5 Elements included in a completion report for an academic, research-oriented audience

The following format, similar to that of the completion report already discussed, is useful for reporting on primary research in a final, formal way. With some modifications, it may also be used for articles intended for publication in professional journals. We have noted, when appropriate, the differences between this format and that of the completion report outlined and explained earlier.

Letter of Transmittal, Cover, Title Page, and Acknowledgments. In general, the front matter for a research report is handled in the same way as it is in the nonacademic completion report. But if the report is being submitted to a journal as an article for possible publication, the cover and the title page would be eliminated, and the letter of transmittal would be altered to reflect the difference in the author's purpose for submitting the report.

Abstract. The rhetorical purpose of abstracts is slightly different from that of executive summaries in that readers do not normally make decisions for subsequent action (except whether they might bother to read the report in the first place) on the basis of the contents of the abstracts. Abstracts, however, still perform a summarizing function, and in that regard, they should still be conclusive, stating the major conclusions of the work and their significance.

If a report is intended for publication in professional journals, authors may also be asked to list in their abstracts any keywords or phrases that will clue readers in to the kind of material covered in the report. These keywords also help libraries index the report in automated databases. Abstracts are often published separately to help researchers narrow down a search for secondary material that will be helpful and pertinent to their own research. In Patricia Wynjenek's case, if she were submitting her report for publication, she might list the following keywords or phrases: *poverty among Mexican Americans along the border; health care among Mexican Americans; economic survival strategies among Mexican Americans.* Researchers in any or all of these areas would then know that her article in some way discusses areas that they are interested in.

Abstracts, like executive summaries, are supposed to be short—some say no more than 5 percent, others no more than 10 percent, of the length of the entire report—but many journals, particularly in the sciences and applied sciences, set arbitrary limits of from 150 to 200 words.

Table of Contents. The table of contents, including lists of figures and appendixes, is presented in the same way as it is for nonacademic completion reports. But the table of contents is not included if the report is being submitted as an article to a journal.

Introduction. The introduction serves the same purpose as it does for the nonacademic completion report. It discusses the purpose and objectives of the research, some background, and the scope or limitations of the report. If the plan of development and the arrangement of the report are unconventional, the introduction explains them.

Review of Literature. The review of literature is the first additional section required of research reports; generally, it is not required of nonacademic completion reports. As is true of the review of literature written for proposals (see Chapter 6, for a detailed discussion of the review of literature), the purpose of the review is to place the writer's work in a larger theoretical context of work already done, or being done, in the field. Readers want to know how the writer's work has been influenced by, or will influence, other similar work done in the field, in order to understand more completely the writer's interpretations of and conclusions about his or her own work.

In reviewing others' work, however, writers shouldn't avoid differences in theoretical approach, methods, findings, interpretations, and conclusions. In fact, such differences are often the meat of meaningful discussion, but writers need to state specifically why other research may differ from their own. In this way, they can prepare readers even more thoroughly than they can in the introduction for what is to come later in the report.

Reviews of literature should also be thought of as expansions of the statement of the problem, the essential element of the introduction. It is important for writers of research reports to demonstrate that problems they address are not formed in a vacuum, that they are a part of a larger set of problems that people are trying to solve. The review of literature demonstrates this larger context.

Finally, writers should keep in mind (a point also emphasized in Chapter 6) that the review is not merely an annotated bibliography. It is a *discussion* that must adhere to all the elements of development, coherence, and interpretation expected of any discussion. It shows ultimately that researchers have done their homework and that they are aware of what is going in the field and how their own work fits into it. Research is now so specialized that writers must provide this context of previous research for their readers. Without the review of literature, it would be difficult for other specialists to appreciate what the writer has done.

Methods and Materials. The second additional section typical of a research report, but not of the nonacademic completion report, is a discussion of methods and materials. This section allows readers to match methods and materials, or how the research was performed, with interpretations and conclusions to see whether they are reasonable. In some cases, readers may want to replicate the methods and materials to see whether they come up with the same interpretations and conclusions.

To satisfy this need, writers must describe the conditions under which data were collected: the kind, setup, and operation of the equipment; the way results were recorded and analyzed; and the peculiar constraints of the method and operation used in the research. This kind of description must thoroughly cover deviations from the conventional, explaining what they were and, of more importance, why they were used.

Results. The results section in reports on primary research is very clear-cut. It lays out the recorded data, often in tabular form, of the investigation in question, accompanied by all necessary written explanations: prefaces, labels, and footnotes. When anomalies or peculiarities occur, those, too, are noted, but they are not usually explained in any detail, since it is still common practice to reserve interpretation for the next section.

Discussion. As one of the most important sections of the primary research report, the discussion reviews, analyzes, and draws conclusions about the data presented in the results section. In essence, the discussion explains what the research has meant, given the context of which it is a part: it suggests, through the researcher's interpretation, what is now known that wasn't known before the research was done and assesses the significance of the new knowledge.

The discussion also explains any questions that the results may pose. Are there anomalies, irregularities, or blanks in the data? Were there recognized flaws in measuring and recording the data? If so, how do these problems affect the conclusions that have been drawn? It is important for the researcher to discuss these problems openly, rather than have a reviewer call them out, as if the researcher didn't know about them. Since professionals understand that problem solving often yields unanticipated new problems, there should be no need to try to hide research problems. In fact, to do so not only is a missed opportunity to do meaningful work but also amounts to unethical professional behavior.

Summary and Conclusion. Often, but not always, a separate section is reserved at the end of the report for writers to assess the significance and worth of their research. It is important here, as with any good conclusion, that there be some speculation about what the research implies for the future. Is more work of the same kind necessary? What more can be learned from additional research? Or should research in the same area take a different path? If so, why, and how should that path be defined?

Answers to questions like these help summarize and justify the work that was undertaken and suggest directions for related work.

Works Cited and Appendixes. Treatment of works cited and appendixes is virtually the same here as it is in the earlier discussion of completion reports. The only inhibiting factor concerns space if the report is being submitted to a journal for publication. Often journals set strict word or page limits, and appended material would be the first to go in order to meet those limits.

SAMPLE PROFESSIONAL COMPLETION REPORT

Figure 7.6 is the rough draft of Patricia Wynjenek's completion report to Las Almas Buenas Foundation on the activities of her work in Indigo on economics and health care. She wrote the draft in the basic format of the first completion report discussed before, but one that could also serve for an article for possible publication in a professional anthropology journal.

Because precedent did not call for it in Las Almas Buenas's case, Patricia did not have the report specially printed and bound. After the letter of transmittal, the report begins with the title page and ends with conclusions and works cited. The report does not contain appendixes.

Patricia N. Wynjenek
Department of Anthropology
Hastings University
Las Palmas, NM 70812

January 16, 2001

Dolores Sanchez, Associate Director
Las Almas Buenas Foundation
1752 Alamo Drive
San Antonio, TX 68201
Re: LAB 990-30-01-15-90

Dear Ms. Sanchez:

Enclosed is my final report on health care practices among low-income residents of Indigo, Texas. I have called it POVERTY, TRADITION, AND HEALTH CARE: THE EXAMPLE OF A TEXAS BORDER CITY. It is the last of three reports I have written for LAB, and it represents my final ideas and conclusions on part of the work I did in Indigo from January to December, 2000.

I have found generally that the case of Indigo, in contrast to reports of other Mexican-American communities, does much to reaffirm the diversity of the Mexican-American population as a whole. Family size has decreased dramatically in the last 20 years; modern or "scientific" medical care is sought when available, but access to it is still restricted by economic, bureaucratic, and professional barriers; and traditional healing practices are well integrated with scientific ones. Decisions about health care are based on perceived acuteness of condition, expense, and failure of previously employed therapies.

I hope you find the report helpful and informative. and that perhaps I can be of further service to LAB in the future. Please let me know if you have any questions.

Sincerely,

Patricia N. Wynjenek

Patricia N. Wynjenek
Associate Professor

Encl.

cc: Gunther T. Eliot, Chair, Department of Anthropology; Lois M. Charing, Dean, College of Arts and Sciences; David J. Price, Provost

The front matter of Wynjenek's completion report is conventional. The letter of transmittal identifies the report (including the contract number in the reference line) and sends it to the appropriate person at Las Almas Buenas. It summarizes briefly Wynjenek's main conclusions, preparing the reader for the executive summary.

FIGURE 7.6 Patricia Wynjenek's completion report

The title page includes necessary identifying information: author, contract number, recipient, and date.

POVERTY, TRADITION, AND HEALTH CARE:
THE EXAMPLE OF A TEXAS BORDER CITY

Submitted to

Las Almas Buenas Foundation
1752 Alamo Drive
San Antonio, Texas

Submitted by

Patricia N. Wynjenek
Department of Anthropology
Hastings University
Las Palmas. NM 70812
(LAB 990-30-01-15-90)

January 16, 2001

FIGURE 7.6 ***(Continued)***

CONTENTS

FIGURE 7.6 (*Continued*)

The table of contents lists major and minor headings, so that readers can turn quickly to particular areas they may have interest in knowing more about. In fact, the main purpose of all front matter (with the possible exception of the acknowledgments) is to identify, summarize, and generally make it easier for busy readers to select parts of the report they consider of greatest interest and importance. This is one of the primary purposes of the kind of formatting we find in technical and professional writing: it allows efficient consideration of a document without requiring a complete linear reading from beginning to end.

The acknowledgments section mentions the people who were of special help to Wynjenek in her work, and she makes sure to include those who supported her financially.

ACKNOWLEDGMENTS

For their enduring patience and good will, I thank first of all the people of La Fabrica. They put up with my intrusive presence and questions, because they are generous people and they thought that some good might come of my interruption of their lives. I sincerely hope it does.

I also have to thank personally Jose Vela and his family for their help, care, and friendship. Without them, I could not have survived, and without people like them, anthropologists like me could never complete the work they believe they are supposed to do. Along with Jose Vela and his family, I also need to thank, for their personal help, Maria Cortez, Magdelena Moreno, and Eduardo de Playa.

For their professional help, and also good friendship, I want to thank my colleagues at Hastings University, James Bowdle and Alan Littleton.

Finally, for their financial support and their intelligent professional assessment of my work, I thank Las Almas Buenas Foundation, the National Anthropological Society, and the Arts and Sciences Research Committee of Hastings University.

FIGURE 7.6 ***(Continued)***

Executive Summary

An investigation of the family planning and other health care practices of the residents of an impoverished barrio in the border city of Indigo, Texas, carried out via traditional ethnographic methods and the administration of a standardized, open-ended interview, yielded the following results:

1. Family size has decreased dramatically in the last 20 years. Despite unaltered adherence to the Catholic Church, women of childbearing age cite economic considerations for their choice to limit the size of their families.
2. The increased availability of modern, "scientific" medical care has greatly reduced reliance on *parteras,* or lay midwives, with generally, though not universally, beneficial results, and increased the frequency and timeliness of medical care generally. Access to medical care is still restricted by poverty and by governmental and professional bureaucracies.
3. Traditional healing practices continue to be employed in definable circumstances. They are well integrated with the resort to scientific medicine, and both traditional and scientific concepts of illness appear to be held without conflict. Decisions about health care are based on perceived acuteness of condition, expense, and failure of previously employed therapies.
4. These findings for Indigo, in their variation from the findings reported for other Mexican-American communities, reaffirm the diversity of the Mexican-American population as a whole.

The executive summary is conclusive, not descriptive. Wynjenek mentions the four areas about which she was able to draw significant conclusions, but more important, she is sure to mention, or allude to, the reasons they are significant. This clues the readers to the fact that Wynjenek's work has turned up new or different information about Mexican-American culture, and suggests those areas of the report readers may want to turn to for a full discussion.

FIGURE 7.6 (***Continued***)

Wynjenek's introduction does what a good introduction should do. It provides the necessary background of what has been done in the area of study, who has done it, when, and what it means for the basis and purpose of her study. It essentially sets the scene and prepares the reader for an intelligent reading of the rest of the report.

Since Wynjenek does not include a separate review of literature (specified as appropriate for research reports in the second format discussed above), her specific references to older and more recent studies others have done help supply the kind of information a review of literature normally supplies.

POVERTY, TRADITION, AND HEALTH CARE:
THE EXAMPLE OF A TEXAS BORDER CITY

Introduction

Early Ethnographic Studies and Their Successors

Studies of health and health care among Mexican-American populations in the United States generally fall into two categories. The first, and by far the older, focuses on the culturally distinctive features of Mexican-American life, particularly the practice of *curanderismo* and other aspects of traditional healing. These studies include not only infuential ethnographic works of the sixties and seventies, such as those of Madsen (1964, revised 1973), and Rubel (1966), Warner (1977), and Achor (1978), but also more recent writings, some of them with an applied focus, that derive from the same tradition. Thus, Maduro (1983) in his discussion of *curanderismo* stresses the significance of "folk illness" in Latino cultures and the beliefs held by Latinos about the causes and treatments. McKenna (1989), in an analysis of the health care needs of elderly Mexican Americans that relies on studies dating from as early as 1951, refers to the "passivity," "passivistic-collectivistic value orientation" and "fondness for traditionalism" of the group under study. And in *Transcultural Concepts in Nursing Care* (1987), a textbook designed to introduce nurses and nursing students to the medical implications of cultural variation, Boyle and Andrews urges nurses to respect the culture-specific needs of many Mexican-American patients who seek "spiritual treatment" from *curanderos* and *espiritualistas*.

More Recent Studies

In contrast to the focus of the traditional and traditionally based studies cited above, new foci in studies of Mexican-American health and health care have emerged in the last decade. Scheper-Hughes and Stewart manifest at least two of these in their 1983 article, "Curanderismo in Taos County, New Mexico—A Possible Case of Anthropological Romanticism." Drawing on their own work among Spanish-Americans in rural New Mexico first studied by Saunders in 1954, Scheper-Hughs and Stewart found that in the 30 years since Saunder's report, health care practices of at least this population have changed greatly. In contrast to Saunders's findings of pervasive reliance on traditional concepts of health and illness and on traditional healers, Scheper-Hughs and Stewart conclude that "*most* Spanish-Americans of Taos County rely primarily on the same explanatory model as Anglos with regard to the definition of health and illness," and that "*curanderismo* in northern New Mexico today has moved from being a primary and important source of medical care to an alternative and very occasional source" (Scheper-Hughes and Stewart 1983:80).

FIGURE 7.6 ***(Continued)***

In addition to these specific findings, Scheper-Hughes and Stewart stress the need to acknowledge and explore the great diversity that exists within the Mexican-American population, rather than to lump together the rural, the urban, the economically secure, the poor, the minority and the majority populations. Indeed, calls for the recognition of Mexican-American diversity have become increasingly common and urgent in a variety of academic disciplines (see Winkler 1990).

An important trend in recent students of Mexican-American health and health care is the move away from a focus on distinctive subcultural variation to a more general demographic and epidemiological focus. Such studies do not aim at ignoring cultural variation where it has a demonstrable effect on the perception of illness or on health care decisions, but they point out that the health of Mexican Americans increasingly presents a profile that is not dramatically culturally distinctive. Rather, it is typical of the social class to which they belong. However, "since a large proportion of the Mexican origin population is poor, health problems which are associated with poverty are particularly important in the study of the health of this group" (Angel 1985:413).

A final topic, less frequently addressed, is what Foster (1987), in his examination of international health agencies, refers to as the "bureaucratic" issue. That is, to what extent are the perceptions of health care and the decisions relating to it the result, not only of subculture, social class, biology, and epidemiological risk factors, but also of the governmental and professional bureaucracies that deliver medical care?

Background of This Study

The following discussion explores these issues as they affect the low-income Mexican-American population of the border city of Indigo, Texas. Data are derived from ethnographic fieldwork (supported by Las Almas Buenas Foundation, the National Anthropological Society, and the Arts and Sciences Committee of Hastings University) conducted from May through December, 2000. Ethnographic fieldwork involved both traditional participant observation in a neighborhood in which I had lived and worked off and on over a period of 14 years, as well as the tape-recorded administration of an open-ended interview schedule by four local residents, who acted as research assistants. One hundred household heads, most of them women, were interviewed in Spanish. Addresses were chosen at random from a list of the slightly more than 300 houses in the neighborhood, which is referred to by the pseudonym of *La Fabrica*.

Indigo is an interesting choice for the exploration of factors involved in Mexican-American family size and other medical choices because of its extreme cultural conservatism. A city of roughly 100,000, Indigo is 93% Mexican-American, with an almost unbroken history of 250 years of economic solidarity and self-direction. Though the city has the lowest per capita income of any city in the United States, it also has an old, entrenched Mexican-American political, commercial, and professional elite. Except for its much larger sister city, Pollo Blanco, Naranjas, Mexico, directly across the international bridges

She also mentions under the heading of the background of her study the basic methods she used: participant observation and interviewing. She says that the interview schedule (survey) was open-ended (questions not limited to a choice of possible answers) and specifies the way the interview was administered, including how many interviewers there were, how many people were interviewed, and how those people were chosen. She does not mention, perhaps because her audience would assume it, that not only the interviewers but also the interviewees were paid. Her introduction thus satisfies the need for more information on methods and materials, suggested above as appropriate for reports on primary research.

FIGURE 7.6 (*Continued*)

The discussion is not headed "Discussion" but rather is divided from the beginning into four major areas of investigation with appropriate headings: "Availability of Health Care," "Family Size," "Childbirth," and "Traditional Healing." Each major area is then subdivided into smaller areas of interest and given a minor heading: "Indigo's Medical Establishment," "Mexican Alternatives to the Medical Establishment," and so on. In this way, Wynjenek has organized her discussion to present the overall problem as one that is naturally made up of distinct parts that need to be analyzed and commented upon as they contribute to an understanding of the whole.

Using the first major area of discussion, Availability of Health Care," as an example, we can see the three basic ways Wynjenek has researched and analyzed the problem. We see these ways repeated, more or less consistently, throughout the report.

1. She uses purely secondary material, often in the form of Census reports, to establish some of the basic facts of the problem: how many doctors and dentists practice in this city of 100,000, where and how they have been educated and trained, and how these compare with other figures.

Thus, readers find out that the number of doctors, a rate of 64.5 per 100,000 persons, is "respectable" for Texas as a whole.

over the Rio Grande, Indigo is isolated from any other settlement of more than a few hundred by 150 miles of scrub desert in every direction, and by the Texas system of independent county governments.

Availability of Health Care

Indigo's Medical Establishment

For a border community in Texas, Indigo enjoys good medical services. Though the city has only ten dentists for 100,000 inhabitants, it has sixty-five doctors, for a rate of 64.5 per 100,000 persons, a respectable showing in Texas. Indeed. Indigo has something of a local reputation as a medical center, and patients from rural towns as far as 80 miles away, as well as from Pollo Blanco and other, smaller Mexican settlements often go to Indigo for medical care. The city has two hospitals, one of them a proprietary hospital in the new, wealthy suburb of Del Mar, which is never patronized by Indigo's poor. Faith Hospital, located in a central, impoverished barrio, is by far the larger of the two hospitals, and accepts medically indigent patients for emergency and urgent care.

In addition to private physicians, dentists, and hospitals, residents of Indigo have access to several neighborhood clinics. La Fabrica, the barrio in which I conducted the interviews cited in this paper, has its own modern clinic within easy walking distance of all houses in this awkwardly situated barrio, isolated from other parts of the city by a large web of railway tracks. The clinic offers routine well baby care, including immunizations; family planning services; and a drop-in clinic for general health care and referrals. It does not handle prenatal care, but it does offer some dental services. The residents of La Fabrica agitated for such a clinic for many years, and its realization, in Mexican colonial-style stucco, has been a source of pride and gratification to the whole barrio.

Aside from the official medical outlets described above, people in La Fabrica can patronize several other sources of health care. There are doctors, dentists and pharmacies in Pollo Blanco; parteras, or lay midwives, in Indigo: and *curanderas,* or traditional healers, on both sides of the river.

Only 18 of the 100 barrio residents surveyed said that anyone in their households had ever visited a dentist in Indigo, and most of them noted that these dental visits had taken place at the clinic. Dental care has an extremely low priority in La Fabrica. Throughout the city it is common to see middle-aged and older people with missing or visibly decayed teeth.

Sixty-three respondents reported that someone in the household had visited a private doctor in Indigo. This group included all those respondents with Medicare (19%), Medicaid (12%), or medical insurance (11%), since these programs pay for office visits that many barrio residents cannot otherwise afford. Unless they have a longstanding relationship with a doctor, most barrios residents believe that they will be turned away if they cannot pay at once for physicians' office visits. While this is certainly true in some cases, it may not be true in others. It is, however, a pervasive view throughout South Texas, and is often responsible for the delaying of medical care.

FIGURE 7.6 ***(Continued)***

Most of Indigo's physicians are Mexican-American and have been trained in Mexican medical schools. Many preserve a reserved, sometimes distant manner with their patients, speaking little to them and explaining less. Patients are often held to be too ignorant to understand details of their conditions and treatments and too irresponsible to arrive at the office at a specific time. Thus most doctors make no appointments, and patients often wait 2 or 3 hours or longer to see a doctor. After such long waits patients are angry and frustrated, feelings that are often intensified after they leave the office with diagnoses and advice they often fail to understand completely but are too intimidated to ask about. Rich or poor, Indigoans almost universally respect modern medicine, but many barrio residents are antagonistic toward Indigo's physicians. Thus even people whose economic resources are limited often seek medical care in San Antonio or Corpus Christi, each 150 miles away. Both cities have large Mexican-American populations, although the cities themselves are much more oriented to Anglo-American culture than Indigo. Still, many Indigoans believe that doctors outside Indigo, whether Anglo-American or Mexican-American, are likely to treat them with greater consideration and respect. One resident of La Fabrica, a middle-aged woman whose eye injury was misdiagnosed in Indigo as permanently incapacitating, now seeks all medical care in Corpus Christi. This not only represents a substantial investment in time and money, but also puts the woman, a dynamic and articulate force for barrio improvement, in a difficult and demeaning position. Since she cannot drive, speak any English, or read or write more than her name, any visit to the doctor now requires the services of her husband or daughter as a driver and translator. Her story, which is only one of many I have heard in Indigo, is powerful testimony of the enormous effort to which some barrio residents will go to avoid having to deal with the Indigo medical establishment.

Most people in La Fabrica, of course, cannot travel 150 miles outside the city for medical care, and must make do with the services Indigo has to offer. This exacerbates rather than tempers their general dislike and mistrust of Indigo's doctors. The most commonly voiced complaint was that "they make you wait and wait, and then it's $30." But despite its 100,000 people, Indigo seems in many respects a much smaller place, and the reputations of individual physicians, as well as stories of their rudeness, favoritism, and maltreatment of patients, are well known throughout the city. Whether or not the stories are apocryphal, the dislike is real.

Because doctors are expensive and time-consuming, and because they are considered arrogant and unsympathetic, few residents of La Fabrica have a regular doctor (the "family physician" of American folklore). Instead they, like many other impoverished Americans, often turn to the emergency room at Faith Hospital. Sixty-seven percent of those interviewed said they or someone in their family had used the emergency room at least once. Often these visits are occasioned by children's illnesses that have worsened during the night, when the clinic is closed, or by acute conditions in adults for which the clinic does not provide services and for which families cannot afford a visit to a private physician, who may turn them away.

More and more people in Indigo are making use of neighborhood clinics: 73% of those surveyed in La Fabrica said they or their

2. She uses primary information gleaned from interviews about the following topics: how many people (usually in raw numbers and percentages) use established medical facilities—doctors, dentists, hospitals, and neighborhood clinics; how often they use the facilities; how those who use them pay for them (Medicare, Medicaid, private insurance, unassisted means). Wynjenek implies that those who must use unassisted means to pay for medical care rarely do, simply because they don't have the means.

3. She uses participant observation, reported anecdotally, to determine how various residents respond to the medical care available to them. Thus readers are told that the private medical establishment, particularly, holds itself aloof from most of its patients; that patients have to wait long periods of time to see doctors; and that they often leave doctors' offices angry and frustrated, knowing very little more about what is wrong with them, or how they should go about making themselves better, than they did before their visits.

Specific anecdotal evidence, like the case of the woman with the eye injury, illustrates and particularizes the experiences of the people in their dealings with the medical establishment.

FIGURE 7.6 ***(Continued)***

As mentioned earlier, we see this pattern of analysis and reporting repeated. Under the major heading of "Family Size," Wynjenek provides Census data on the birth rate in Indigo and how it has decreased dramatically; primary data from respondents on how and why they limit family size; and finally, interpretive notions, based on participant observation of limited family size (for one male respondent, the births of only two children in thirteen years of marriage) that have to do with how people think they are responding to culturally approved norms, and how they actually appear to be responding to them.

families used the neighborhood clinic regularly. The proximity of the clinic and the fact that its services are available free or at very low cost have undoubtedly improved health care for those in the barrio, especially children. One barrio resident who worked tirelessly to bring the clinic to La Fabrica (the woman with the eye injury mentioned earlier) said that her greatest sense of achievement came from the realization that "now people don't wait when their children are sick until they are almost dead" to seek medical care for them.

Mexican Alternatives to the Medical Establishment

Sixteen residents of La Fabrica told interviewers that they patronized doctors or dentists in Pollo Blanco, and sixteen (an overlapping but not identical group) said they bought medicines in Pollo Blanco's pharmacies. The decision to seek medical or dental care across the river is usually based on cost: most office visits and procedures are less than half as expensive as they are in Indigo. In addition, some people find Pollo Blanco's physicians more sympathetic and less distant. This is difficult to understand, since most of Indigo's doctors (and virtually all those patronized by barrio residents) are Mexican-American, some are Mexicanborn, and many are graduates of Mexican medical schools. One would expect that their demeanor would be similar to that of their counterparts across the river. Finally, Mexican pharmacies offer not only several kinds of patent medicines believed by many Indigoans to be extremely effective, but also some kinds of drugs that can be bought over the counter in Mexico but only by prescription or not at all in the United States.

The use of Mexican health care facilities correlates neither with low income nor with advanced age nor with briefness of residence in the United States. It seems to be an unpredictable, idiosyncratic choice. Some respondents said brusquely, "No, *nada de eso*" (No, none of that), when asked if they went across the river for health care. Others responded, "Yes, sometimes, for reasons of economy" (*por economia*). My landlady provides a good example of the mixture of elements involved in choosing where to seek medical care. She is a strong, almost passionate, believer in American medicine; she has a good income by barrio standards, as well as medical insurance, thanks to her husband's rare union job: she is an American citizen, having lived 40 years in Indigo, although she was born in rural Naranjas. She would not consider going to a doctor in Pollo Blanco, but all her dental work is done there, at fees that are roughly a third of U.S. dental fees for comparable work. And she is very partial to some Mexican patent medicines, which she considers superior to anything available in the United States.

Family Size

Birth Rate

In 1968 the birth rate in Johnson County, Texas, of which Indigo is the county seat and only city, was 43.8 per thousand. Twenty years later the birth rate had declined to 26.6 per thousand, nearly 40% lower. It is interesting to note that, although Johnson County has the second

FIGURE 7.6 *(Continued)*

highest concentration of Mexican Americans of any county in the United States, its rank in birth rate is only seventieth. It should also be remembered that figures for Johnson County are inflated by the large numbers of residents of Mexico who give birth in Indigo so that their children may claim U.S. citizenship. Indeed, the first thirty-seven places in the birth rate sweepstakes go, not to counties with high Hispanic populations, but primarily to counties in Kansas, South Dakota, North Dakota, and Utah, with high Hutterite, Mennonite, and Mormon populations.

A look at Indigo's 5-year age cohorts in the 1990 Census report gives graphic evidence of the city's declining birth rate:

under 5 years: 10,350; 5–9 years: 10,755; 10–14 years: 11,154; 15–19 years: 11,165. These figures are all the more remarkable because of the constant emigration of Indigoans who cannot find work in the city (whose official unemployment rate has recently ranged from 12 percent to 36 percent) and the constant immigration of Mexican nationals, who may be even more culturally conservative than lifetime or longtime residents of Indigo.

Reasons for Limiting Family Size

Johnson County's birth rate is nearly twice as high as that of the state of Texas as a whole, which is 19.2 per thousand. Its dramatic decline over the last two decades, however, seems to indicate that Mexican-American birth rates, even in such a traditional and isolated enclave as Indigo, are not mere cultural epiphenomena, but that they respond to the same kinds of conditions as do the birth rates of other populations. Shrinking opportunities for migrant farm workers, the need for longer schooling, the increased acceptability of women's working outside the household, and necessity for women to do so have all had an effect on family size. These phenomena are consistent with the patterns that not only prevail in developed, industrial countries, but that have emerged in the developing world as well. They are also, and at least equally significantly, specifically expressed in the words of Indigo residents as well.

Feelings about family planning in the barrios of Indigo are both interesting and fairly clear-cut, and they provide insight into the human realities of family planning that is not available from Census statistics. In the survey of households in La Fabrica, answers to questions about family size fell into three categories. Just over half of those questioned favored birth control outright. Nearly another quarter believed that birth control was essential in contemporary life, but they regretted its necessity. Only a quarter said they believed in having "all the children God sends."

People with an apparently unalloyed preference for family planning included men and women of all ages, and their reasons were always economic. "You have to think, how many can I support, feed, educate, protect, and defend against everything," said one young woman with energy, while an eighty-three-year-old mother of five sons responded urgently, "I tell my grandchildren that they must limit their children to one or two—or three." The different options for rich and poor informed many responses. "If you make good money you can have a hundred children, but if you're poor, you have to plan," one man stated firmly.

FIGURE 7.6 *(Continued)*

A common response of barrio residents was that family planning had not always been necessary, as it now was. One great-grandmother represented this view in her response: "You used to be able to have all the children God sent; everyone had five. But in these times, no. Now you can have only one." A middle-aged woman offered a similar opinion: "Before, you could have twelve, fifteen, twenty children. Now everything is so expensive you can't maintain them." People who mentioned an earlier era nearly always regretted the necessity of birth control, as, indeed, did those who compared the rich to the poor in this respect. The desire for numerous children is still present for many people in Indigo's impoverished barrios, although most have decided that they cannot afford to indulge this desire.

The makeup of the group of barrio residents who said that they favored having all the children God sent is extremely interesting. It includes primarily men, older women who had never been married, and women over forty, some of whom had had as many as ten or fifteen children. Some of the responses in this category were religious in form. One of the never-married women gave the most doctrinaire response: "According to my religion, marriage is for having a family, and to limit one's family would be a sin." But a seventy-year-old mother of four answered similarly: "I think you should have all the children God sends you. To limit them would be a mortal sin." Most of the reasons married women gave for preferring not to limit their families, however, were references to their own children. "No, I don't like family planning," said one woman, "because I had fifteen, myself." None of the men gave any reason for not liking birth control, nor did any of the interviewers, all of whom were women, press them. An interesting feature of their responses, however, was that their professed desire for "all the children God sends" seldom seemed to accord with the number of children they actually had. In 13 years of marriage one of these men had had two children, who were eleven and twelve years old. Another had had two in 7 years, and a third had had three in 17 years. It seems clear that either these men were responding according to a culturally approved model while acting according to economic self interest, or they believed they were acting according to cultural rules while their wives were acting according to economic self-interest.

While there is still a strongly voiced minority sentiment in Indigo's barrios that opposes limiting family size, it is not the sentiment of women of childbearing age. All but five female respondents aged forty or under said they favored birth control, as of course, did many men and older women. Whatever the motivation of the men opposed to birth control, the size of their families seem to be determined not by their stated views, but by those of their wives.

Only one of the household heads interviewed in Indigo identified herself as anything but Catholic, and she was a former Catholic, now converted to the Jehovah's Witnesses. Though there is a great range of piety and degree of formal religious observance, virtually all Indigoans consider themselves Catholic, and all adults are well aware of the prohibition of the Catholic church on artificial methods of birth control. For a small minority, like the two women quoted earlier, this prohibition was given as the reason for which they did not approve of birth control. It should be noted, however, that several of those interviewed who disapproved of birth control on religious grounds were

FIGURE 7.6 (*Continued*)

women who had never married (and who had no children). Those persons who favored birth control and who also mentioned the position of the Church were divided as to their feelings about not complying with Church teachings. One mother of two in her early twenties said, "If a person knows that he is poor, he can't have too many children, if you accept what God sends, he could send a bunch (*un bonche!*)." Though this woman is remarkably cheerful and self-reliant and apparently unmoved by Church teachings on birth control, some of her neighbors were less insouciant. "They say, 'Have faith,'" one older woman remarked bitterly, "but it's too many." And the father of ten children under sixteen, a man who now favors birth control, responded similarly: "They say it's a sin, but I say you can't support them."

Childbirth

Access to Prenatal Care

The fundamentally economic basis for birth control in Indigo is mirrored in the way other medically related decisions are made. One of these concerns is childbirth, which, because of Indigo's location on the U.S. Mexican border, is marked by some unusual features. Not only do Mexican nationals living in Pollo Blanco come across the river, sometimes while in labor, to give birth in Indigo and thus assure their children of U.S. citizenship, but both residents of Pollo Blanco and Indigoans themselves make use of the services of parteras, or lay midwives.

Pregnant women who do not plan to give birth in a hospital often spend the last week or so of their pregnancies at the house of a partera in one of Indigo's barrios. Here the woman's vital signs are monitored, and if obvious problems arise, the partera can usually send her patient to Faith Hospital for care and delivery. Parteras are usually shrewd, cautious, and experienced women, who much prefer to lose a patient to the hospital than to death, should complications arise.

A woman in labor, with acute preeclampsia or toxemia, or with severe postpartum complications will not be refused admission to Faith Hospital, whatever her economic situation, residence, or nationality. In this, Indigo is very different from many other settlements along the border or elsewhere in South Texas, where there are no medical facilities at all, or only privately funded hospitals without a federally mandated obligation to accept indigent patients. Although the hospital provides inpatient maternity care to virtually anyone who requires it, it does not provide prenatal care, nor does any clinic or physician provide this service at low or no cost. The only alternatives available to pregnant women are a private physician (most of whom require a fee of roughly $700, payable in cash by the fourth or fifth month of pregnancy), a partera (whose fee, usually $200 to $300, must also be paid in advance) or no prenatal care at all, culminating in arrival at Faith Hospital in labor. Because of the high unemployment and underemployment in Indigo, common unreported employment (involving payment "off the books" or "under the table," to benefit the employer), as well as thousands of long-term jobs that simply offer no employee benefits, only eleven of the households surveyed re-

The same happens with "Childbirth" and "Traditional Healing," although purely secondary material, like Census data, is noticeably lacking, mainly because respondents are even more reluctant to report such information to Census takers than they are to people like Wynjenek and her research assistants. Thus information about access to and use of prenatal care, traditional midwives (parteras), and curanderas, can be obtained only through primary means (answers to interviewers' questions.)

Primary evidence seems to take over in these sections of the report. Readers learn about how many women take advantage of prenatal care, how many use parteras, how many go to professional physicians to find out about complications of their pregnancies, and how much they have to pay for such services. Anecdotally, readers also learn about how midwives work, how often their services are used, and what they change.

FIGURE 7.6 ***(Continued)***

ported any coverage by employment-related insurance. Twelve households reported coverage for some family members by Medicaid, and nineteen elderly householders reported that they or their spouses received Medicare, a program that is of little assistance in defraying the costs of childbirth.

Traditional Midwives

In the households I surveyed, 37% of the female household heads had given birth with the assistance of a partera, 50% had been attended by a doctor, and 2% had employed both during the course of their reproductive lives. The rest had had only the assistance of their families. Of those who had employed parteras, only two said they preferred them to doctors. All the others cited a lack of money to pay a doctor, or, in the case of two elderly women who had lived on remote ranches, the unavailability of doctors. Those who had been delivered by doctors were almost universal in citing the greater security or safety of physician-attended births. Five women added that they had been forced to turn to doctors because of complications surrounding their pregnancies or deliveries: hemorrhaging, preeclampsia, or the need for a caesarean delivery. Another three women volunteered that they had been afraid to go to a partera. Formal interviews in La Fabrica did not request information about the complications of pregnancy or delivery, although some women volunteered them. Among my friends in other barrios in Indigo I have heard, however, several stories of problems accompanying delivery by partera. One involved the death, shortly after delivery, of a young woman who lived in Pollo Blanco and worked in Indigo (a "commuter"). The woman had been staying at the house of an Indigo midwife awaiting delivery. The baby was born healthy, and the mother appeared healthy as well. Shortly after delivery, however, complications arose in the mother, who was immediately sent to Faith Hospital. She died shortly after admission. Without more information it is impossible to know whether delivery in a hospital would have saved the mother's life, but it seems likely. Another story, more typical, involves complications that, while less tragic, would certainly never have occurred in a hospital delivery. The midwife in this case performed an episiotomy on her patient, and the baby was born healthy. The episiotomy, however, was never sutured. It healed slowly and painfully, with infection and the development of awkward scar tissue that necessitated caesarean delivery for the next pregnancy.

One of the interviewers, a former migrant farm worker, all of whose children had been born in hospitals, some of them in northern states, occasionally asked respondents who had had doctor-assisted births whether they had done so because they had been "afraid." Though I have no further evidence than this, it seems as though there is some vestigial virtue that can be attached to the courage involved in foregoing a doctor. Whether or not this is true, few Indigo residents are choosing delivery by partera these days. No respondent in her twenties had employed a partera, and only five in their thirties at the time of the interview had done so. This is in great measure the result of Faith Hospital's relatively recent policy of accepting all patients for delivery and emergency inpatient prenatal and postnatal care. It cannot necessarily be seen, however, as unalloyed progress, since many women

FIGURE 7.6 ***(Continued)***

who would once have had some prenatal oversight from a partera now have none at all, their only goal being hospital delivery.

The story of Maria Esmeralda Diaz (a pseudonym) illustrates the difficult position of impoverished pregnant women in Indigo, as well as the style of the city's medical community. Maria was born 25 years ago in Chicago to parents who came from Indigo. She lived in Chicago off and on until she was married, at 23. She then returned to Indigo with her Puerto Rican husband so that she could be near her parents, who lived in La Fabrica. She wanted to be near her family when her first child was born, and she wanted to escape Chicago's crime. But her husband had trouble finding a job in Indigo's crisis economy, and the couple had little money. Meanwhile, Maria's pregnancy was advancing, but since she could not put together either $200 for a midwife or $700 for a doctor, she relied on her mother's care and advice and put off medical assistance. When it began to seem as though Maria's pregnancy was continuing too long, she made an appointment with a prominent Indigo general practitioner. This woman refused to accept Maria as a patient because she had come to her only in the ninth month of pregnancy. Now more concerned than ever, Maria immediately took a bus to Chicago, where her aunt still lived. The morning after she arrived, she went straight to a hospital, where (despite her indigence) she was examined, the fetus was examined via ultrasound, and for reasons of which Maria was not entirely sure, doctors decided to induce labor immediately. It is chilling to consider what would have happened if Maria had lacked such energy, experience, and family connections.

The growing conviction that modern medicine, however and whenever acquired, is superior to traditional care has created new difficulties for impoverished pregnant women in Indigo. In addition, time spent in the maternity ward of Faith Hospital is for many women a rare chance for respite from their strenuous household duties. One informant, who had developed preeclampsia during her last two pregnancies and had spent several weeks in the hospital before each birth, told her interviewer with enthusiasm, "*Me encanta el hospital* (I adore the hospital)." It is thus clear that in Indigo, as Scheper-Hughes and Stewart report for Taos, the use of *parteras* does not represent reactionary adherence to a dangerously maladaptive cultural practice, but rather a locally available optional strategy employed primarily for economic reasons and eagerly discarded as soon as possible from economic and/or medical motivations, even when it might prove beneficial.

Traditional Healing

Home Remedies

It remains to consider the issue of traditional medicine, including both household remedies *(remedios caseros)* and traditional healers *(curanderas)*. Two-thirds of the household heads interviewed in La Fabrica said they resorted to home remedies first when family members were ill. These remedies are usually herb teas of various leaves, flowers, or barks, many of which can be bought in small cellophane bags at the grocery store. Indeed, Indigo is the home of La India, a company

FIGURE 7.6 ***(Continued)***

Further, readers learn about midwives, home remedies, traditional healers, and the ailments they treat.

Wynjenek is sure to point out that although there is a certain embarrassment (she more significantly calls it "opprobrium" attached to traditional healing methods, the people of La Fabrica take part in them. The evidence Wynjenek presents in these sections is more clearly and dramatically anecdotal, even more convincing because it comes from individual sources, who have decided for any number of reasons to be forthcoming in their responses. One example involves the death of a young woman who lived in Pollo Blanco, who had come to a partera in Indigo for the delivery of her child. Another example tells about a woman who traveled back to Chicago, where she had an aunt living, who could help her in the last hours of her pregnancy, because she could not get the help she needed in Indigo. A third example deals with the problem of one of Wynjenek's informants and friends who had been suffering from a stomach ailment, who had first used established medical means to diagnose and treat the ailment; then when none of that treatment seemed to work, had gone to a curandera in Pollo Blanco; and finally, when that treatment had not worked, had gone back to established medical care in Indigo. Her review of the steps that the man took to get better indicates that although the established

whose packaged medicinal and culinary herbs can be found in stores serving Mexican-American customers all over the western and southwestern United States. For other, more recherché ingredients, some barrio residents go to special herb shops in Indigo or Pollo Blanco. In addition to teas to bring down fevers, settle queasy stomachs, or relieve insomnia, some women can make poultices for cuts, burns, pains, and fevers, and soothing syrups for coughs. Among those interviewed who said they could do this were both old and young women, including two nurse's aides.

Traditional Healers

Besides the ability to cure everyday discomforts with home remedies, some women can cure such disorders as *mal de ojo* (usually translated as "evil eye," though this distorts the nature of the illness), *empacho* (intestinal blockage), and *susto* (fright). Though social scientists usually treat these disorders as a separate category of "folk" or supernaturally caused illnesses, they do not seem to be regarded universally as a distinct category by Indigoans. One woman, for example, said she takes her young son to the neighborhood clinic when she suspects him of having *susto,* while at least one respondent seemed to distinguish *mal de ojo,* as supernaturally caused, from *susto* and *empacho,* which had natural causes (and the latter of which was curable by Pepto Bismol). Similarly, respondents do not use the term *"curandera"* to designate all women who can cure *ojo, susto,* or *empacho.* I did not hear of a single man who engaged in traditional curing, although there are some in the region, including some who are extremely well known. Some women can cure these ailments in their own children, while others have sisters or friends who are particularly good at it. Only when an adult is suspected of having *empacho* (usually only children or adolescents are afflicted with *ojo* or *susto*) and home remedies and patent medicines have failed, would most families consider looking for an official *curandera* whom they do not know personally.

Ambivalence and Integration

An extremely unfortunate flaw in the interview schedule resulted in unreliable responses to questions about *curanderas.* Naively, I had constructed the questions using the term *"curandera."* In response to all questions using this term, every informant responded that she never consulted such persons. One interviewer, however, an extremely bright and insightful young woman, simply rephrased questions involving traditional healers. Instead of asking whether the informant ever consulted *curanderas,* she asked, "When someone in your family is ill with *susto* or *empacho* or something like that, do you know how to cure him, or do you know someone who does?" To such a question, virtually everyone responded affirmatively, and then proceeded to discuss such techniques as rolling an egg across the forehead of a child to remove a fever, or rubbing the body with holy oil to cure *empacho.*

Clearly, even in Indigo's impoverished barrios, people are well aware of the opprobrium attached to *curanderismo,* especially when it is clearly identified with the traditional lexical label. They are embarrassed to admit that they resort to it when asked directly, but readily describe their use of the system when the interviewer both assumes

FIGURE 7.6 ***(Continued)***

means of his cure eventually worked, he couldn't rule out the traditional means, mainly because he perhaps hadn't believed strongly enough in them.

that they use it and refrains from labeling it. I have had a number of discussions with Indigo's barrio residents about traditional concepts of health, traditional healing, and supernatural power generally, and only one person, my monolingual, nonliterate, but very modern landlady, resolutely maintained disbelief. Indeed, her dislike for or discomfort with such topics was so great that it took days of pleading to discover why an old woman and her great-grandchild came regularly to gather leaves that had fallen from a particular tree in the neighborhood. It was a pirul tree, the leaves of which are used in many traditional curing rituals, but having finally told me that, my landlady burst out, "You will think we are savages. You will think we are crazy. It's nothing but witchcraft" (*pura brujeria*).

It is safe to say that most improverished Indigo residents continue to believe in traditional concepts of medicine together with modern western concepts. Many accounts of Mexican-American *curanderismo* stress its conflict with the practice of contemporary western medicine. The results of my experience and surveys in Indigo, however, paint a picture of traditional and scientific medicine as an integrated system, in which there is little conflict for most barrio residents. Home remedies, simple homeopathic and sympathetic curing practices, and patent medicines, such as aspirin and Pepto Bismol, are the first choice for curing sick or injured family members. They are cheap and easy to procure, and they disrupt the family very little. If these attempts fail, and the illness or injury does not improve, the patient is usually taken to the clinic, the emergency room, or a private doctor. Only if these latter measures fail after some time is a patient likely to look for outside help in traditional curing. Most barrio residents seem to consider the majority of maladies susceptible of cure by either traditional or scientific medicine, depending on circumstance, severity, and the individual. Certainly no Indigoan of my acquaintance would hesitate to consult a doctor for an obvious acute condition, whether he or she thought it was *susto* or heart trouble.

A medical anecdote from the life of a friend of mine of twenty years' standing, a man I will call Jose Vela, illustrates the ways in which traditional and scientific medicine may be combined in Indigo's barrios. In the spring Jose's stomach, which had been troublesome for some time, began to give him genuine pain. Jose, the thirty-eight-year-old son of uneducated Mexican migrant farm workers, who himself had earned a B.A., had been a medical corpsman in the Air Force, and the pattern of his discomfort seemed to him to indicate an ulcer. Since over-the-counter remedies and his mother's teas had no effect, Jose made an appointment with a general practitioner. The doctor examined him, ordered several tests, and ultimately concluded that Jose did not have an ulcer. Seeing that Jose's job was very demanding, he decided that the stomach pains were caused by "tension," and prescribed a mild tranquilizer. When, after several weeks, Jose's stomach had not improved, his mother, who does not like or trust doctors in Indigo, persuaded him to consult a physician in Pollo Blanco. This doctor agreed that Jose did not have an ulcer, and prescribed something to "calm the nerves," presumably another tranquilizer.

A month later, when Jose felt no better, his mother and his wife conferred and came to the conclusion that Jose must have *empacho*. Since he was by now miserable and living on Rolaids, Jose allowed his

FIGURE 7.6 ***(Continued)***

Wynjenek's conclusions, although implied throughout the report, are separately laid out for readers to consider immediately if they want to. They expand upon and develop the conclusions she has offered in both the letter of transmittal and the executive summary, and again they suggest the significance that the study has had, not only for the time in which it was done, but also for the future.

Readers can decide from her conclusions whether what she has done and what she thinks it means is significant enough to ask her for specific recommendations about what should be done in the future, or whether projects like hers should be supported in similar ways in the future.

Patricia Wynjenek has written a typical completion report. With a few modifications, it is also in good shape to submit to a professional journal for possible publication.

mother to take him to a *curandera* in Indigo's central barrio of La Virgen (a pseudonym). The *curandera* was a pregnant middle-aged woman, whose unemployed husband was often at home doing chores around the house, a situation that made Jose extremely uncomfortable. What made him more uncomfortable was having to take his clothes off, not only in front of the *curandera,* but in front of his mother as well!

The treatment, repeated several times a week, consisted of a vigorous massage, preceded by prayerful invocations and succeeded by downward brushing with leaves of the *pirul* tree, also accompanied by prayer. In between treatments Jose was to drink a bottle (about half a pint) of Pompeiian olive oil a day. The *curandera* asked for no payment: her skills were a gift from God. But if her patient was cured through her efforts, he was expected to give her a monetary gift.

Jose gave the *curandera* a gift, but he was not cured. In desperation he made an appointment with an Indigo internist, who diagnosed and treated him for diverticulitis. Within a week Jose's pain had subsided.

When I asked Jose, a thoughtful, insightful, educated, and intelligent man, what he had thought about the *curandera,* he replied that although he had disliked the treatments, he knew of many cases in which *curanderas* had effected cures for people who believed in the process. "I guess I didn't believe strongly enough," he said.

Conclusions

Most of Indigo's population is extremely poor, and nearly all of it participates in the most conservative Mexican-American culture anywhere in the United States, with the possible exception of the culturally distinctive area of highland New Mexico. Indigo's cultural system offers its participants traditional medicinal and healing options not open to most other Americans who participate more fully in the majority culture. And the city's location on the Mexican border offers scientific western medicine in a slightly different setting and at a much lower cost. But it would be a mistake to believe either that (1) Indigoans do not themselves participate to a great extent in Anglo-American culture, or that (2) they make decisions informed primarily by considerations shaped by their own distinctive subculture. In most matters that directly affect their health, well-being, and family size, Indigo's barrio residents appear to make choices based on the same values as any other Americans: they want to maximize their well-being and minimize the cost wherever possible—cost being of great importance in Indigo. Thus, as child and infant mortality declines, and as the raising of children becomes more and more expensive, considering urban life, prolonged schooling, and the cessation of income derived from child labor, people in Indigo, like people everywhere for whom this is true, simply have fewer children. They are aware of the teachings of the Church, but with more or less pain they, like most other Catholics, and certainly like most other American Catholics, have decided that they must ignore those teachings.

In the cases of prenatal care and delivery, Indigoans, like most other Americans, are convinced of the superiority of doctors and

FIGURE 7.6 ***(Continued)***

hospitals over *parteras*. When they could not afford doctors or hospitals, or could not reach them, Indigoans have turned to *parteras* simply "por economia," as most respondents stated, with evident amazement that anyone would wonder why they had done so. They have not feared or mistrusted modern medicine: they have not been moved by modesty: they have not been put off by Anglo-American medical personnel (most doctors and virtually all nurses in Indigo are Mexican-American, in any case). They have simply not had any other economic option.

As for *remedios caseros* and *curanderismo,* these again represent primarily a low-cost alternative to very expensive scientific medical care. As Anglo Americans turn to Tylenol, Indigoans turn to *yerbas* (as well as to Tylenol) as a first line of attack. And as Anglo-Americans turn to psychiatry, chiropractic, and various alternative strategies when mainstream modern medicine fails, Indigoans turn to *curanderismo*. It is not a substitute for scientific medicine, but an adjunct to it. And if impoverished Mexican Americans rely more heavily on *yerbas* and *curanderas* than Anglo Americans do on their alternative strategies (and I am not sure that this is the case), this is far less because they are Mexican-American than because they are impoverished. In the end, Indigoans' decisions about health care are based on essentially the same considerations addressed by other Americans: perceived acuteness of condition, expense, and failure of previously employed therapies.

Works Cited

Achor, Shirley
1978 Mexican Americans in a Dallas Barrio. Tucson: University of Arizona Press.

Angel, Ronald
1958 The Health of the Mexican Origin Population. In The Mexican American Experience. Rodolfo O. de la Garza et al., eds. Austin: University of Texas Press.

Boyle, Joyceen S. and Margaret M. Andrews
1989 Transcultural Concepts in Nursing Care. Glenview, Illinois: Scott Foresman.

Clark, Mark
1970 Health in the Mexican Culture. Berkeley: University of California Press.

Foster, George M.
1987 Bureaucratic Aspects of International Health Agencies. Social Sciences and Medicine 25 (9):1039–1048.

Madsen, William
1964, revised 1973. Mexican-Americans of South Texas. New York: Holt, Rinehart and Winston.

Maduro, Renal do
1983 *Curanderismo* and Latino Views of Disease and Curing. The Western Journal of Medicine 139:868–874.

FIGURE 7.6 ***(Continued)***

McKenna, Margaret
1989 Twice in Need of Care: A Transcultural Nursing Analysis of Elderly Mexican Americans. Journal of Transcultural Nursing I (1):46–52.

Rubel, Arthur J.
1966 Across the Tracks: Mexican-Americans in a Texas City. Austin: University of Texas Press.

Scheper-Hughes, Nancy and David Stewart
1983 *Curanderismo* in Taos County, New Mexico—A Possible Case of Anthropological Romanticism? Western Journal of Medicine 139:875–884.

Sounders, Lyle
1954 Cultural Differences and Medical Care: The Case of the Spanish Speaking People. New York: Russell Sage.

Winkler, Karen J.
1990 Scholars Say Issues of Diversity Have "Revolutionized" Field of Chicano Studies. Chronicle of Higher Education, 26 September, 1990:A4–A8.

FIGURE 7.6 ***(Continued)***

EXERCISES

1. Patricia Wynjenek's report to Las Almas Buenas is a combination of the two formats for completion reports outlined in this chapter. It does not take the exact form of the nonacademic completion report for government, corporate, or industrial groups; neither does it take the exact form of the report of primary research for academic groups. And yet the commentary has claimed the formatting and reporting successful. Do you agree?

   Given Wynjenek's obligations to Las Almas Buenas and to another, later academic audience, analyze specifically how the report is set up and whether it is successful in meeting the needs of both audiences and of Wynjenek herself. You will have to define what those needs are in each case.

   Do this task as a class or in small groups, and be specific in your analysis.

2. The kinds of completion reports discussed in this chapter are by no means the only ways of presenting the results of research. Several others, such as manuals, brochures, and pamphlets, can be the appropriate products of legitimate research; it all depends on the purpose of the research and the audience who will benefit from reading about it.

If your research calls for a different kind of final presentation, you should plan its formatting before the work is actually completed. To aid in this planning, go to the library or to the Internet, and find at least two samples of final presentations that would be appropriate for the kind of research you are doing. Analyze them for how well they meet the needs of both audiences and writers. Consider not only what has been written but also how it has been presented, and then decide how the presentations may be adapted to fit your needs.

You may do this task with others who are working on similar kinds of projects; the final product should be a detailed outline of your plan.

ASSIGNMENT

1. Following the "Assignment" section is a completion report written by a student in a technical and professional writing class (see Figure 7.7). The student proposed her own project in her major, set up the audience she wanted to address, and decided on the form she thought would best address her readers. The report gives you an idea of what students can accomplish in their majors, as they put together their first formal completion report. We don't intend it as a model you should copy, but as a demonstration of work done by a fellow traveler.

   Evaluate the student report. Your comments should address not only audience considerations, but also such issues as consistency with the proposal work and the extent to which the report adheres to the principles discussed in this chapter. What is admirable about the report? What is weak?

TERM PROJECT OPTION

Write a completion report (or whatever final product is appropriate for the work you formally proposed in Chapter 6). Besides adhering to the principles discussed in this chapter and elsewhere in the text, you should edit your project for the problems of usage discussed in the Appendix and for problems of basic grammar and sentence structure.

Whatever form your final project takes, write an executive summary that serves two functions. First, it should do the conclusive summarizing required of an executive summary. Second, it should explain to your instructor exactly what decisions you made about audience and form and how you adapted your choice of words, level of abstraction, and the like, on the basis of those decisions. You'll also want to discuss the purpose, context, and ethical considerations that you saw for this project.

SAMPLE STUDENT COMPLETION REPORT

40 Oakwood Dr. #114
Pocatello, ID 83204

April 23, 1991

Dr. Tracy Montgomery
LA #212
Idaho State University
Pocatello, ID 83209

Dear Dr. Montgomery:

I am submitting my major project, an article for the journal Sociology and Social Research. I have also attached my graded proposal per your request. This article is primarily based upon an original survey administered to ISU students in the Fall of 1990.

I have made several changes since submitting my proposal. First, I changed the title to "AIDS Education and Variables Affecting Safer Sex Behavior in a Conservative State University." This reflects a slight change in emphasis to how AIDS education should proceed and also the addition of two variables in my examination of what affects behavior. My interview with Dr. Farrell and my further review of previous studies influenced my decision to incorporate these additional two variables. My five variables now include behavior, knowledge, concern, attitude, and contact with a person living with AIDS (PLWA). All five of these variables were tested on the original survey instrument.

Second, although many of the schools where previous studies have been done are of similar size to ISU, most are probably not as conservative. My article (and the title) reflect this difference in emphasis. Third, I chose to include only works cited as my references because of the length of the list. I have used three works that I had not reviewed at the time of my proposal and I have dropped several sources which did not fit in my article. Fourth, in my proposal I said that I would turn in this project on April 22, 1991. However, I have extended this self-imposed deadline by one day to finalize my project.

This survey found that ISU students are sexually active and do put themselves at risk of contracting HIV. Further, no one variable is effective in promoting all the safer sex behaviors measured in this survey. Recommendations call for a comprehensive education plan incorporating all of these variables because they all have some influence in changing some behaviors.

Sincerely,

Sheryl Soderstrom

Sheryl Soderstrom

FIGURE 7.7 **Student completion report**

AIDS Education and Variables Affecting Safer Sex Behavior
in a Conservative State University

Sheryl Soderstrom

Submitted to
Dr. Tracy Montgomery
Instructor
Professional Writing
Idaho State University

April 23, 1991

FIGURE 7.7 (*Continued*)

AIDS EDUCATION AND VARIABLES AFFECTING SAFER SEX BEHAVIORS IN A CONSERVATIVE STATE UNIVERSITY

Executive Summary

This paper reports the results of a survey of ISU students (n - 245) during the fall semester 1990. The survey tested five variables: behavior, knowledge, concern, attitude, and contact with a person living with AIDS (PLWA). Results indicate that students are sexually active (68%) and are placing themselves at risk of contracting AIDS because of sexual behavior. Other findings include the following: 90% gave accurate answers to knowledge questions, 76% expressed a high level of concern, 92% indicated a positive attitude about AIDS, 34% have had contact with a PLWA, and 78% claim that they are practicing safer sex because of AIDS. Basic transmittal knowledge was high (95%), personal concern was low (47%), 92% supported condoms on campus, and 95% supported the right of an HIV+ child to go to public school.

Respondents were more willing to discuss AIDS/safer sex with their children (98%/97%) than to take positive action in their own sexual behavior. Remaining celibate until marriage (only 44% Strongly Agree/Agree), and giving up oral sex (only 49% SA/A) were the least popular of the safer sex behaviors on the survey. This was followed by dating longer before having sex (only 76% SA/A). None of the variables can be correlated to changes in all eight behaviors tested, but accurate knowledge, high concern, and positive attitude do affect some behaviors. Contact with a PLWA had no effect on behavior, but did have an effect on concern which in turn affects some behaviors. Recommendations include a comprehensive format that enhances all these variables for the individual (especially personal concern) to facilitate behavior changes, and modules on self esteem, decision-making skills, negotiation skills, and condom use skills. Education at ISU needs to reach males, LOS students, and freshmen in particular. AIDS education for everyone must move from the general and basic to the specific and detailed, and it must move from the statistical "other" to personal concern and the realization that no one is immune.

The audience for this article are the readers of <u>Sociology and Social Research</u>. These are the professionals in the field of Sociology (and perhaps other social sciences) as well as Professors and students in this field. They should be familiar with the research format and the use of tables, cross-tabulations, and surveys and will thus understand my methods and results as presented. They should also be familiar with AIDS and the various studies conducted in the past to the extent of their personal interest in this area.

To appeal to this audience I followed the typical social research format. I reviewed past research in this area and tried to do something different that would pique their interest. For instance, I looked at five variables instead of just two or three, and I did my survey at a conservative school. I used language and style appropriate to this particular journal and comparable to other articles already published in this

FIGURE 7.7 (*Continued*)

journal. I used a modified "Scientific Articles" format which includes Abstract, Introduction (not identified as such), Literature Review (not identified as such and basically incorporated into the Introduction), Methods, Results, Discussion, and Works Cited. These modifications reflect the pattern typically seen in articles published in this journal. I attached four tables of information (which are discussed in the text) at the back of the article to meet with the convention of this journal.

Within the section on Methods, I divided my report into two sections: Sample and Survey Instrument. The section on sample describes the demographic information on my respondents. The section on Survey Instrument defines the variables in terms of the survey items measuring them and gives a brief recounting of how the data was gathered and analyzed. In the Results section, I chose to begin with information about students' sexual activity, followed by the overall percentages for each variable. Next, I reported on the variables broken down by demographic information. Then I went into some detail about each of the variables, and finally I addressed the influence each of the variables wields on safer sex behavior. All discussion of results is confined to the section labeled Discussion and is limited to the major points. Also in this section, I address limitations of the study and recommendations for future AIDS education.

FIGURE 7.7 (*Continued*)

AIDS EDUCATION AND VARIABLES AFFECTING SAFER SEX BEHAVIOR IN A CONSERVATIVE STATE UNIVERSITY

Abstract

Past research has focused on four variables relating to AIDS: behavior, knowledge, concern, and attitude. However, most previous studies did not look at all four of these variables. This study of students at Idaho State University (n-245) examined all four of the variables listed above as well as contact with a person living with AIDS. Results indicate that students at this conservative school are at risk of contracting HIV sexually and that no one of the variables examined is sufficient to cause a change in all the safer sex behaviors examined. Different behaviors were affected by different variables, leading to the conclusion that AIDS education must be comprehensive to include all these variables and to cause a change in sexual behaviors.

FIGURE 7.7 ***(Continued)***

AIDS EDUCATION AND VARIABLES AFFECTING SAFER SEX BEHAVIOR IN A CONSERVATIVE STATE UNIVERSITY

Today, having unprotected sex is like playing Russian Roulette because of a small, fragile, but deadly virus known as Human Immunodeficiency Virus (HIV). While HIV is transmitted in ways other than just semen and vaginal fluid (blood, shared needles, mother's milk), this study only looks at sexual transmission—perhaps the primary route of transmission on college campuses. This study of 245 students at Idaho State University found that students are indeed sexually active (68%) and are putting themselves at risk of contracting HIV. General knowledge of AIDS was high, but personal concern was low. Further, it was determined that a comprehensive system of education that addresses myths and brings concern down to a personal level will be the most effective in bringing about safer behavior.

For years, college campuses have had a reputation of sexual freedom and experimentation, and coupled with the relatively high density of single people, one would expect this population to be particularly vulnerable. However, one might question the validity of this observation in a small public university set in an extremely conservative community and state. But if the observation and reputation hold true at the small school, then they can be expected to be true (and perhaps multiplied) at larger, more liberal universities. One would also expect a university population to be better educated, and better informed on a variety of current events and issues than the general public. Given these factors, it would, therefore, be an appropriate place to measure HIV/AIDS awareness and concerns in relation to sexual behavior.

The presence of HIV in our society and its fatal consequences should cause responsible people to change their sexual practices to protect themselves and others. However, articles in <u>Mademoiselle</u> (Hacinnis, 1988), <u>U.S. News & World Report</u> (-1988), and <u>Newsweek</u> (Leslie et al., 1988) indicate that AIDS is spreading into the heterosexual population, and misinformation about AIDS still exists. Results of a Centers for Disease Control (CDC) study of student blood samples from universities across the country show that 1 in 300 students are HIV positive (Bennett, 1991). Most researchers believe that the threat of AIDS has done little to curb sexual activity among heterosexuals and especially among college students (Smilgis, 1987; Hirschorn, 1987; Gayle, 1990; Greatorex, 1989; Masters et al., 1988).

In a quest to prevent unnecessary deaths, AIDS educators are searching for "the key" to producing sexual behavior changes in people. Many previous studies have looked at three factors—knowledge, concern, and attitude. While knowledge about HIV/AIDS—its transmission and its prevention—has increased (Winslow, 1988; -1990a: Ward & Ault, 1990), it has not been sufficient to produce behavior changes (Ward & Ault, 1990; Ishii-Kuntz, 1988; Fennell, 1990; -1988; -1990c). Most agree that knowledge alone is not the answer in behavior modification (Fennell, 1990). Researchers note that concern about HIV/AIDS has led to some reported changes in sexual behavior on some campuses (Carroll, 1988; Ishii-Kuntz, 1988; Bruno, 1985), but on

FIGURE 7.7 (*Continued*)

others the effect was nil (Ward & Ault, 1990; Ahia, 1991; -1990c; Hirschorn, 1987; Smilgis, 1987). And Carroll noticed some discrepancy between reported and actual behavior changes (1988). Fennell reported that attitudes improved over a 30-hour AIDS class that included critical thinking skills, ethical discussions, and group activities, but did not investigate the relationship between attitude and behavior (1991). A study done with high school students found that 61% overall had positive attitudes toward HIV/AIDS and PLWAs, but again did not check for the effect attitudes had on behavior (Jones et al., 1991). Others found a change in attitude, but not behavior (-1990b; Hirschorn, 1987; Smilgis, 1987).

Overall, most agree that sexual behavior has not changed over the last 15 years. Young people are still as sexually active as always, and they continue to engage in risky behavior (Newman, 1988; Hanna, 1989; -1990b; -1990c; -1988; Hirschorn, 1987; Katz & Cronin, 1980; Smilgis, 1987). The current study looks at the variables "knowledge," "concern," "attitude," and "contact with a PLWA," and their effect on reported sexual behavior.

Methods

Sample. The 49-item survey was administered to a non-random sample of undergraduate students in Sociology and Social Work classes at Idaho State University (ISU) during Fall semester 1990 (n-245). These lower division classes provided a cross-section of ISU's students, although not in the same proportion as the school's actual population. Most of the survey respondents were female (62%) (See Table 1, p. 15). More than 10 majors were identified with the largest field, social sciences, accounting for 27% of the total. Freshmen represented the largest class (35%) and almost half (44%) of the respondents were LOS (Mormons). Respondents ranged in age from 18 to 50 years old. For ease of handling, they were grouped into five age categories. Almost half (46%) were single, while a little more than a third (37%) were married. Married respondents reported marriages ranging from 1–35 years long, while respondents living with a significant other in a sexual relationship reported that those relationships had been ongoing for up to 5 years. Sixty-eight percent of the respondents reported being sexually active.

Survey Instrument. This survey measured five variables: behavior, knowledge, concern, attitude, and contact. Questions measuring behavior ranged from discussing AIDS/safer sex with children to six behavior questions borrowed directly from the Ishii-Kuntz study (1988). These six items began with the phrase "Because of AIDS" to isolate that the threat of AIDS was the motivation behind these reported behaviors. Responses were analyzed and cross-tabulated with the four other concepts, and then grouped to indicate safer sex and not safe behaviors. Six items testing awareness of transmission methods, prevention, and outcome of an AIDS diagnosis determined level of knowledge. Responses were then analyzed and grouped as accurate or inaccurate knowledge. Concern was measured by five items testing the individual's perception of the magnitude of the AIDS problem

FIGURE 7.7 ***(Continued)***

both in the U.S. and in the State of Idaho, and the extent of the individual's concern: everyone should be concerned. I am personally concerned. I am concerned about my teenagers. Responses were grouped to reflect high and low concern. Attitude reflects opinion of AIDS and persons living with AIDS (PLWAs), as well as two important social issues: condom dispensers on campus including in the dorms, and the right of HIV infected children to attend public schools. Responses to the five attitude items were grouped as either positive or negative. Two items measured contact: attendance at one (or more) of a PLWA's (Richard Carper) lectures two weeks prior to taking the survey, and "Do you personally know anyone who is HIV+ or who has AIDS?"

Most survey questions were offered on a Likert scale—strongly agree (SA), agree (A), disagree (D), strongly disagree (SD), and not applicable (NA). Some items gave the choice of yes, no, or undecided, some were mark-all-that-apply, and some were fill-in-the-blank. Collected data was analyzed for frequency and then cross-tabulated with the Absurv survey analysis program. Percentages for each item testing a particular variable were grouped, adjusted to remove those who thought the particular item did not apply to them, and a mean was obtained. Answers to individual questions were also adjusted, polarized by variable, and then a mean was obtained for the polarized ends of that variable.

Results

Forty-five percent of the singles reported they are sexually active, while 52% of the divorced respondents, and 75% of the separated respondents also reported that they were sexually active. To compound their risk factor, of those reporting the number of partners they've had in the last year, 7% of the married reported 2–3 partners: 75% of divorced reported 2–5 partners: 50% of separated reported 2–3 partners; 61% of singles reported 2–20 partners: and 21% of those living with someone reported 2–5 partners.

Overall, 92% of respondents indicated they had a positive attitude about AIDS, 34% have had contact with a PLWA, 76.4% expressed a high level of concern, 90.3% gave accurate answers to knowledge questions, and 78% claim that they are practicing safer sex because of AIDS (See Table 2, p. 16).

More females have accurate knowledge (91%) than males (86%), and more females also have a positive attitude (91%) compared to 81% of males (See Table 3, p. 17). There is little difference between males and females on concern or behavior. Fewer people with no religion had accurate knowledge (82%) than either Catholics (91%), Protestants (91%), or LDS (90%). More Catholics expressed high concern, positive attitudes, and safer behavior than the other religion groups. Fewer LDS people expressed a positive attitude (82%), although that is still a majority, and less than half (46%) reported safer sex behavior because of AIDS. There is a negative relationship between class level and reported safer sex behavior. However, more seniors expressed accurate knowledge (94%), high concern (72%), and a positive attitude (94%), than did their lower class mates.

FIGURE 7.7 (*Continued*)

A high percentage of respondents (See Table 2, p. 16), gave accurate answers on the questions testing knowledge of AIDS (90%), but 39 people still believe AIDS is transmitted by giving blood. More respondents were unsure of the correct answer to a question about whether AIDS is fatal or not than other knowledge questions, but still more than half (62%) gave the correct answer. Seventy-eight percent of respondents reported that they have, will, or are more likely to engage in safer sex behaviors. This ranged from a high of 98% reporting that they have or will discuss AIDS with their children, to a low of 44% who SA/A they have decided to remain celibate prior to marriage. Also unpopular was giving up oral sex—only 49% SA/A they are less likely to engage in oral sex. Regarding the variable "concern," the more general the concern statement, the more people SA/A. For example, 97% SA/A that AIDS is something everyone should be concerned about, but only 47% admitted that they were personally concerned about contracting AIDS. In spite of an overall positive attitude reported by the majority, 16% indicated they SA/A with the statement "AIDS is a punishment sent by God because of people's immoral behavior."

Accurate knowledge appears to heavily influence the decision to remain celibate prior to marriage (See Table 4, p. 18). Forty-five percent of those with accurate knowledge SA/A, while only 23% of those with inaccurate knowledge SA/A to remain celibate. All other differences between accurate and inaccurate knowledge, regarding any of the behaviors, was small, but it is interesting to note that the safer behaviors of using condoms, and not engaging in oral or anal sex were more popular among those with inaccurate knowledge. Regardless of knowledge, a high percentage (> 70%) say they are engaging in safer sex, with the exception of being celibate and giving up oral sex. A positive attitude seems to have an important effect on six of the eight behaviors. However, 56% of those with a negative attitude SA/A to remain celibate until marriage because of AIDS as opposed to only 42% of those with a positive attitude. The biggest difference between positive and negative attitudes in relation to behavior is on using condoms: 88% of those with a positive attitude SA/A, while only 67% of those with a negative attitude SA/A. High concern seems particularly effective in getting people to date longer before sex, and to not engage in oral sex. It is only moderately effective in getting people to ask for a potential partner's sexual history, to use condoms, and to not engage in anal sex. Having had contact with a PLWA had no major effect on any behavior; however, it does have a positive effect on concern, which in turn has an effect on behavior.

Discussion

It appears that at least at ISU, conservative morals have little effect on college students. If the behavior indicated on the survey is true, ISU students are indeed similar to their cohorts across the country. They are sexually active and they are engaging in risky sexual behavior. And if people in this morally conservative area are sexually active to the point of putting themselves at risk, it seems reasonable to assume that people in more liberal areas are doing the same.

FIGURE 7.7 (*Continued*)

This risky behavior cannot be attributed to ignorance, lack of concern, or a poor attitude because the majority of respondents expressed accurate knowledge, high concern, and a positive attitude. While no one variable was effective in influencing safer behavior across the board, knowledge, concern, and attitude all had a major effect on at least one behavior. In particular, increasing personal concern seems to be effective in getting people to practice safer sex. One way to increase personal concern is to increase contact with PLWAs. Dr. Larry Farrell explained that, "the most effective thing I've seen so far is for people who are diagnosed or infected to make themselves known . . . . That establishes some real immediacy . . . . It's no longer an abstract idea. It's no longer an invisible statistical entity. It's a person" (1991). Personal stories can open some people's eyes to the reality and risk of their own actions. All these variables, then, must be encouraged and nurtured in the individual to increase safer sex behavior.

Consistently throughout the various cross-tabulations and in spite of one's position with regard to any variable, people were not agreeable to remaining celibate or to giving up oral sex. The other behavior viewed with less favor was dating longer before having sex. This should send a clear message to the school, the community, and the state that people are sexually active in spite of AIDS, in spite of religious influences, in spite of moral conservatism, and in spite of "knowing better." Although we know that the only safe sex is no sex, those groups who insist that that is all that should be taught are fooling themselves.

This small study is subject to the same limitations that any survey research is subject to, i.e., are respondents telling the truth or are they marking answers to meet assumed expectations? Additionally, the following limitations are noted: first, this survey did not ask which safer sex behaviors people actually did or how often they did them, but instead focused on what they were likely to do. Second, this survey tied all safer sex behavior to the motivation or threat of AIDS; however, at this school people may be agreeing with these statements for other reasons. Parallel statements without the "Because of AIDS" qualifier may have made this distinction clearer.

AIDS education to date has been effective in providing basic knowledge about AIDS and the transmission of HIV. But as Fennell reminds us, "Changes in knowledge without subsequent changes in behavior are not sufficient when dealing with HIV because of its potentially fatal consequences." The next step is to bring that knowledge down to a personal level. We must remove the notion that AIDS can be handled in a coldly statistical manner, and we must change people's perception that AIDS only happens to other people so different, so far removed from themselves that they are not at risk.

Future education should incorporate more detailed knowledge of AIDS to deal with the myths and misconceptions. It should also incorporate modules on self-esteem, decision-making skills, negotiation skills, and familiarity with condoms (Fennell, 1990; Farrell, 1991). Education at ISO should be particularly aimed at males, LDS students, and freshmen. As educators we need to boost the personal concern level of the individual, and we need to rid people of the negative attitudes associated with AIDS because negative attitudes act as a barrier to hearing the message and thus protecting oneself.

FIGURE 7.7 ***(Continued)***

This study finds that none of the variables examined is, by itself, sufficient to cause a change in sexual behavior. However, different variables seem to be related to different safer sex behaviors. Therefore, a comprehensive education plan incorporating all these variables in a format that requires the student to think, to role play, to participate in activities, and to interact with PLWAs and HIV+ people may be our best hope to accomplish sexual behavior changes in this population which considers itself "not at risk."

Works Cited

____1988. A Scary Little Survey of AIDS on Campus. U.S. News & World Report. 14 November: 12.

____1990a. AIDS Survey. AIDS Task Force. Lewis-Clark State College, Lewiston, Idaho.

____1990b. College Students only 'Paying Lip Service' to Safe Sex Message. Pediatric News. Orlando. Florida. May.

____1990c. Threat of AIDS having little impact on sex habits of college students. (AP) Idaho State Journal.

Ahia, Ruth N. 1991. Compliance with Safer-Sex Guidelines Among Adolescent Males: Application of the Health Belief Model and Protection Motivation Theory. Journal of Health Education. Jan./Feb. 22: 49–52.

Bennett, Jack. 1991. Lecture notes. Guest speaker in Dr. Larry Farrell's Biology 469 Microtopics: AIDS class. Idaho State University. Pocatello, ID. 29 January.

Bruno, Mary, et al. 1985. Campus Sex: New Fears—AIDS is the latest STD to hit the nation's colleges. Newsweek. 28 October: 81–82.

Carroll, Leo. 1988. Concern with AIDS and the sexual behavior of college students. Journal of Marriage and the Family. 50:405–411.

Farrell, Larry D. 1991. Personal Interview. Idaho State University. Pocatello, ID. 8 March.

Fennell, Reginald. 1990. Knowledge, Attitudes and Beliefs of Students regarding AIDS: A Review. Health Education. July/Aug. 21:20–25.

1991. Evaluating the Effectiveness of a Credit Semester Course on AIDS Among College Students. Journal of Health Education. Jan./Feb. 22:35–41.

Gayle, Helene D., et al. 1990. Prevalence of the human immunodeficiency virus among university students. New England Journal of Medicine. 29 November. 323:1538–41.

Greatorex, Ian Frederick & John Michael Valentine Packer. 1989. Sexual Behaviour in University Students: Report of a Postal Survey. Public Health. 103:199–203.

Hacinli, Cynthia. 1988. AIDS, straight: A Heterosexual-risk update. Mademoiselle. August:138.

Hanna, Jack. 1989. Sexual Abandon: The Condom is unpopular on the campus. Maclean's. 25 September:48.

Hirschorn, Michael W. 1987. AIDS is Not Seen as a Major Threat by Many Heterosexuals on Campus. The Chronicle of Higher Education. 29 April. 33:1:32–34.

FIGURE 7.7 (*Continued*)

Ishii-Kuntz, Masako. 1988. Acquired Immune Deficiency Syndrome and Sexual Behavior Changes in a College Student Sample. Sociology and Social Research. October. 73:13–18.

Jones, Herb, et al. 1991. HIV Related Beliefs, Knowledge and Behaviors of Ninth and Eleventh Grade Public School Students. Journal of Health Education. Jan./Feb. 22:12–18.

Katz, Joseph and Denise M. Cronin. 1980. Sexuality and College Life. Change. F / M:44–49.

Leslie, Connie, et al. 1988. Amid the Ivy, Cases of AIDS—College campuses offer students no sanctuary. Newsweek. 14 November:65.

Newman, Frank. 1988. AIDS, Youth, and the University—An interview with Admiral James D. Watkins, the Chairman of the AIDS Commission. Change. September/October:39–44.

Smilgis, Martha. 1987. The Big Chill: Fear of AIDS—How heterosexuals are coping with a disease that can make sex deadly. *Time*. 16 February:50–53.

Ward, Sally K. and Susan Ault. 1990. Fraternity and Sorority Membership, AIDS Knowledge and Safe-sex practices. Sociology and Social Research. April. 74:158–161.

Winslow, Robert W. 1988. Student knowledge of AIDS transmission. Sociology and Social Research. 72:110–113.

FIGURE 7.7 (***Continued***)

TABLE 1 Demographics

	#	%
SEX		
MALE	93	38
FEMALE	151	62
AGE		
18–20	84	35
21–25	60	25
26–30	32	13
31–35	26	11
36–50	39	16
RELIGION		
CATHOLIC	25	11
PROTESTANT	63	29
LDS	97	44
NONE	34	16
CLASS		
FRESHMEN	86	35
SOPHOMORE	60	25
JUNIOR	64	26
SENIOR	33	14
MARITAL STATUS		
MARRIED	88	37
DIVORCED	23	10
SEPARATED	4	2
SINGLE	110	46
LIVING WITH	15	6
SEXUALLY ACTIVE		
YES	159	68
NO	74	32
MAJOR		
SOC. SCIENCES	65	27
BUSINESS	40	17
HEALTH RELATED	34	14
EDUCATION	37	15
COUNSELING	16	7
SCIENCES/MATH	8	3
ENGL/PHIL/MC	6	3
MISC	35	15
PARTNERS		
1	90	66
2–3	32	24
4–5	9	7
6–10	3	2
11–15	1	1
16–20	1	1

TABLE 2 Polarized Variables By Survey Questions

	ACCURATE	INACCURATE
KNOWLEDGE		
AIDS is transmitted by: (All-that-apply)	95%	5%
There is a cure for AIDS.	93%	7%
People can catch the AIDS virus from public toilet seats.	96%	4%
AIDS can be transmitted from females as well as from males.	99%	1%
Most people who become infected with the AIDS virus will not get fully developed, fatal AIDS.	62%	38%
There is little that can be done to prevent the spread of AIDS.	97%	3%
MEAN	90.3%	9.6%
	SAFER	**NOT SAFE**
BEHAVIOR		
I have or will discuss AIDS with my children.	98%	2%
I have or will discuss safer sex practices with my children.	97%	3%
Because of AIDS I have increased the length of time I date someone before engaging in sexual intercourse.	76%	24%
Because of AIDS I am now likely to ask a potential sex partner about his/her sexual history.	87%	13%
Because of AIDS I have made the decision to remain celibate prior to marriage.	44%	56%
Because of AIDS I am more more likely to use condoms when having intercourse.	86%	14%
Because of AIDS I am less likely to engage in oral sex.	49%	51%
Because of AIDS I am less likely to engage in anal sex.	83%	17%
MEAN	78%	22%
	HIGH	**LOW**
CONCERN		
AIDS is something most people should be concerned about.	97%	3%
AIDS is the most urgent health problem facing the U.S.	89%	11%
AIDS is the most urgent health problem facing Idaho.	67%	33%
I am personally concerned about contracting AIDS.	47%	53%
I am concerned about my teenage children getting AIDS.	82%	18%
MEAN	76.4%	23.6%
	POSITIVE	**NEGATIVE**
ATTITUDE		
Would you stop being friends with a person if you discovered that he or she had AIDS?	96%	4%
AIDS affects only limited, specific groups of people.	95%	5%
Condoms should be available in the bathrooms on campus including in the dorms.	92%	8%
HIV infected children have a right to go to public school.	95%	5%
AIDS is a punishment sent by God because of people's immoral behavior.	84%	16%
MEAN	92%	8%
	YES	**NO**
CONTACT		
Did you attend a lecture by Richard Carper during the week of Oct. 22–26, 1990?	54%	46%
Do you personally know anyone who is HIV+ or who has AIDS?	14%	86%
MEAN	34%	66%

TABLE 3 **Variables by Demographics (Means)**

	KNOWLEDGE ACCURATE	KNOWLEDGE INACCURATE	CONCERN HIGH	CONCERN LOW	ATTITUDE POSITIVE	ATTITUDE NEGATIVE	CONTACT YES	CONTACT NO	BEHAVIOR SAFER	BEHAVIOR NOT SAFER
Male	86%	14%	64%	36%	81%	19%	29%	71%	49%	51%
Female	91%	9%	66%	34%	91%	9%	38%	62%	51%	49%
18–20	87%	13%	63%	37%	85%	15%	25%	75%	57%	43%
21–25	87%	13%	61%	39%	85%	15%	30%	70%	45%	55%
26–30	93%	7%	66%	34%	91%	9%	42%	58%	47%	53%
31–35	92%	8%	72%	28%	93%	7%	33%	67%	57%	43%
36–50	91%	9%	70%	30%	89%	11%	53%	47%	47%	53%
Cath	91%	9%	75%	25%	97%	3%	46%	54%	57%	43%
Prot	91%	9%	63%	37%	90%	10%	34%	66%	54%	46%
LDS	90%	10%	63%	37%	82%	18%	28%	72%	46%	54%
None	82%	18%	64%	36%	88%	12%	35%	65%	47%	53%
Fresh	88%	12%	63%	37%	82%	18%	20%	80%	55%	45%
Soph	89%	11%	63%	37%	87%	13%	33%	67%	51%	49%
Jun	88%	12%	65%	35%	92%	8%	44%	56%	47%	53%
Sen	94%	6%	72%	28%	94%	6%	53%	47%	48%	52%
Married	93%	7%	67%	33%	88%	12%	42%	58%	40%	60%
Divorced	91%	9%	73%	27%	97%	3%	37%	63%	61%	39%
Separat.	75%	25%	50%	50%	80%	20%	50%	50%	41%	59%
Single	87%	13%	64%	36%	86%	14%	27%	73%	58%	42%
Liv. with	85%	15%	64%	36%	88%	12%	37%	63%	48%	52%
Sex. Act.	89%	11%	67%	33%	90%	10%	36%	64%	50%	50%
Not Sex. Act.	89%	11%	60%	40%	82%	18%	27%	73%	53%	47%
1 Partner	98%	2%	58%	42%	91%	9%	37%	63%	47%	53%
2–3 Part.	91%	9%	63%	37%	95%	5%	30%	70%	59%	41%
4–5 Part.	89%	11%	69%	31%	95%	5%	45%	55%	55%	45%
6–20 Part.	100%	0%	75%	25%	100%	0%	50%	50%	68%	32%
Not At All	95%	5%	62%	38%	93%	7%	35%	65%	52%	48%
Somewhat	96%	4%	57%	43%	92%	8%	31%	69%	57%	43%
Very Much	98%	2%	53%	47%	80%	20%	35%	65%	43%	57%

TABLE 4 **Variables X Safer Sex Behavior (SA/A)**

	DISCUSS AIDS	DISCUSS SAFER SEX	DATE LONGER	SEXUAL HISTORY	REMAIN CELIBATE	MORE LIKELY TO USE CONDOMS	LESS LIKELY ORAL SEX	LESS LIKELY ANAL SEX	MEAN
Knowledge									
Accurate	98%	97%	76%	88%	45%	86%	49%	82%	78%
Inaccurate	88%	93%	71%	73%	23%	90%	56%	88%	73%
Concern									
High	99%	98%	80%	89%	44%	89%	52%	85%	80%
Low	95%	92%	45%	76%	56%	74%	29%	78%	68%
Contact									
Yes	99%	98%	65%	83%	47%	82%	52%	77%	75%
No	98%	98%	80%	88%	43%	88%	48%	84%	78%
Attitude									
Positive	99%	98%	78%	87%	42%	88%	48%	83%	78%
Negative	87%	90%	58%	75%	56%	67%	41%	84%	70%

Solving Problems Through Trip Reports, Feasibility Studies, and Scientific Reports

CASE STUDY

BRUCE KENNEWICK'S OFFICE

The Brighton House ▪ 1530 Shore Boulevard ▪ Ozark Lake, Missouri

Bruce Kennewick had just gotten back from his task force assignment in Mexico. Brighton Hotels' head office had sent him to Puerto Vallarta to straighten out the Catering Department at its new hotel, Brighton Playa–Puerto Vallarta, that had opened at the beginning of November, the start of the winter tourist season. Although all the departments of the hotel had gone through the usual task force treatment before a new hotel opens, which involves having a team of experts in various areas of a hotel's business coming to the new property to establish procedures and to train personnel, the catering department had met with unusual reverses. Fifty percent of the personnel hired had quit or been fired, and the Director of Catering, a Mexican national, had quit suddenly to take a better job in Guadalajara with a competing hotel chain.

PREVIEW OF THE CHAPTER

(continued)

Brighton Playa–Puerto Vallarta was certainly in business but needed some help—especially in the Catering Department. That's where Bruce Kennewick came in. After some twelve years with Brighton Hotels, in various capacities of providing food, beverage, banquet, and meeting and convention services, Bruce was a good person to send to Brighton Playa. He could assess the problems quickly and begin to solve them, or at least suggest how they might be solved. And so on November 22, he got his assignment to be in Puerto Vallarta on December 3, to run another mini-task force, specifically for the Catering Department.

Bruce accepted the assignment gladly. He knew a lot about running a Catering Department, he was willing to take an interim assignment if the company wanted him to, and he had never been to Mexico. What could be better?

Although it seemed ideal, the job turned out to be sometimes troublesome and frustrating. He had difficulty with language and cultural differences that created some tensions and misunderstandings between him and the people he worked with. He also had been away from home and his family for a prolonged period—four months—and that absence had created other problems. In fact, Bruce Kennewick was bone tired. He thought, in the end, that he had done well on his special task force assignment, but there it was, sitting on his desk: the form for reporting company activities off of home site. He would have to fill out the form, but it would mean at this point a major effort, one that he would have to put a good deal of thought into, and one that could not reflect how tired he was.

Despite his fatigue, he decided he would have to put the best face on the situation. His task force effort, in fact, had worked pretty well. He left Brighton Playa–Puerto Vallarta with a better staff, one that could work well with the new management that showed promise in developing new business along Brighton lines. Bruce had done basically what he was supposed to do, but he knew also that there were still some problems he would have to discuss, problems he felt sure that he had addressed, and in some cases had solved, but ones that Brighton needed to know about so that it could plan for the future.

Pushing aside for now the other paperwork demands on his desk, Bruce turned his attention to this important trip report he had to write.

FUNDAMENTALS OF A TRIP REPORT

Purpose and Rationale

Trip reports have a very specific purpose. They are written to record the activities of someone away from the usual work site. Organizations send their employees on trips for any number of reasons: to drum up new business, to check the progress of business that was already under way, and to assess and solve problems that have cropped up with business already under way. Management needs to know what has gone on, what is going on, or what needs to be done, so that it can plan for the future. In this way, trip reports are very much like periodic reports, because they answer the same kinds of questions.

Bruce Kennewick's case is typical of organizations like Brighton Hotels. They regularly send out people to start up new hotels, a process called task forcing—an effort by veteran employees to establish local standard operating procedures (LSOPs) at a new site consistent with the general standard operating procedures (GSOPs) throughout the chain. The work involves analyzing the needs and setup of the new site, defining the operations and procedures of the chain in light of the situation at the new site, training new employees, and working through the usual functions of the hotel to anticipate and forestall problems.

Bruce had been a member of numerous task forces, which usually worked for a month or so away from their own home assignments. But he had never been on a follow-up or troubleshooting task force. He had also never gone to an international property, where there were differences in language, culture, and delivery of services that had to be coped with. This assignment was special in those ways, requiring him to report more thoroughly and thoughtfully on the outcome of his trip.

All trip reports, whether special or routine, can be organized under the following five general topics:

1. Introduction
2. Overall assessment of the work done
3. Problems
4. Solutions
5. Conclusions and, perhaps, recommendations

ANALYZING THE PROCESS OF WRITING A TRIP REPORT

Again, the answers to several questions help define the purposes for writing and reading trip reports, and therefore help determine their basic contents (see Figure 8.1).

Introduction

As is true of periodic reports, trip reports must establish the basics to orient the reader. The audience—again middle to upper management—needs to sort out one project from another and make decisions on the basis of the answers to several other questions. So at the outset, readers need to know where the writer went, whether the writer went singly or as part of a team, when the

Questions That Trip Reports Answer

Introduction

- Where was the work done?
- With whom?
- When?
- For how long?
- Why?

Overall Assessment of the Work Done

- What work was done?
- How did it meet the goals of the assignment?

Problems and Solutions

- What problems were encountered?
- Were they solved?
- How?
- If they were not solved, how might they be solved?
- By whom?
- When?

Conclusions and Recommendations

- What work has to be done in the future?
- When should it happen?
- Who should do it?

FIGURE 8.1 **Questions to ask and answer to define the contents of a trip report**

trip happened, who authorized it, how long the trip lasted, and finally, why the trip was taken—what purpose it was to serve.

To this requirement, Bruce was able to respond fairly easily because Brighton Hotels, typical of many organizations, had provided a format to fill in. The first page of Bruce's trip report shows this format (see Figure 8.2, pp. 280–285).

Overall Assessment of the Work Done

As we have already suggested, management calls for trips to be taken for a number of reasons. A new sales territory may need to be established, an undertaking that could call for market testing or contacts with potential customers to be made; or perhaps a new product needs to be introduced in an already established market, in which case specific local marketing strategies need to be devised. Also, as in the case of Brighton Hotels, management needs to know whether operations are being conducted according to plan, whether expectations are being met, including how and to what extent. Management wants to know whether its short- and long-term goals are being served, and if not, why not. If an organization spends money sending someone to another place to find out what is going on there, then clearly a full report of what happened on the trip needs to be written.

Although these needs can often be met on any individual site by those who work there, sometimes it takes a fresh and practiced eye, someone not connected with the day-to-day business, accurately to assess the situation. Bruce Kennewick fills this need for the Brighton Hotels. His years of experience in the food and beverage areas of the hotel business, but particularly in catering for Brighton, have enabled him to analyze and evaluate a property's operations and to make certain changes that will further the effective and smooth running of the business. Thus, the work he had to do in Puerto Vallarta—general reorganizing of the Catering Department, soliciting business, and training personnel—becomes the necessary substance of this section of his trip report.

Problems and Solutions

This part of the trip report mirrors the same section of the periodic report. If problems exist that have implications for the future, they need to be analyzed in this section of the report. Readers want to know what is the nature of the problems, how they were solved, and how and when they may happen again. If the problems cannot be solved within the trip period, writers need to explain why and to suggest how they might be solved in the future, and by whom.

Besides the usual problems of finding reliable personnel, Bruce Kennewick had the additional problems of working in a foreign culture and speaking a different language. Although he knew some Spanish, it was limited pretty much to a basic "getting around" vocabulary, not a facility for carrying on business easily. And even though the hotel had hired some bilingual employees, they were often in key positions and had their own work to do. At the beginning of his assignment, therefore, he found that he had to spend a good deal of time learning the language and finding people who could translate and interpret for him.

But the bigger problem was the cultural differences, especially in the way of conducting business. Bruce's task force experience had been limited to properties in the United States, where he could usually rely on a large pool of skilled labor that delivered services quickly and efficiently. And even though Puerto Vallarta is a resort town that is geared in many ways to the service and hospitality industry, its pool of skilled labor is nevertheless small and in great demand. Therefore, services cannot be rendered in the same way as they might be in the United States. The need for electricians, carpenters, and people versed in audiovisual and computer technology is great and sometimes is met haltingly.

Problems like these have significant impact over the long term. They need to be addressed so that management can plan for the future, so that future task force operations can be made more efficient, and so that follow-up task force assignments, like the one Bruce was on, could either be reduced in frequency or be eliminated entirely.

Conclusions and Recommendations

Once again, this section is very similar to the section that answers the same questions in the periodic report. The difference is that in the periodic report, writers and readers are reporting on the general, ongoing activity of a project over time, whereas in the trip report, the interest lies in an activity that is specified at a particular time. In the periodic report, concern about the future is normal as a project evolves; in the trip report, concern about the future is more specifically directed toward follow-up activities or toward predictions about what may happen.

Thus, Bruce Kennewick will reserve this section to suggest how follow-up task force actions may be made less frequent or eliminated entirely. Making sure, for instance, that the initial task force has at least one bilingual member and someone who is familiar with the cultural differences of the country in question would make initial training and contacts with service personnel more efficient.

The future work section of a trip report, then, requires specific recommendations to management, rather than a review of work already projected.

Format and Substance of a Trip Report

Bruce Kennewick wrote his report on a form that was provided by Brighton Hotels. When no company form exists, however, the basic format of his report can be followed.

Sample Professional Trip Report

Figure 8.2 presents the rough draft of Bruce's trip report to T. P. Simpson on the activities of his follow-up task force assignment.

BRIGHTON HOTELS, INC.

3200 LAKESIDE DRIVE, SALT LAKE CITY, UTAH 70121
806-326-9800 FAX: 806-243-1861

OFF-PROPERTY REPORT

Primary Reporter: Bruce Kennewick, Director of Catering. 044-34-4739

Home Property of Primary Reporter: The Brighton House, Ozark Beach, MO 1530

Authorization: T. P. Simpson, Vice President, International Division, Salt Lake City, Utah, November 10, 2000

Reporting Property: Brighton Playa, Puerto Vallarta, Mexico, 1720

Other Members of the Reporting Team: None

Dates of Assignment: December 1, 2000–April 1, 2001

Purpose of the Assignment: Reestablish and refine GSOPs. Catering Department, in light of difficulties with personnel changes. Specifically, train new Assistant Catering Director to take over as Catering Director.

FIGURE 8.2 **Bruce Kennewick's trip report**

Work Accomplished

Summary

General Activities

This assignment fell within the usual guidelines and expectations of a follow-up task force. I identified four general areas of concern: personnel, administration, sales, and support services. The problems with the four areas were either solved or substantially diminished to the point that productivity has increased steadily over the last 2 months. I spent the first 10 days assessing the local operations to see where strengths and weaknesses lay and then started to decide which areas needed the most improvement, in line with GSOPs, and how that improvement could be brought about quickly and be sustained over a longer period of time.

Following are the significant accomplishments in each area:

Personnel:

1. Trained the Assistant Catering Director to take over as Director of Catering, with ultimate recommendation that he be promoted
2. Hired and trained a new banquet manager and coordinated his efforts with the Assistant Director of Catering's, according to GSOPs
3. Hired and trained four new waiters

Administration:

Set up the following on the property's computer system:

1. A reorganization of the Catering Department's central filing system
2. A separate trace and file system on local groups and businesses
3. A new database of menu information

Sales:

With the help of the Director of Sales and the Food and Beverage Director:

1. Developed yacht sales (sales to private yachts docked at Puerto Vallarta)
2. Planned weekly theme parties

Support Services:

Finding skilled, reliable vendors of support services proved to be something of a problem, but we established at least the beginnings of a vendor network.

The first paragraph summarizes briefly the work done on site during the time spent there. It identifies the four areas of concern and concludes what Kennewick did to solve the problems of each, the problems he encountered in solving them, and what has to be done to solve those not yet solved. As in a periodic report, this introductory section serves as an executive summary, which lays out the substance of the report so that busy readers can find out immediately what happened and then return to the report to study any details.

FIGURE 8.2 (*Continued*)

The "Personnel" section identifies the work that Kennewick had to do with regard to personnel: hiring and training new waiters and a banquet manager, and continued training of the assistant catering director. He is specific about who the principal parties are and where they come from, but does not waste time going into detail about their qualifications, since that kind of information is available more completely and efficiently through the personnel office.

Kennewick speaks of the specifics of his work with Moreno and Vasquez, keeping in mind the purposes of the assignment to reestablish, and refine GSOPs and specifically to train the new Assistant Catering Director to take over as Catering Director. And he specifies the activities he and his trainees took part in to show how he was meeting those purposes. He comments briefly on how the training for the two went generally.

In the fourth paragraph Kennewick writes specifically about his training, of Moreno in sales procedures, a crucial element in the operations of the Catering Department. He shows how he involved the Banquet Manager and refers the readers to a later part of the report for greater detail.

Work Accomplished

Problems

Differences in language and culture initially slowed down the reorganization effort, but by the end of the period both problems had been significantly diminished.

Recommendations

1. Provide at least one bilingual member who is also versed in the mores of the host culture on task force teams to international properties.
2. Establish a pool of reliable, competent off-property service personnel as soon as possible at international sites.
3. Hire an audiovisual manager at Brighton Playa.

Details of the Report

Personnel

On the property when I arrived were an assistant catering director, Sr. Jose Moreno, who had been hired only 3 weeks before; a banquet manager; a catering manager; and eight waiters, three full-time, five half-time. Within 3 weeks the banquet manager and four of the waiters, one full-time, three half-time, had quit. We advertised the positions in the local press and through local contacts, and at the end of the first 2 weeks, we were able to hire four new waiters and a new banquet manager, Sr. Pedro Vasquez.

The most important work of the period was to prepare Sr. Moreno to take over as Director of Catering and to train the new banquet manager, Sr. Vasquez. The two men have combined experience of 7 years with Hilton and Stratton in Mexico City and Guadalajara, but both are new to Puerto Vallarta. Their complete résumés are available through Central Personnel.

It was essential to coordinate the efforts of these two people to meet the standards of Brighton's GSOPs. I therefore reviewed with them, over a 3-week period, the paperwork involved in ordering, preparation, and presentation, and was able to take advantage of a short lull to supervise a dry-run rehearsal for a large convention at the end of the month. The two men worked well with me and soon began to understand their relationship in planning and coordinating clearly. They also worked well with waiters, bringing them up to standard performance in about 3 weeks.

During this time I also gave special attention to Sr. Moreno's training in sales procedures, pointing out and clarifying his responsibilities for coordinating local bookings with those of our own booking centers in the United States. We also worked together on plans for increasing local sales (discussed in greater detail below). We involved Srs. Vasquez and Castorena (Catering Manager) in several of our meetings so that they could better understand the relationship of sales to their parts of the operation.

In general, work with personnel went smoothly, though a bit slowly, and I felt confident at the end of the period that the Catering Department was working well as a team.

FIGURE 8.2 (*Continued*)

Administration

A number of important administrative chores has to be completed before the department could run smoothly. The existing filing system for such things as sales orders, purchase orders, and disbursements had to be reorganized to conform to Brighton's central system. A good deal of miscommunication had resulted from using the ad hoc, local system the original Director of Catering had established, and that apparently only he understood.

I also established a separate trace and file system for soliciting sales from local groups. As a subset of the main filing system, the new database now provides quick and easy access to information on local groups and businesses in order to solicit new business and to encourage repeat business. Although 4 months was not enough time to test the system fully, we were able to book four new meetings and two repeat meetings for the 2000 Christmas season, an encouraging indication of the new file's future utility.

The last administrative chore was to establish a database of menu information based on several considerations: time of year (accounting for seasonal differences, especially in fish, fruits, and vegetables), local service availability, kind of meeting, size of meeting, and costs (including customer's budget). Having this information, and being able to update it according to changes in production and the economy, will make our sales efforts both more efficient and convincing.

In the three paragraphs under "Administration," Kennewick refers to three administrative chores he accomplished: reorganizing the central filing system, establishing a trace and file system, and setting up a database of menu information. Since the latter two have a direct influence on sales, he is sure to point that out, given the primary importance of sales to any operation of this kind. He is also specific as to the effects the changes will most likely have.

Sales

With the help of the Director of Sales and the Food and Beverage Director, the Catering Department was able to suggest two local initiatives for local sales. The first is yacht food sales, something new to our business, but that Hilton has tried already with some success. The procedure is fairly simple. A "grocery list" of foods and meals available is delivered by uniformed personnel to private yachts docked in the port, and the customers are invited to decide what they might want delivered to them, ranging from uncooked foods to complete dinners, including full service.

Customers may order uncooked foods to be prepared on ship by their own staffs or a complete banquet delivered and served by Brighton personnel. There are of course variations of service available between the two extremes.

The initiative capitalizes on the leisurely resort atmosphere of Puerto Vallarta. And even though this initiative may require some additional outlay, such as more delivery equipment and additional servers, the move could also prove very profitable. People with the means to take private cruises are also willing to pay for courteous, efficient, and tasteful service.

The modest beginnings of yacht sales when I left—ten or so orders, mostly of small dinners of six to seven people—were profitable. We were able to use existing hotel equipment and personnel, without hindering operations, and the rewards, both in revenue and customer satisfaction, were substantial. A more detailed report on this initiative will follow from the Director of Sales.

The other local initiative, which will prove beneficial for extra-local sales, is the theme party. Based on successful specials, happy hours, and seasonal parties at hotels in the United States, the Catering Department, with help from the Food and Beverage Director, has decided on weekly theme parties that will be held in the Aztec Room (a

The "Sales" portion of the report is the longest, again because of the central importance sales have to the Catering Department. Kennewick explains in some detail the two local initiatives he was able to put into place: yacht sales and theme parries. He explains how the initiatives will work and offers in paragraphs three and five the underlying business rationale, besides increasing profits, for the moves.

Finally, Kennewick is quick to mention the help he received from the Director of Sales and the Food and Beverage Director. This step is impotant because it demonstrates the interconnectedness of the operation, and also because it shows he is willing to draw on other in-house resources, a central part of the notion of team playing that Brighton Hotels strongly fosters.

FIGURE 8.2 (*Continued*)

Grade B banquet room) on each Wednesday. The themes will vary from the mystery of the sea to the culture and mythology of the Aztecs, but we plan an Aztecan theme at least twice a month. The kind of food and beverages will attempt to reflect the theme, although choices from the bar and menu will also include standard, nontheme items; waiters will wear appropriate costumes in keeping with the theme. Service will be mostly buffet style, but waiters will serve cocktails, coffee, and dessert, as requested. Music will also be provided, mostly by recording, but sometimes live. Theme parties will start at 7:00 p.m. and food service will conclude at 11:00 p.m. Bar service, including hors d'oeuvres and music, will continue until 2:00 a.m.

Kennewick ends the 'Work Accomplished' section with talk of support services. Although he found the work in this area problematic, Kennewick is able to point to some success with the physical therapists' meeting and to suggest a reasonable way of solving the problems he came across.

Support Services

Support services were in some ways the most troublesome of all the work for the period. Although Food and Beverages has trouble with timely delivery of some of its commodities, Catering is able nevertheless to piggy-back on its orders for most of our functions. But Catering requires other support services that F and B doesn't require, such as stage construction and audiovisual services, and finding vendors for these services was a problem (to be discussed in some greater detail below).

We were able, however, to find some local plumbers, carpenters, and electricians who could serve our needs in a fairly reasonable time. For the first meeting of the American Association of Physical Therapists at Brighton Playa, we were able to section off areas by collapsible partition and in some cases construct performing stages for various demonstrations of new techniques in physical therapy. We also provided minimum services for audiovisual requirements, such as overhead and slide projectors. The meeting was successful, although we got some complaints about the immediate availability of these services. Given some time, and an established network, these services can be as easily rendered as any other. But it will take time and practice. More of this later under "Problems."

Kennewick speaks of two basic problems typical of any international property: differences in language and culture. He is specific about how language was a problem, especially in training personnel, since he is not bilingual, and those who are were not always available. This made for some misunderstanding and confusion and in general tended to slow down the operation. But he also points out that time and practice in Spanish made him at least functionally fluent toward the end of his stay.

Problems Encountered

Differences in language and culture were the biggest problems I had to confront for most of the time on the property. Although I spoke some Spanish when I arrived, I soon realized that I did not speak enough to get along easily. The GM and all other department heads are bilingual, but the manager's level has only a few who are bilingual, and Srs. Moreno and Vasquez are not among them. Although they speak better English than I do Spanish, their facility in English is halting at times, which at the beginning of the training period caused some difficulties. They seemed to understand the basic concepts of the GSOPs, for instance, but when I attempted to alter them for specific local needs, they frequently misunderstood what I was trying to explain.

The problem was even greater in the training of waiters and setup personnel. At that level everyone speaks only Spanish. And the same is true for off-property service personnel: carpenters, electricians, and the like. Even though Srs. Moreno and Vasquez took a willing hand in training waiters, there were times when they were either unavailable or themselves had trouble understanding and explaining the kinds of procedures I wanted waiters to follow. At the times I had to deal with service personnel, Moreno and Vasquez would have to do al-

FIGURE 8.2 (*Continued*)

most continuous translation, and if they were not around, I would have to find someone else to translate for me.

Over the course of 4 months I was able to learn much more Spanish and became at least functionally fluent in the needs of day-to-day business. Nevertheless, much of the slowness at the beginning of the reorganization efforts I must attribute to problems with language.

The other main problem has to do with cultural differences, especially in the general area of punctuality. North Americans in business usually operate on the notion that time is money, that meeting a fairly rigid schedule of operations will in the end be profitable. Mexican nationals in business, at least at the mid to lower levels, however, tend to operate on more flexible schedules, those that allow for more relaxed interactions between people, interactions they believe in the end will also be profitable. The result of the two is a clash of expectations: North Americans expecting a more rigid conformity to schedules, Mexicans expecting a more flexible schedule. The two notions can eventually come together or be compromised at some point, but I found that it takes time, and neither group is immediately comfortable with the other's expectations.

I tended to complain about lateness to meetings and missed appointments by both hotel and off-property personnel: they tended to show confusion about why I was complaining. It took us both a while until such difficulties could be compromised, which meant essentially changes in planning and schedules on both of our parts: but eventually a much smoother working relationship was worked out.

The other problem, cultural differences, is more delicate and difficult to solve. But Kennewick does a good job of defining the differences as they affected him and his operation, without making qualitative judgments. Referring to the differences as a "clash of expectations," he is careful not to diminish the mores of the host culture, and in speaking of the "compromise," he gives credit to both sides for their willingness to take part in that activity.

Recommendations for the Future

Several recommendations for the future grow immediately out of the problems I encountered, although others might simply move business along more expeditiously and profitably.

1. Initial Task Forces on international properties should have at least one member who is fluent in the native language and who has clear ideas of cultural differences that will affect the conduct of business. The two usually go together, but not always. This will make conduct of business much easier and more efficient.
2. A pool of reliable, competent off-property service personnel should be established as soon as possible. This involves some research: finding vendors who have worked well on other properties in the area and determining their availability or perhaps getting recommendations for vendors from the local Chamber of Commerce.
3. At Brighton Playa in particular the Catering Director needs to hire someone to manage audiovisual services.

The recommendations for the future are brief and straightforward and grow directly out of problems Kennewick came across. When possible, as in the case of item 2, some idea of how the recommen-dation could be carried out is included.

FIGURE 8.2 ***(Continued)***

EXERCISES

1. Taking a trip on a company's business obviously costs something, in both time and money. Trips, therefore, have to be planned in advance, so that time and money are spent wisely. Bruce Kennewick, for example, spent several hours of several days in planning what he would do when he got to Puerto Vallarta and in anticipating as well as he could the problems he might confront. Taking these steps made his work more efficient when he got on site.

   Using a problem that you and your instructor have agreed is worthy of attention and that would require a trip to help address, plan what you would do, how you would do it, and why. An example might be the decision of the city in which your college or university is located to establish a comprehensive solid waste recycling program in response to new state mandates for such programs in all municipalities of 10,000 people or more. The city council has turned to the college or university for help in planning and implementing the program.

   The problem would require a review and assessment of existing recycling efforts and facilities. When planning this work, you might consider a visit to the mayor, city manager, or city planner to ask questions about the city's current recycling efforts and the way that they might be changed to meet the new state mandates. You would also want to visit the current recycling facilities to see how they operate.

   Imagine yourself as a member of an assessment team, and decide what it is you need to find out from the visits just suggested. In addition to deciding on the kinds of questions you would ask and the kinds of operations you should observe, anticipate any problems you might have that would interfere with your plans.

   Remember that you will eventually have to report to someone regarding your visits. Define that audience carefully, and keep it in mind as you make your plans.

   A specific outline of your plans for the audience intended should be the result of this exercise.

2. Plan a trip to gather data for any other problem that may exist on your campus or in the community: access to facilities for handicapped students and visitors; the extent of and access to computer facilities for students; a safety program for women traveling alone on campus; or programs for reeducation and reemployment of workers who have been laid off at a local or regional manufacturing facility.

ASSIGNMENT

In accordance with the principles discussed in this chapter and elsewhere in the text, write a report on a trip that you have taken to collect data on the research problem you are attempting to address or solve.

CASE STUDY

AL ROBINSON

Somewhere on the Llano Estacado in New Mexico ▪ Temperature 115°F

Al Robinson took another swig from his canteen. He and his crew had been running pilot holes into the desert for five months now to find a spot that could support the project that his company was interested in: burying nuclear waste. This exploratory work had turned up some interesting things; the pilot holes had first of all looked like diagonals instead of verticals because of the unexpected resistance they had encountered from various soils. A number of methodological changes were then instituted. Another problem, expected but not in its current frequency, was the volume of brine infiltration that the crews encountered.

After consulting with the hydrologists and mining engineers, Al made some significant adjustments to the pilot hole rationale. After those changes, the crew started getting some reliable numbers for depth, time, and costs. Unfortunately, these numbers were not exactly what Al's management seemed to be looking for, from what Al could perceive. As a result of the memos and preliminary reports he had seen, Al had the impression that management was interested in a simple project, not one that would require greater depths and, therefore, greater costs than a standard potash mine would require.

Al felt that he had a dilemma. Was he obliged to report all the work they had done to conduct the feasibility study? Should he try to do his managers' thinking for them by leading them to what he thought was a

reasonable answer, or should he just lay down the facts and let them form their own opinions? He didn't think it was his job to place undue influence on his higher-ups. Exactly what should he do to write up this feasibility study?

As a last resort, Al called Tammy, the technical writer he had worked with on procedures. She was not knowledgeable in the technical aspects, but she certainly understood how to write up technical subjects. She had some good advice for him. She was a cool breeze in the middle of the desert.

FUNDAMENTALS OF A FEASIBILITY STUDY

Purpose and Rationale

A feasibility study reports exploratory work performed to assess whether a larger project is advisable. The advisability can be based on any parameters that the authors want considered, like cost, physical feasibility, or schedule. This kind of report used to be called a recommendation report, because that is the essence of the report's importance.

More important than the description of work done to try to determine feasibility is the logic of the conclusion that the authors draw. A feasibility study is not complete without the recommendation; yet, the recommendation cannot be evaluated without the data on which it was based. So a feasibility study will usually contain two distinct sections, a descriptive section and an evaluative section. The descriptive section, like that of a lab report, will review in appropriate detail the tests performed to try to determine feasibility. The evaluative section will pull all the discrete conclusions together into an overall recommendation.

Although feasibility studies are not difficult to write, they may be more difficult to make readable than other types of reports because they have so much descriptive work to do. The temptation for the writer is to get bogged down in descriptive details instead of subordinating these details to the readers' concerns. Different audiences will read feasibility studies for different reasons, but whenever possible, the writer should structure one report for ease of reading by a variety of audiences, rather than write different versions of the same report.

Although feasibility studies may take various forms and may be addressed to varying audiences, the following ground rules always apply:

1. Make the report conclusive.
2. Review reasonable alternatives.

3. Provide sufficient detail and analysis to support conclusions and recommendations.
4. Include specific recommendations.

ANALYZING THE PROCESS OF WRITING A FEASIBILITY STUDY

The essential questions that feasibility studies answer are summarized in Figure 8.3.

A good feasibility study will state its conclusions up front, either in the report's introduction, in an abstract, or in an executive summary. This sort of recommendation report is very tempting to structure like a detective novel instead of like a professional report because one of the purposes of the report is to give a definitive answer to a specific question: Should we take on this project or not? Will this project be worth financing? Is this project doable? The writer gets caught up in the suspense of the enterprise. But the reader wants answers immediately; the suspense wastes their time. So the author must be forthcoming with the conclusions. Responsible audiences will read on. With the suspense gotten out of the way, the reader can immediately read for coherence between test data and conclusions.

Reviewing reasonable alternatives means that the writer will consider choices to originally projected actions in a responsible way. For example, if an author were estimating the feasibility of burying nuclear waste in bedded salt in a specific site, the author would have to review alternatives in good faith, not skipping any alternatives that, for personal reasons, did not appeal. Thus, the writer would have to admit that a site with an extant but defunct potash mine shaft might work, despite the fact that the company would rather test out all kinds of innovative shaft-sinking technology and would, therefore, prefer virgin territory. It is tempting to "load" alternatives into a feasibility study, but it is better to try to cover only the reasonable ones, since lack of coverage of those might be apparent to reviewers and thus might provide them with reason to dismiss the report.

Questions That Feasibility Studies Answer

- What is the problem being studied?
- Briefly, what alternatives are being considered?
- What is the recommendation?
- What evidence supports this recommendation?
- What financial and schedule consequences result from this recommendation?
- What further work needs to be done?

FIGURE 8.3 **Questions to ask and answer to define the contents of a feasibility study**

To establish credibility, the feasibility study must show clear logical connections between the descriptive data it provides and the conclusions and recommendations it reaches. Thus, it is not enough to expect the facts to speak for themselves. Instead, the facts must be analyzed for the reader, and the logic spelled out in clear, simple prose. So, an analysis section is at the heart of a feasibility study. It serves as the connection between the raw data and the conclusions that the data support.

It is important for those who are writing the feasibility study and who, therefore, have the most detailed and current understanding of the possibilities in a situation, to write specific recommendations based on their findings. These findings, although perhaps rejected in their specific language, may provide the model of acceptable choices. They essentially instruct the reader in how to think about the project's implications. If a feasibility study dodges this responsibility, then the report may lose all credibility.

SAMPLE OUTLINE FOR A FEASIBILITY STUDY

Having reviewed these particulars with Tammy, Al had a much better idea of what to include in his report. He decided that some brief mention should be made of their early drilling problems, just to honor the accuracy of the descriptive information section, but he wouldn't need to dwell on these problems. He also decided that the use of defunct potash mine shafts should be clearly explained as an alternative, despite the fact that he didn't think that his own management was very interested in the alternative. In addition, he would walk his readers through the process of elimination that he had gone through to provide what he saw as the most feasible alternative: keep looking. His report thus became a very straightforward job, with the following organization:

I. **Abstract or Executive Summary.** This section of the report contains Al's conclusion that the project is not feasible in any of the sites so far. He also notes here his recommendation to keep looking, but to shift the search to the southwest by 50 miles.

II. **Description of the Problem and Summary of Testing.** This is where Al summarizes the problem of finding new sites and presents a chronological summary-level review of the four months of testing.

III. **Discussion.** Briefly, Al summarizes the alternatives being considered.

IV. **Conclusions and Recommendation.** Al lists these for his readers.

V. **Analysis.** Al walks his readers through the evaluation and process of elimination by which he reached his conclusion and recommendation. This discussion serves as the rationale for his recommendation, and includes financial and schedule consequences that would result from his recommendation.

VI. **Future Needs.** Al lays out what work has to be done from this point on.

Note that short reports could combine sections III and IV, as long as those sections emphasize analysis.

SAMPLE ENGINEERING FEASIBILITY STUDY

We have included a copy of an engineering feasibility study from a professional source. (See Figure 8.4.) You'll note that this feasibility study uses organizational headers and topic headers.

This study is a fine model of explicitness. Note that the executive summary is conclusive; we know what the upshot of this feasibility study is without having to search through the report to find it. The report is also well marked with descriptive headers, topic sentences, and an explicit conclusions section.

A notable element of this report is its graphics. They reinforce central concepts of the study and the authors discuss them in the text. They don't expect the tables and figures to speak for themselves. The drawings, of course, visualize for the reader the sort of technology these authors have in mind. Since feasibility analyses concern areas of speculation, drawings help a great deal to make the speculations seem rational and well grounded in reality.

We think this report is a fine example of professional and technical writing. You may also note that the authors write in the first person plural (we), which allows them to avoid many passive constructions (refer to the Appendix for an explanation of passives). The authors also express several opinions in this study. Thus the persuasive purpose of so much technical and professional writing is acknowledged and taken up in this study.

DISCLAIMER

This document was prepared as an account of work sponsored by an agency of the United States Government. Neither the United States Government nor the University of California nor any of their employees, makes any warranty, express or implied, or assumes any legal liability or responsibility for the accuracy, completeness, or usefulness of any information, apparatus, product, or process disclosed, or represents that its use would not infringe privately owned rights. Reference herein to any specific commercial product, process, or service by trade name, trademark, manufacturer, or otherwise, does not necessarily constitute or imply its endorsement, recommendation, or favoring by the United States Government or the University of California. The views and opinions of authors expressed herein do not necessarily state or reflect those of the United States Government or the University of California, and shall not be used for advertising or product endorsement purposes.

This report has been reproduced
directly from the best available copy.

Available to DOE and DOE contractors from the
Office of Scientific and Technical Information
P.O. Box 62, Oak Ridge, TN 37831
Prices available from (615) 576–8401, FTS 626–8401

Available to the public from the
National Technical Information Service
U.S. Department of Commerce
5285 Port Royal Rd.,
Springfield, VA 22161

FIGURE 8.4

PROOF-OF-CONCEPT DEVELOPMENT OF PXAMS
(PROJECTILE X RAY ACCELERATOR MASS SPECTROMETRY)*

White Paper prepared as part of the final report for the Laboratory Directed Research and Development Program project "Proof-of-Concept Development for Accelerator Mass Spectrometry Inverse PIXE" (tracking code 95–ERP–126).

I.D. Proctor, M. L. Roberts, J. E. McAninch and G. S. Bench
Center for Accelerator Mass Spectrometry
Lawrence Livermore National Laboratory
Livermore, California, 94551–9900, USA

EXECUTIVE SUMMARY

Prior to the current work, accelerator mass spectrometry (AMS) was limited to a set of ~8–10 cosmogenic isotopes. This limitation is caused primarily by the inability to discriminate against stable atomic isobars. An analysis scheme that combines the isotopic sensitivity of AMS with similar isobar selectivity would open a large new class of isotope applications. This project was undertaken to explore the use of characteristic x rays as a method for the detection and identification of ions, and to allow the post-spectrometer rejection of isobaric interferences for isotopes previously inaccessible to AMS.

During the second half of FY94 (with Advanced Concepts funding from the Office of Non-Proliferation and National Security), we examined the feasibility of this technique, which we are referring to as PXAMS (Projectile Xray AMS), to the detection of several isotopes at Lawrence Livermore National Laboratory (LLNL). In our first exploratory work, we measured the x ray yield vs. energy for ^{80}Se ions stopped in a thick Y target. These results, shown in Fig. 1, demonstrated that useful detection efficiencies could be obtained for Se ions at energies accessible with our accelerator, and that the count rate from target x rays is small compared to the Se Kα rate. We followed these measurements with a survey of x ray yields for $Z = 14 - 46$.

With support from this LDRD project during FY95, we successfully developed the PXAMS concept and applied it for the detection of ^{63}Ni, which we will apply in the determination of the fast neutron fluence from the Hiroshima bomb. This first application to ^{63}Ni is in collaboration with Tore Straume and Alfredo Marchetti of the Health and Ecological Assessment Division at LLNL. In a demonstration experiment, ^{63}Ni was measured in Cu wires (2–20 g) which had been exposed to neutrons from a ^{252}Cf source. We successfully measured ^{63}Ni at levels necessary for the measurement of Cu samples exposed near the Hiroshima hypocenter. To our knowledge, this work represents the first attempt to apply AMS to the detection of this long-lived isotope, and was the first application of PXAMS to any isotope at LLNL. We

FIGURE 8.4 (*Continued*)

have also begun to extend the methods developed for ^{63}Ni to the detection of ^{59}Ni as a biomedical tracer in living systems.

Our success in the detection of ^{63}Ni in Hiroshima wire samples—a pathological technical challenge—will have several major impacts on future development of isotopes for AMS. As the first application of PXAMS at CAMS, this work has demonstrated the effectiveness and simplicity of the technique, has illuminated both the advantages and the drawbacks, and has revealed many of the parameters which determine the sensitivity. The work also demonstrates the applicability of analytical chemistry techniques to AMS, and the value of considering novel chemical methods and sampling handling. Our efforts to reduce instrumental Cu background in the ion source will be directly applicable to other isotopes. Primarily however, this work should encourage the AMS community to explore what at first glance appears to be a set of near intractable problems.

PXAMS holds the promise of enabling the application of AMS to long-lived isotopes throughout the periodic table, pointing to a wide range of potential applications. In just the range $Z = 22 - 50$, there are 15 long-lived isotopes which were previously unmeasurable at LLNL. PXAMS can be an effective detection technique for these isotopes and, as a general statement, PXAMS can be expected to provide 100–1000 times better sensitivity for these isotopes than would be possible with any other mass spectrometric technique (for a given sample and a given preparation chemistry). Further feasibility studies for these isotopes will now concentrate on potential applications, funding opportunities, required sensitivities, options for chemical sample purification and preparation, and availability of material for production of standards and tracers.

In the near future, there are plans at LLNL to extend AMS to several new isotopes with applications in biomedical tracing (^{53}Mn, ^{60}Fe, ^{79}Se, ^{126}Sn, ^{154}Hg, $^{202,205}Pb$), effluent migration (^{90}Sr, ^{99}Tc, ^{93}Zx, ^{205}Pb), and non-proliferation (^{90}Sr, ^{99}Tc). The success of the current work has significantly increased the feasibility of detecting these isotopes by AMS. More importantly, it has spawned a new interest in developing new isotopes at LLNL.

FIGURE 8.4 ***(Continued)***

1. INTRODUCTION

Stable atomic isobars are the dominant background limiting the sensitivity of AMS for most isotopes. For isotopes of atomic number $Z < 20$, the post-spectrometer rejection of these backgrounds relies on the energy loss of ions in matter. The rate of energy loss dE/dx is a function of Z. It is therefore possible to distinguish isobars by sampling the energy deposited at different depths in the material. The standard AMS detector is the multiple anode gas counter, in which the ionization of a low pressure (50–200 Torr) gas is used as a measure of energy deposition. The ΔE-E telescope using silicon surface barrier detectors is similar. Another altnerative is the gas-filled magnet, in which a magnetic field combines with energy loss and charge changing interactions to physically separate the isobars.

Isobar rejection techniques that rely on energy loss suffer from two intrinsic features of the energy loss process—charge changing interactions and the statistical nature of multiple collisions—which cause the width of the energy distribution to increase as the ions slow down. At lower Z this spreading, known as energy straggling, is small compared to the difference in energy loss between isobars. With increasing atomic number, the straggling increases relative to the energy loss difference, so that isobar separation becomes progressively less effective, and at some point unworkable. For the energies accessible with larger tandem accelerators, this point is somewhere in the range $Z \approx 25$–30.

We are exploring an alternative method for isobar rejection which does not rely on energy loss. In this method, a variant of PIXE (particle induced x ray emission), projectile ions are identified by the characteristic x rays they emit when slowing down in matter. These x rays are of atomic origin, and can therefore be used to distinguish the ions by atomic number. Artigalas *et al.*[1,2] examined this approach for the AMS detection of ^{36}Cl, ^{59}Ni, and ^{94}Nb at 15–21 MeV, and Wagner *et al.*[3] explored its application in the detection of ^{59}Ni and ^{60}Fe at 55 MeV. In both works, the purpose was to find a method for detecting the radioisotopes given the constraints of the available accelerators.

During the second half of FY94,[4] we examined the applicability of this technique, which we are referring to as PXAMS (Projectile X ray AMS), to the detection of several isotopes at Lawrence Livermore National Laboratory (LLNL). In our first exploratory work, we measured the x ray yield vs. energy for ^{80}Se ions stopped in a thick Y target. These results, shown in Fig. 1, demonstrated that useful detection efficiencies could be obtained for Se ions at energies accessible with our accelerator, and that the count rate from target x rays is small compared to the Se $K\alpha$ rate. We followed these measurements with a survey of x ray yields for $Z = 14$–46.

During the course of this LDRD project in FY95, we successfully developed PXAMS for the detection of ^{63}Ni, which we will apply in the determination of the fast neutron fluence from the Hiroshima bomb, in collaboration with Tore Straume and Alfredo Marchetti of the Health and Ecological Assessment Division at LLNL. To our knowledge, this work represents the first attempt to apply AMS to the detection of this long-lived isotope, and was the first application of PXAMS to any isotope at LLNL. We have also begun to extend the methods developed for ^{63}Ni to the detection of ^{59}Ni.

FIGURE 8.4 (*Continued*)

In this paper we present the results of our x ray yield measurements, describe the experimental arrangement currently used for PXAMS at LLNL, and discuss our work on the detection of ^{63}Ni, ^{59}Ni, and other isotopes.

2. MEASUREMENT OF THE X RAY YIELDS

X ray yields were measured using the 10 MV FN tandem accelerator at the Center for Accelerator Mass Spectrometry (CAMS) at LLNL.[5] Picoamp currents of selected ions were stopped in thick targets, and the resulting x rays were detected with a lithium-drifted silicon detector (Si(Li)) normally used for PIXE analyses. The number of incident ions was determined by collecting the charge deposited on the targets, which were mostly elemental metal foils of 10 mg/cm^2 thickness or greater. Absolute x ray detection efficiencies were measured using thin (~50 $\mu g/cm^2$) PIXE standards. In general, the uncertainties in the measured $K\alpha$ yields were insignificant compared to the ~10% uncertainty in the charge integration. For the $K\beta$ yields, the uncertainties were somewhat larger because of increased statistical uncertainty and increased uncertainty in the peak extraction.

Results of the yield measurements vs. ion energy for Ni and Se ions are shown in Figs. 1 and 2. The Ni results exhibit a trend similar to the Se results, but with significantly higher yields for the same energy per unit mass. Also seen in Fig. 2 is the good agreement of the present Ni data to the x ray yields of Ref. 2.

Projectile and target yields for Se ions on various targets are displayed in Fig. 3. The optimum target is either Y or Zr, and the shape of the yield curves is essentially the same for different energies (compare the 80 and 100 MeV results). The projectile and target $K\alpha$ yield curves illustrate the cross-section resonances which results from the formation of molecular orbitals between the projectile and target atoms when there is an energy overlap between the respective K shells. This atomic physics phenomenon has been studied in detail by Meyerhof *et al.*,[6] and the present data follow the trends in that reference. In addition, we studied the second resonance region (near $Z = 80$), in which there is an overlap between the projectile K shell and the target L shell. The maximum yield in the second resonance region is about 75% of the Y yield, however this is a more difficult region for PXAMS applications because of the intensity and number of target $L\alpha$ lines.

In Table 1, we present additional measured yields which are representative of the efficiencies that will be accessible with our accelerator. Yields as high as 1 x ray per incident ion are accessible for light ions such as Si. The yields are significantly reduced for heavier ions, primarily because of the reduced energy per unit mass. However, even for ^{107}Pd ions, where the best yield is less than 1 x ray per 300 incident ions, we expect that PXAMS will allow more sensitive detection than is possible with other techniques.

We have also examined the use of projectile $L\alpha$ lines, which have significantly higher yields, however with standard x ray detectors these lines are not resolved from the target $L\alpha$ lines. Spectrum fitting is an option, however this would likely require significant intervention and analysis time per sample on the part of the experimenter. We have

FIGURE 8.4 (*Continued*)

considered the possibility of detecting Lα lines using a wavelength dispersive x ray detector, which could provide increased isobar rejection, but the small acceptance of these detectors is expected to be a significant drawback.[7]

3. PXAMS AT LLNL

For further PXAMS development, we have purchased a high resolution high-purity germanium (HPGe) detector† and which is installed on the AMS beamline at CAMS. A schematic of our setup is shown in Figs. 4 and 5. The target and absorbers are placed immediately in front of the Be window of the detector. This design was chosen to maximize the solid angle intercepted by the detector. While the solid angle in this geometry could in principle approach 2π, the solid angle at present is $\sim 0.08 \times 4\pi$ because of the ~10mm distance between the crystal face and the outer edge of the cryostat, plus the combined thickness of the foils.

The 0° detector position leads to significant shifting and broadening of the x ray lines due to Doppler shifts. A somewhat smaller effect (which acts at all angles) is the shifting of the atomic levels caused by the high average charge state of the ions as they slow down in the target. Because of the Doppler broadening, a careful choice of target thickness and accepted angular range is important to find a balance between maximizing the x ray yield and minimizing the width of the peaks to improve for isobar rejection.

Examples of the observed x-ray spectra for a ^{63}Ni standard (^{63}Ni/Ni = 1×10^{-8}) and blank (^{63}Ni/Ni = 0) are shown in Fig. 6. A fraction of the Cu counts extend through the Ni region as a low energy tail, so that subtraction of background Cu counts from the Ni sum is necessary for ultimate sensitivity. Ni and Cu counts are extracted using the indicated summing regions. The amount of isobar rejection, defined as the ratio of Cu counts in the Cu region to Cu counts in the Ni region (i.e., the "peak-to-tail" ratio), is ~300 in this case. For measurements with a signal-to-background of >0.1, ^{63}Cu/^{63}Ni levels can be as high as ~3000 (larger ratios are possible with sufficient statistics). It was found that the Cu peak-to-tail ratio is stable to <0.5% over periods of hours, and for count rates up to ~3 kHz. It is therefore conceivable that ^{63}Ni could be detected at the 3σ level for samples in which the ^{63}Cu/^{63}Ni ratio was as large as ~10000.

4. PXAMS APPLIED TO THE DETECTION OF ^{63}NI AND ^{59}NI

The long-lived isotopes of nickel have current and potential use in a number of applications including cosmic ray studies,[8,9] biomedical tracing,[10] characterization of low-level radioactive wastes,[1,2] and neutron dosimetry. As our first application of PXAMS, we are developing methods for the routine detection of these isotopes by AMS.

†An Iglet-X detector, 100mm^2 active area and 145 eV FWHM resolution at 5.9 keV, purchased from EG&G Ortec, Oak Ridge, TN.

FIGURE 8.4 (*Continued*)

One intended application of ^{63}Ni is in Hiroshima dosimetry, in support of work by Tore Straume and Alfredo Marchetti in the Health and Ecological Assessment Division of LLNL. Currently there is a discrepancy in the neutron fluence emitted by the Hiroshima bomb which may be as large as a factor of ten.[11,12] Resolution of this discrepancy has important implications in studies of the long-term health effects of radiation exposure which follow the health histories of Hiroshima and Nagasaki survivors.[13] The reaction $^{63}Cu(n,p)^{63}Ni$ has been identified as one of a small number of reactions which might be used for the direct determination of the Hiroshima fast neutron fluence.[14] The current (1994) level of ^{63}Ni in copper samples exposed near the Hiroshima hypocenter has been estimated to be ~1.4 fg ^{63}Ni per g Cu.[15]

The detection of ^{63}Ni in copper samples from Hiroshima is in many ways a pathological, worst-case challenge for AMS. The sample consists of only femtograms of analyte in grams of the isobaric interference, the isobar easily produces high negative ion currents in a Cs sputter source, the isobar is relatively ubiquitous in nature, and a cursory look at the chemistry of Cu and Ni would indicate that their separation at ultratrace levels is difficult if not impossible.

AMS measurement of ^{63}Ni ($t_{1/2}$ = 100 y) requires the chemical removal of ^{63}Cu (natural isotopic abundance 68%), which is a stable isobar of ^{63}Ni. PXAMS will allow a post-spectrometer rejection of ^{63}Cu by a factor of ~3000, so that, for the measurement of a 10 g copper sample (~14 fg ^{63}Ni), the Cu must be reduced to <50 pg. Strict demands are therefore placed on the chemical separation technique, which must provide a reduction of Cu by ~10^{12}, with quantitative retention of the ^{63}Ni.

For sample preparation, two chemical steps were used: electrodeposition of bulk Cu, and reaction of Ni with carbon monoxide. The first step relies on the large difference in electropotential between Cu and Ni to preconcentrate the trace Ni from gram-sized copper samples.[16] Samples are dissolved in an electrolytic cell, and the Cu is simultaneously electrodeposited on a platinum electrode, leaving Ni in solution. Cu concentrations are reduced from grams to sub-microgram levels in this step.

Following the electrochemical separation, the Cu concentration is further lowered using the reaction of Ni with carbon monoxide to form the gas nickel tetracarbonyl ($Ni(CO)_4$). This reaction, suggested by an external collaborator (Prof. P. Jones, Chemistry Department, University of the Pacific), is highly selective for Ni, and has been exploited by other researchers for the detection of sub-nanogram quantities of Ni by graphite furnace-atomic absorption spectrometry.[17,18,19,20] We have adapted this technique for the preparation of AMS samples (see Fig. 7). Samples containing 1 mg Ni (used as a chemical carrier for the ^{63}Ni) in weak nitric acid solution are adjusted to pH ≈ 10 using concentrated ammonia. A mixture of carbon monoxide and helium is bubbled through the solution. A solution of sodium borohydride is added, reducing Ni^{11} ions to Ni^0, which then react with the CO to form $Ni(CO)_4$. The $Ni(CO)_4$ is carried in the gas flow to a cold trap immersed in liquid nitrogen. Following completion of the reaction, the trap is allowed to warm to room temperature and the $Ni(CO)_4$ is transferred in a He flow to the AMS sample holder, where it is thermally decomposed to Ni metal at ~160°C (Fig. 8).

FIGURE 8.4 (*Continued*)

The $Ni(CO)_4$ reaction has proved to be an exceedingly effective method for the removal of Cu. For samples containing 1 mg Ni carrier, the Cu concentration is consistently reduced to $<2 \times 10^{-8}$ (Cu/Ni), or <20 pg Cu. This Cu level appears to be the background limit of the AMS ion source. These levels are regularly achievable even for samples which have been spiked with 1 mg Cu, demonstrating a reduction in Cu by $>10^8$. Processing times are ~15–30 min per sample. Decomposition of the $Ni(CO)_4$ directly in the sample holders obviated the need for any additional handling of the sample following purification. Any such handling would have caused additional Cu contamination, and would likely have required clean-room conditions.

For AMS measurements, the terminal of the accelerator was set to +9.0 MV, and the high energy spectrometer was tuned to select the 99 MeV $^{63}Ni^{10+}$ ions. Using a 2.0 mg/cm^2 Zr foil and a 10 mm foil to detector distance, the total ion detection efficiency for this arrangement is $\sim 3.4 \times 10^{-3}$ Kα counts per incident $^{63}Ni^{10+}$ ion.

To fully demonstrate our ability to measure ^{63}Ni induced in Cu samples, we prepared a set of demonstration samples for measurement by AMS. This experiment was performed in collaboration with T. Straume and A.A. Marchetti. Cu wires (2–20 g) were exposed to neutrons from a ^{252}Cf source at LLNL. Fast neutron fluences were ~2 and 20 times the estimated fluence near the hypocenter of the Hiroshima bomb. The results of this experiment are listed in Table II, and are presented graphically in Fig. 9. We successfully measured ^{63}Ni at levels necessary for the measurement of Cu samples exposed near the Hiroshima hypocenter. For the demonstration samples, the Cu content was chemically reduced by a factor of 10^{12} with quantitative retention of ^{63}Ni. Detection sensitivity (3σ) was ~24 fg ^{63}Ni in 1 mg Ni carrier ($^{63}Ni/Ni \approx 2 \times 10^{-11}$). The linearity of the results for wires of different sizes and different neutron fluences illustrates the accuracy of the technique. Significant improvements in sensitivity are expected with planned incremental changes in the methods. Because of precious, historical nature of the archived Hiroshima samples, measurements of the actual samples will be delayed until these improvements have been made.

We have begun to modify our methods for the detection of ^{59}Ni ($t_{1/2} = 1.0 \times 10^5$ y) which we plan to use as a biomedical tracer in living systems. Initial results indicate that only minor changes will be required. In a first experiment using a series of ^{59}Ni standards, we demonstrated a detection sensitivity (3σ) of <10 fg ^{59}Ni in 1 mg Ni carrier ($^{59}Ni/Ni \approx 1 \times 10^{-11}$). This corresponds to a ^{59}Ni activity of ~20 μBq (~0.6 fCi), well below the limits of decay counting techniques.

5. CONCLUSIONS

Prior to the present work, AMS at LLNL was limited to isotopes of the elements with atomic number $Z \leq 20$ (Ca), allowing a set of ~8 long-lived isotopes to be accessed.‡ Even with the largest accelerators suit-

‡ This count includes ^{129}I, Z = 53, which is an exception because there is no stable isobar with mass 129.

FIGURE 8.4 (*Continued*)

able for AMS, there are <15 available isotopes. PXAMS holds the promise of enabling the application of AMS to long-lived isotopes throughout the periodic table, pointing to a wide range of potential applications. In just the range Z = 22 – 50, there are 15 long-lived isotopes which were previously unmeasurable at LLNL. We now know that PXAMS can be an effective detection technique for these isotopes. As a general statement, PXAMS can be expected to provide 100–1000 times better sensitivity for these isotopes than would be possible with any other mass spectrometric technique (for a given sample and a given preparation chemistry). Further feasibility studies for these isotopes will now concentrate on potential applications, funding opportunities, required sensitivities, options for chemical sample purification and preparation, and availability of material for production of standards and tracers.

Our success in the detection of ^{63}Ni in Hiroshima wire samples—a pathological technical challenge—will have several major impacts on future development of isotopes for AMS. As the first application of PXAMS at CAMS, this work has demonstrated the effectiveness and simplicity of the technique, has illuminated both the advantages and the drawbacks, and has revealed many of the parameters which determine the sensitivity. The work also demonstrates the applicability of analytical chemistry techniques to AMS, and the value of considering novel chemical methods and sampling handling. Our efforts to reduce instrumental Cu background in the ion source will be directly applicable to other isotopes. Primarily however, this work will encourage the AMS community to explore what at first glance appears to be nearly intractable problems.

In the near future, our plans at LLNL are to extend AMS to several new isotopes with applications in biomedical tracing (^{53}Mn, ^{60}Fe, ^{79}Se, ^{126}Sn, ^{194}Hg, $^{202,205}Pb$), effluent migration (^{90}Sr, ^{99}Tc, ^{93}Zr, ^{205}Pb), and non-proliferation[21] (^{90}Sr, ^{99}Tc). The success of the current work has significantly increased the feasibility of detecting these isotopes by AMS. More importantly, it has spawned a new interest in developing new isotopes at LLNL.

REFERENCES

[1]H. Artigalas, M.F. Barthe, J. Gomez, J.L. Debrun, L. Kilius, X.L. Zhao, A.E. Litherland, J.L. Pinault, Ch. Fouillac, P. Caravatti, G. Kruppa and C. Maggiore, *Nucl. Inst. and Meth. B* 79, 617 (1993).

[2]H. Artigalas, J.L. Debrun, L. Kilius, X.L. Zhao, A.E. Litherland, J.L. Pinault, C. Fouillac and C. Maggiore, *Nucl. Inst. and Meth. B* 92, 227 (1994).

[3]M.J.M. Wagner, H. Synal and M. Suter, *Nucl. Inst. and Meth. B* 89, 266 (1994).

[4]J.E. McAninch, G.S. Bench, S.P.H.T. Freeman, M.L. Roberts, J.R. Southon, J.S. Vogel and I.D. Proctor, *Nucl. Inst. and Meth. B* 99, 541 (1995).

[5]J.R. Southon, M.W. Caffee, J.C. Davis, T.L. Moore, I.D. Proctor, B. Schumacher and J.S. Vogel, *Nucl. Inst. and Meth. B* 52, 301 (1990).

FIGURE 8.4 (*Continued*)

[6]W.E. Meyerhof, R. Anholt, T.K. Saylor, S.M. Lazarus and A. Little, *Phys. Rev.* A5, 1653 (1976).

[7]D. Morse, G.S. Bench, A. Pontau, and S.P.H.T. Freeman, *Nucl. Inst. and Meth.* B 99, 427 (1995).

[8]M. Paul, L. K. Fifield, D. Fink, A. Albrecht, G.L. Allan, G. Herzog and C. Tuniz, *Nucl. Inst. and Meth.* B 83, 275 (1993).

[9]W. Kutschera, I. Ahmad, B.G. Glagola, R.C. Pardo, K.E. Rehm, D. Berkovitz, M. Paul, J.R. Arnold and K. Nishiizumi, *Nucl. Inst. and Meth.* B 73, 403 (1993).

[10]*Nickel and Human Health: Current Perspectives,* E. Nieboer and J.O. Nriagu, eds., John Wiley and Sons, Inc. (1992).

[11]T. Straume, S.D. Egbert, W.A. Woolson, R.C. Finkel, P.W. Kubik, H.E. Gove, P. Sharma, and M. Hoshi, *Health Physics* 4, 421 (1992).

[12]T. Straume, L.J. Harris, A.A. Marchetti and S.D. Egbert, *Radiat. Res.* 138, 193 (1994).

[13]*US-Japan Joint Reassessment of Atomic Bomb Radiation Dosimetry in Hiroshima and Nagasaki, Final Report,* Vol. 1 and Vol. 2, Roesch W. C., ed., Radiation Effects Research Foundation, Hiroshima (1987).

[14]A.A. Marchetti and T. Straume, *Appl. Radiat. Isotop.* 47, 97 (1996).

[15]T. Shibata, M. Imamura, S. Shibata, Y. Uwamino, T. Ohkuba, S. Satoh, N. Nogawa, H. Hasai, K. Shizuma, K. Iwatani, M. Hoshi and T. Oka, *J. Phys. Soc. Jap.* 63, 3546 (1994).

[16]A.A. Marchetti, personal communication (1995).

[17]D.S. Lee, *Anal. Chem.* 54, 1182 (1982).

[18]J. Alary, J. Vandaele, C. Eserieut, and R. Haran, *Talanta* 33, 748 (1986).

[19]W. Drews, G. Weber, G. Tölg. *Fresenius Z. Anal. Chem.* 332, 862 (1989).

[20]R.E. Sturgeon, S.N. Willie, and S.S. Berman, *J. Anal. At. Spectro.* 4, 443 (1989).

[21]J.E. McAninch and I.D. Proctor, Lawrence Livermore National Laboratory Internal Report, UCRL-ID-122643 (1996).

FIGURE 8.4 ***(Continued)***

ISOTOPE	HALF-LIFE (Y)	COMPETING ISOBAR		MEASURED ION		TARGET	ENERGY (MeV)	Kα X RAYS PER INCIDENT ION
^{33}Si	100	^{33}S	(Z + 2)	^{38}Sl	7+	$CaCO^3$	56	1.0
					7+	Ti	56	0.3
^{60}Fe	1.5×10^6	^{60}Ni	(Z + 2)	^{56}Fe	11+	Cu	102	0.8
^{39}Ni	1×10^5	^{59}Co	(z − 1)	^{58}Nl	11+	Zn	102	0.5
						Ge	102	0.4
^{79}Se	$\sim 10^2 – 10^4$	^{71}Br	(Z + 1)	^{60}Se	10+	Y	100	0.05
					12+	Y	120	0.08
^{93}Mo	4×10^3	^{93}Nb	(Z − 1)	^{92}Mo	11+	Rh	106	0.008
^{107}Pd	7×10^6	^{107}Ag	(Z + 1)	^{104}Pd	11+	Ag	106	0.003

TABLE I. **Representative measured x ray yields for various ions and targets**

IRRADIATION TIME (d)	EXPECTED ^{63}NI/CU (fg/g)	DISSOLVED WIRE MASS (g)	TOTAL[a] NICKEL (mg)	MEASUREMENT TIME (min)	MEASURED ^{63}NI (fg)	STATISTICAL[b] UNCERTAINTY (fg)	SYSTEMATIC[c] UNCERTAINTY (fg)	EXPECTED ^{63}NI (fg)
0	2.7	2.92	1.11	24	−9	±14	±14	0
1.00	27	19.81	1.17	22	59	±9	±8	54
10.0	27	2.24	1.11	30	65	±14	±10	60
10.0	27	4.33	1.11	24	133	±10	±7	116
10.0	27	8.73	1.13	17	243	±13	±7	233

[a]Includes 1.1 mg Ni added as carrier and trace nickel from the wire sample (3.4 ± 0.5 μg Ni/gCu).
[b]Uncertainty resulting from counting statistics.
[c]Uncertainty resulting from uncertainty in the calibration.

TABLE II. **Results of AMS measurement of ^{63}Ni in irradiated copper wires.**

FIGURE CAPTIONS

FIG. 1. X ray yields vs. incident ion energy for ^{80}Se ions on a thick Y target. Shown are the Se Kα (squares) and Kβ (diamonds) yields, and the Y Kα yields (crosses). For the Kα yields, error bars are smaller than the symbols. The curves are guides to the eye.

FIG. 2. X ray yields vs. incident ion energy for ^{58}Ni ions on a thick Zn target. Shown are the Ni (squares) and Zn (crosses) Kα yields of the present work, and the Ni (inverted triangles) Kα yields of Ref. 2. For the present data, error bars are smaller than the symbols. The curves are guides to the eye.

FIG. 3. X ray yields vs. target atomic number for 80 and 100 MeV ^{83}Se ions. Shown are the Se Kα yields at 80 MeV (squares) and 100 MeV (open triangles). Also shown for 80 MeV are the Se Kβ yields (diamonds), and the Y Kα (crosses) and Lα (plusses) yields. For the Kα and Lα yields, error bars are smaller than the symbols. The curves are guides to the eye.

FIG. 4. Schematic of the AMS spectrometer at LLNL. For PXAMS measurements, the ions are detected and identified via characteristic x rays (see also Fig. 5).

FIG. 5. Detection and identification of ions via characteristic projectile x-rays (PXAMS). Following analysis in the AMS spectrometer (Fig. 4), the ions are incident on a Zr foil, including characteristic x-rays which are detected with a HPGe detector. Additional absorbers (Be and mylar) are included to stop the ions and attenuate low-energy x-rays.

FIG. 6. ^{43}Ni PXAMS x-ray spectra for a ^{63}Ni standard (^{63}Ni/Ni = 1×10^4) and a blank. 99 MeV 10+ ions were incident on a 2.0 mg/cm^2 Zr foil. For these samples, the copper concentration was ^{63}Cu/Ni = 1×10^2.

FIG. 7. Apparatus for generation of $Ni(CO)_4$. Sample solution containing Ni is placed in the reactor. A mixture of He and CO is bubbled through the reactor. BH_4^- is added, reducing Ni^{1+} to Ni^0, which reacts with CO producing $Ni(CO)_4$. The $Ni(CO)_4$ condenses on allanized quartz wool in the LN_2 trap. After completion of the reaction, the trap

FIGURE 8.4 ***(Continued)***

is warmed and the $Ni(CO)_4$ is transported with He to the AMS sample holder for thermal decomposition.

FIG. 8. Thermal decomposition of $Ni(CO)_4$ in target holder. $Ni(CO)_4$ is carried in helium to the sample holder, where it is thermally decomposed to Ni.

FIG. 9. Measured vs. expected ^{43}Ni contends in the demonstration experiment. The results demonstrate the accuracy and sensitivity of the methods. Wires were irradiated with fast neutrons from a ^{252}Cf neuron. Fast neutron fluences corresponded to ~1.4 and ~14 times the estimated fast neutron fluence near the hypocenter of the Hiroshima blast. The estimated fluence is from Ref. 15, which is based on the D586 evaluation (Ref. 13). Crosses: ^{43}Ni standards and blanks. Open square: unirradiated wire. Filled square: 20 g wire irradiated with ~1.4 times the estimated Hiroshima fluence. Open triangles: wires of various masses irradiated with ~14 times the estimated Hiroshima fluence.

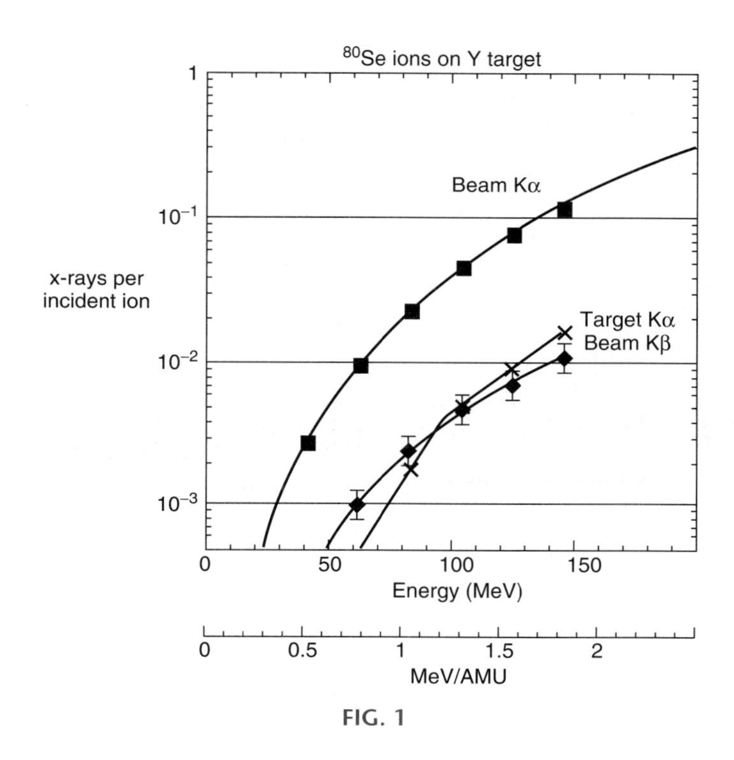

FIG. 1

FIGURE 8.4 (*Continued*)

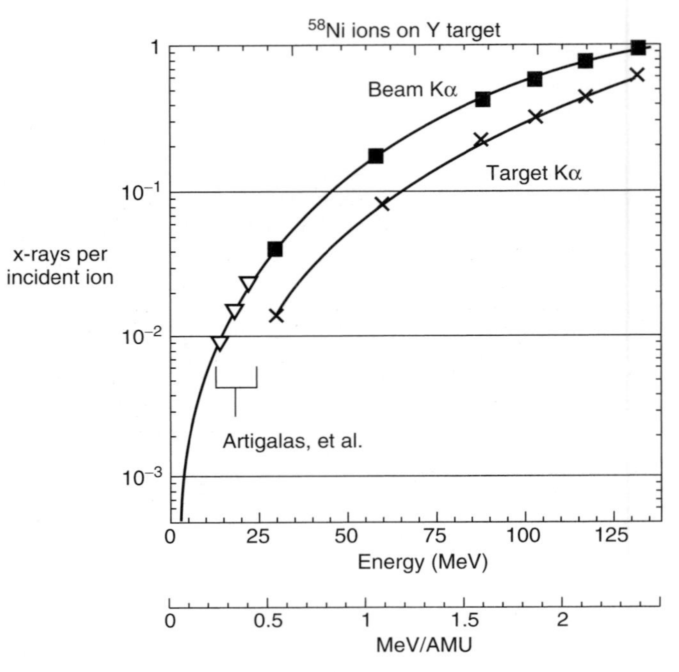

FIG. 2

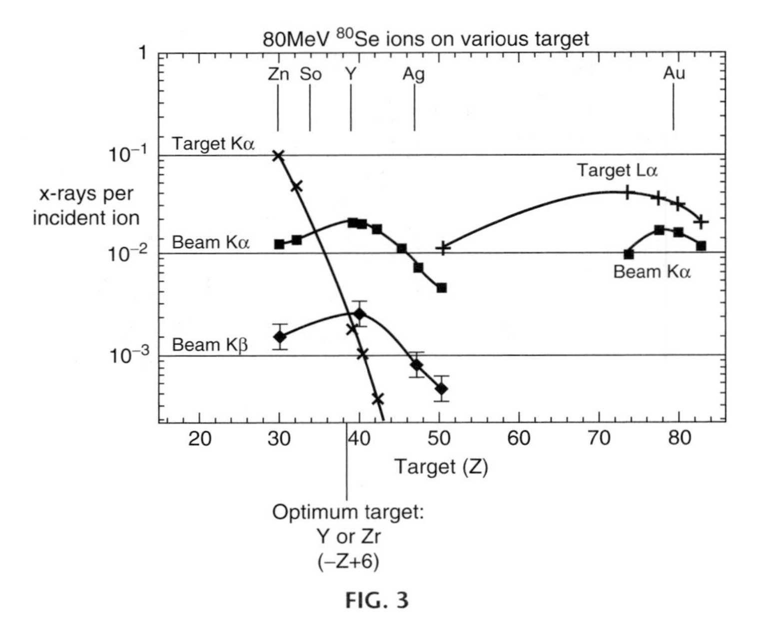

FIG. 3

FIGURE 8.4 (*Continued*)

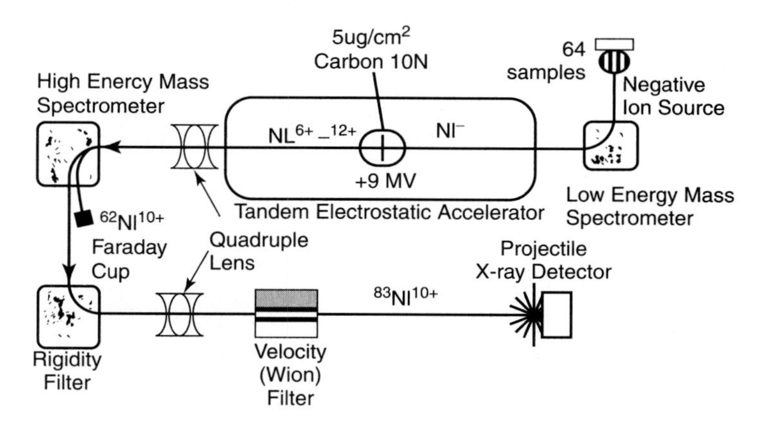

FIG. 4

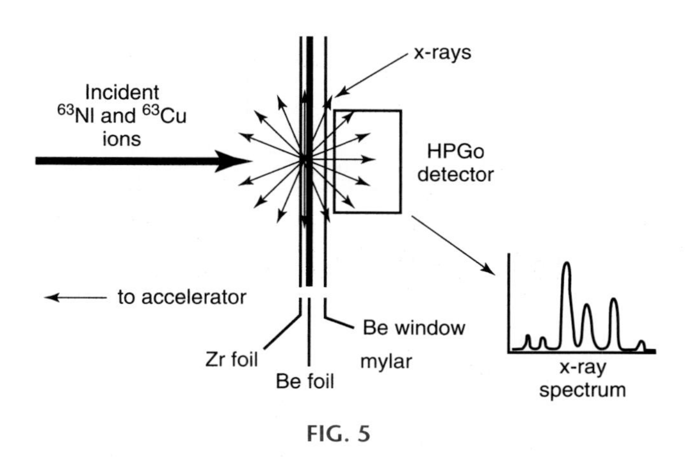

FIG. 5

FIGURE 8.4 (*Continued*)

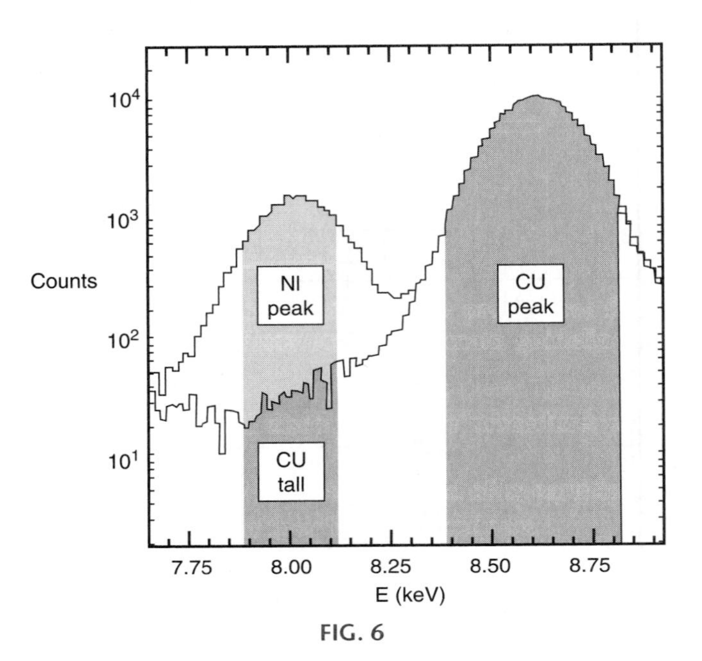

FIG. 6

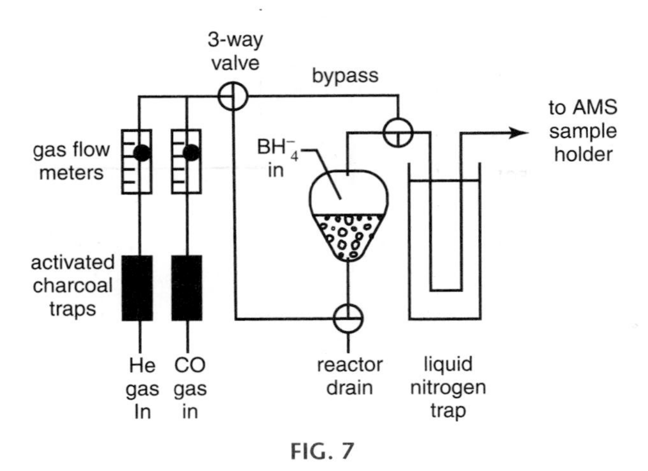

FIG. 7

FIGURE 8.4 (*Continued*)

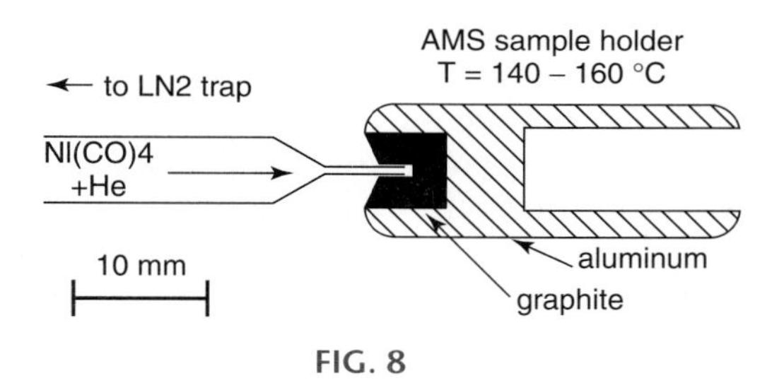

FIG. 8

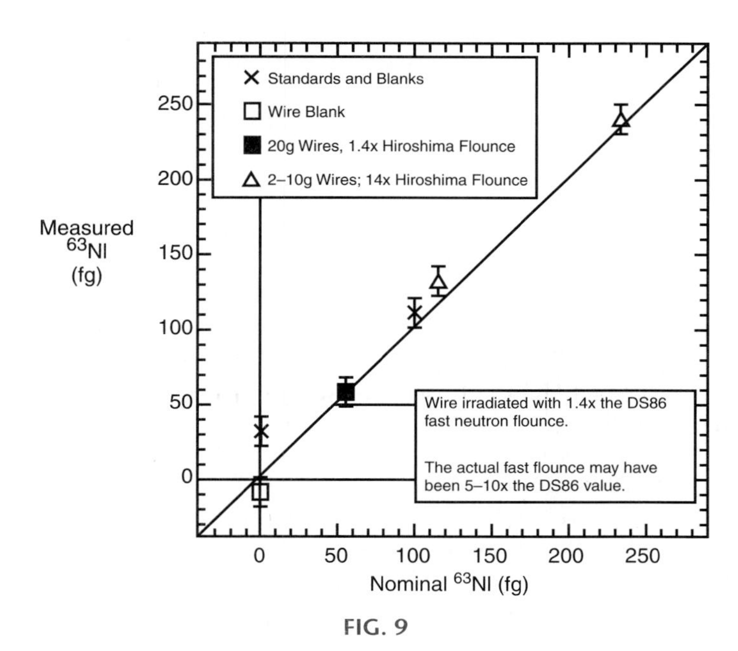

FIG. 9

FIGURE 8.4 (*Continued*)

EXERCISES

1. Investigate the possibility of adding a special interest class to the curriculum in your field. First, you'll need to obtain the necessary data on the requirements to add a new course in your department. Then, you'll need to do some exploratory work on the interest other students might have in such a course. You might write up a brief questionnaire to assess interest. Finally, after putting the procedure and interest assessment together, write a brief report to your English instructor on the feasibility of your idea. Consider your instructor a first-level reviewer; if you get the go-ahead from her or him, you might then submit your report to your department chair.

2. Investigate the possibility of establishing a new student service or facility, or enhancing an existing one: an after-dark transit or escort service for women, a counseling service for minorities, an up-to-date computer lab. As in exercise 1, you'll have to obtain the necessary data on the need for such a service or facility and then determine whether existing resources can provide or be expanded to accommodate the change you think is necessary. Your investigation can be preliminary, but it should be specific about what needs to be done and how it should be done, although the work may have to begin in the future. Keeping these points in mind, prepare a short report to an appropriate audience of students (perhaps members of the student government) or administrators.

FUNDAMENTALS OF SCIENTIFIC WRITING

Scientific reports, including lab reports, fall into the category of professional writing that we call science writing or scientific writing. There are many fine texts devoted solely to science writing, for example, Michael Alley's *The Craft of Scientific Writing*. Although the rhetorical choices the author has to make are the same as those in other types of professional writing, the conventions are very different. Scientific reports are dictated almost entirely by convention. Beyond any obvious advice for writing lab reports, such as to use precise information and provide all the data requested, there is not much general advice we can give. To explain the format-driven nature of lab reports, we will consider the evolution of science writing and its place in the field of professional writing.

Evolution of Scientific Writing

In the early days of scientific writing, the scientists told each other about their work through stories, little narratives that described the experiment chronologically. In the early volumes of *The Philosophical Transactions of the Royal*

Society, Sir Isaac Newton tells the story of his research on Opticks. He says, in effect, "I took me a prism and set it up in such and such a place to get such and such an angle of light."

By the nineteenth century, scientific discourse looked very different. Especially in the discourse of scientists addressing other scientists, the form became rigidly conventional. No longer could experimental work be reviewed in the leisurely form of a story, because there was so much more work than there had been two hundred years before. Instead, scientists used standard categories to describe their work. These categories, which predominate in science writing today, include method, conclusions, previous work, and so on. The use of a conventional format allows other scientists to skim work quickly so that they can stay current in their field.

Contemporary scientific writing is even more rigid than its nineteenth-century counterpart. One factor is the great increase in scientific research today. Another factor is the fierce competition for funds and reputation, which has made reproducibility, which was so important to Newton, secondary to other concerns, such as the review of literature and speculation about other necessary work. Such changes mean that science writers must familiarize themselves with the conventions that dominate both their field of expertise and the particular journal they want to publish in.

The effect of such rigid convention in science writing is not altogether constructive. A great deal of the writing is mindlessly passive. And increasingly, illustrations, photographs, and other graphics are expected to carry the weight of meaning. There is little attempt in some scientific publications to incorporate into the text the significant aspects of the graphics. Instead, the text seems to exist to orient the reader to the graphics. Although not all science journals have gone this way, others certainly have.

The Professional Writing Continuum

A look at the continuum of types of professional writing (see Table 8.1) shows the place of scientific writing. Some interesting things about professional writing as a whole are reflected in this continuum. First, as one moves from scientific writing to administrative writing, the audience becomes more varied and less expert. The audience for administrative writing is even more generalized than that for technical and professional writing. Although many different users must interpret most technical and professional reports, they are generally "insiders," whereas correspondence, the primary type of administrative writing, can also go to "outsiders": people who have nothing at all to do with the authoring organization.

Where style and format are concerned, science writing is by far the most conventional. One can find technical and professional reports in many arrangements and formats, but such variety cannot be found in scientific writing, for the reasons just discussed. If any change is to occur in science writing to challenge established conventions, it will probably come only from those scientists who have come to write well, who take an interest in the workings of language, and who will then understand the advantage of some flexibility in approach for rhetorical reasons. In technical and professional writing, there is

	SCIENTIFIC WRITING	TECHNICAL AND PROFESSIONAL WRITING	ADMINISTRATIVE WRITING
Document types	journal articles lab reports	technical reports procedures manuals feasibility studies proposals	letters memoranda brochures
Style Tone	conventionalized "objective"	varying styles somewhat objective	varying styles subjective

TABLE 8.1 **Continuum of Professional Writing**

some flexibility of arrangement and style, whereas in administrative writing, the only standards are the purely arbitrary ones of setting up letters and memos.

The level of objectivity also changes across the continuum. On the scientific writing end, objectivity is at a premium, the notion being that the investigator will draw conclusions from objective data instead of from personal opinion or prejudice. Thus, there is much passive construction in science writing; the scientist is trying to emphasize the objective nature of the work, performed by accepted protocols. Yet all scientists acknowledge the importance of the observer, who reads measuring instruments, sees reality from an individual orientation, and interprets, no matter how objectively, as a human being.

The continuum shows that objectivity is partly in evidence in technical and professional writing but is not at all the goal of administrative writing. In other words, as one moves across the continuum, the writing becomes no longer a matter of presenting facts to an audience, such as one might find through experimentation; instead, the writer is trying to persuade the reader to accept the view of reality presented in the particular report. Proposals are at the more obviously persuasive end of technical and professional documents, and cover letters and brochures are obviously persuasive. In fact, at the administrative end of the continuum, subjectivity—a focus on the person behind the message—is a strength, something to be sought for in the writing. Thus, a job application that paints a vivid picture of the writer is admirable, not unprofessional.

Technical and professional writing as a whole used to be defined as we, in this text, define science writing. Generally, people thought that all the documents produced in the workplace were about the serious, objective work of engineers, businesspeople, bankers—all the standard professions associated with the work that professionals do. This work was all about numbers and costs and schedules and other objective parameters, that is, measurable parameters. The current notion is more sophisticated. Professionals understand that the objective data do not speak for themselves. They must be framed in a context, within a purpose, and for an audience with expectations and logical requirements. In other words, these data must be conveyed effectively; to accomplish this goal, the writers use rhetorical techniques to persuade the audience that certain "facts" make sense and certain conclusions are more or less reasonable. One can see, then, that it is a mistake to equate technical and professional writing with total objectivity or with rigid format conventions. As a matter of fact, the writing that best fits into this old-fashioned notion of technical and professional writing is scientific writing, a small part of the whole professional writing field.

FORMAT AND SUBSTANCE OF A SCIENTIFIC REPORT

For the scientific report, then, our discussion is descriptive instead of prescriptive. One type of scientific report is the lab report, usually an in-house document that reports interim findings informally, before they are written up as a formal scientific report. Thus, an engineer will record in a lab report the preliminary data from an experiment in progress. But this report makes few demands on the writer, since it usually is a sort of fill-in-the-blanks report that requires the writer to transfer essential information from personal notes.

The scientific report is simply a more thorough, formal documentation of findings arising from a number of related lab reports. The scientific report adds the review of literature section, to situate the work in significant other work in the field; an abstract (since it would be addressed to outsiders, other scientists at other facilities); formal citations; and suggestions for further investigation. One pitfall of the scientific report's abstract is that it may simply be a tease, telling the reader what the article is supposed to be about, instead of reporting very concisely what conclusion the researchers came to. So, again, the concern is to be conclusive, to know how to write the main report clearly, knowing that the point is not to string the reader along, but to respect the harried reader's limited time and attention span where professional reading is concerned.

The best general advice we can give to science writers is to follow the rhetorical principles in this book, while writing within the conventions that dominate the field, understanding that these conventions are not written in stone; many of them are relatively new rules but are subscribed to as if they were the only way one could write well about science. In the range of professional documentation, science documentation is the most rigidly prescribed, partly because the audience wants certain essential information from the writing and partly because the most appropriate language for relaying scientific information is not discursive but symbolic. Mathematicians and musicians are not expected to convert all their symbols into prose any more than scientists are expected to convert their symbols into prose. Such symbolism is appropriate only for an expert audience, and most scientists understand this matter.

SAMPLE STUDENT-AUTHORED SCIENTIFIC ARTICLE

Authored by a chemistry student in a technical and professional writing class, the accompanying sample (see Figure 8.5) is an award-winner that translates the student's field notes into a professional-level article that was submitted to a competition for chemistry undergraduates. This student had published before in his field and so was accustomed to writing to professional conventions.

Jon Sutter
2534 S. Fairway
Pocatello, Id. 83201
April 24, 1991

Professor Montgomery
Department of English
Idaho State University
Pocatello, Id. 83207

Dear Professor Montgomery:

Enclosed is a copy of my final project for the Spring semester of English 307. As you may recall, the title of my project is "Convergence of Generalized Simulated Annealing with a Variable Step Size for Function Optimization." The paper was written for a group of PhDs in Chemistry (namely the Kent State Undergraduate Research Symposium reviewers). The paper will discuss two major topics. These topics are generalized simulated annealing (GSA) and generalized simulated annealing with a variable step size (GSAVS).

It is necessary to discuss GSA since GSAVS is a modification of this well known optimization technique. Dr. Kalivas and I created GSAVS and it is my intention to show that GSAVS is able to converge to exact global conditions, whereas GSA is not. A comparison will be drawn between GSA and GSAVS to show that this is true. Also, GSAVS will be used to perform a chemical analysis (predict concentrations of an unknown sample) to present some of the possible uses.

The data used in this paper were all taken from experiments performed by me last summer. I also did a great deal of research on this topic to support claims that were not achieved through my research. I consulted Dr. John Kalivas, an expert in the field of chemometrics, to explain the importance of this research and the long term goals of GSAVS. A combination of all three are represented in this paper.

Sincerely,

Jon Sutter

Jon Sutter

Enclosure JS/js

FIGURE 8.5 **Student scientific article**

Executive Summary:

The need to optimize chemical processes is often a must in the field of chemistry. Over the years many approaches have been investigated, some with limiting features that render them impractical. After much research generalized simulated annealing (GSA) was found to be the best in that it is able to find the area of the best conditions; therefore generalized simulated annealing with a variable step size was created to find the best conditions, which is the discovery through research presented in the paper.

GSA is a computer program that requires user assistance. First, the user must pick a set of parameters s (the initial conditions such as time, temperature and concentration of reactants of a chemical reaction). GSA uses the parameters to calculate the cost function (the numerical answer of a mathematical equation that represents the situation being optimized, using the parameters as the variables). GSA then randomly selects a new set of parameters in the neighborhood of the previous parameters, calculates a new cost function, and compares the two cost functions. If the new cost function is better than the old cost function, GSA accepts this new step and continues by selecting new parameters. If the new cost function is worse than the old, GSA may accept the step based on a probabi1ity number P. GSA is able to find the near global (best) conditions due to the fact that it accepts some detrimental (bad) steps. The equation of the probability of accepting a detrimental step is based on Boltzmann's distribution and is represented as:

$$P = \exp(-\beta\Delta(crs)/C(s) - C(o))$$

This equation is based on the differences in the last two cost functions and also the difference between the new cost function and the best cost function (predicted by the user). Beta is a constant adjusted by the user to keep P in a range of values that will allow GSA to work for different experiments (p must be between 0.5 and 0.9 at first). If P is greater than a random number from 0 to I then the detrimental step is accepted. Theoretically, the P values should go to zero when GSA is close to the best conditions since the difference between the current cost function and the predicted best cost function will be small. This will cause detrimental steps to be rejected and GSA to converge to global conditions. Unfortunately, the step size is often too large and GSA steps over the exact global. It is at this point that GSAVS will automatically shrink the step size and converge to the exact global conditions. GSAVS adjusts step sizes based on a number resulting from the number of steps accepted divided by the total number of steps in a certain block of steps. Since this number is small when the P values are at zero, GSAVS shrinks the steps.

Three major things had to be determined in order to make GSAVS work. First, through trial and error, a block of 20 steps was used to calculate the acceptance ratio. Second, beta was adjusted until the P values were between 0.2 and 0.7 so that a forced increase of the step size was not induced. Third, beta was adjusted after every step size change to keep the P values consistent. The adjustment is based on a

FIGURE 8.5 (*Continued*)

ratio of the old beta and the corresponding difference in cost functions and the current difference in the cost functions.

GSAVS was used to optimize two situations, a mathematical equation and a chemical analysis. First, GSA and GSAVS were both used on a mathematical equation starting from points that should give no difficulties for either technique. Here it was evident that GSAVS was able to converge to exact conditions where GSA was not. A figure of the response surface was supplied to show the readers the difficult starting conditions (close to global conditions and a local optimum). GSAVS was started at these difficult areas and was still able to converge to global conditions. Second, GSAVS was used to perform a chemical analysis. A figure of the curves are presented in a figure. The assumption here was that the reader would know that spectral overlap would cause difficult selection of concentrations by any technique. This is not an unreasonable assumption since the readers are all chemistry PhDs (Kent State Undergraduate Research Symposium reviewers). Despite the spectral overlap, GSAVS was able to find the concentrations of the unknowns. Noise was then added to this spectra and GSAVS was compared to a well known technique of predicting concentrations (the K-matrix approach). Here it was shown that GSAVS did a better job at predicting concentrations in areas of spectral overlap than the K-matrix approach.

In conclusion, when global conditions are absolutely needed, GSAVS is a technique that can find them.

This paper was written under the assumption that the readers were well versed in chemical procedures, but not in the field of chemometrics. Therefore, a great deal of emphasis was placed on the description of how the optimization techniques worked. The paper was written in accordance to the stipulations created by the reviewers of the Kent State Undergraduate Symposium and also technical papers written in the field of chemistry.

Each entry had to include a short abstract, an introduction section, an experimental section, a results and discussion section, and a works cited section which had to be presented in order of appearance in the paper. Since the reviewers were interested in the student's contribution to the project, I wanted to make this come through clearly in the paper. Therefore, I tried to use simple language that would make the paper clear.

Also, in regards to the experimental section, the paragraphs don't contain topic sentences since this is not standard procedure of the papers written in chemistry. The experimental sections are usually short sentences that simply state what kinds of instruments were used and other technical details that would help to recreate data. Tables and figures were included so that the reviewers could more easily visualize details of GSAVS. I tried not to use passive tense, but it was difficult not to in some areas without using first person.

FIGURE 8.5 (*Continued*)

CONVERGENCE OF GENERALIZED SIMULATED ANNEALING WITH A VARIABLE STEP SIZE FOR FUNCTION OPTIMIZATION

Jon Sutter
Dr. Montgomery
Technical Writing

FIGURE 8.5 ***(Continued)***

Introduction

The need to optimize chemical processes is often a must in the field of chemistry. Over the years many different optimization techniques have been developed. Unfortunately, research has shown that many of them have limited uses (1-4). The limiting factor can be the amount of time required to perform the optimization or simply that the technique converges to a local optimum rather than the global optimum. Whatever the reason, often the optimization technique is considered useless for certain situations.

After much research in the area of Chemometrics. an optimization technique called generalized simulated annealing (GSA) has been found to be the best (4-8). GSA does not require much time to perform an optimization, and it is almost guaranteed to find the area of the global optimum.

This paper is concerned with a modification of GSA that will allow convergence to the exact global optimum rather than the area of the global optimum. The modification is the addition of a variable step size. Thus the technique will be referred to as generalized simulated annealing with a variable step size (GSAVS). Since the paper is concerned with GSAVS, only a brief description of GSA will be offered. For a more detailed description of GSA refer to reference 5. For a comparison of GSA and another optimization technique (simplex) refer to reference 4.

For the sake of clarity, it is necessary to give a qualitative description of how GSA works and how it differs from other optimization techniques before describing the quantitative features of GSA. GSA has been suitably called a semi-random walk, because it is able to walk off of local optima and onto the global optimum due to the fact that, unlike other optimization techniques, it will accept some detrimental steps. To begin GSA the user must pick a set of parameters; GSA then computes a cost function C(s) using these parameters, takes a random step from s and obtains the set of parameters r, computes a cost function C(r) based on the set r, and finally compares the two cost functions C(s) and C(r). In the case of a minimization, GSA accepts C(r) if it is less than C(s), and it accepts C(r) if it is greater than C(s) based on a calculated probability P. This procedure is repeated until the area of the global optimum is discovered. Whenever cost functions are compared it will be referred to as a step or iteration in this paper. A more detailed description of GSA will now be given.

Quantitative Description of GSA

Through research, GSA was developed by Bohachevsky (5). It is a variation of simulated annealing which is related to the annealing process of a solid. The reason this technique is able to walk off of a local optimum and locate the area of the global optimum stems from a statistical equation that is derived from the Boltzmann's distribution. This equation is:

$$P = \exp(-\beta\Delta C(rs)/C(s) - C(o))$$

FIGURE 8.5 (*Continued*)

where β is a constant adjusted by the user, ΔC(rs) – C(r) – C(s), C(r) is a randomly selected cost function, C(s) is the last computed cost function, and C(o) is the predicted optimal cost function selected by the user.

This equation is used to determine whether a detrimental step will be accepted or not. In the case of a minimization, when C(r) (the selection of the set r will be discussed later) is greater than C(s). GSA has taken a detrimental step. Most optimization techniques would not accept this step, which will allow them to converge to a local optimum in certain situations. GSA, however, will accept this step if P is greater than a random number that is taken from a uniform distribution from zero to one. This makes GSA a semi-random optimization technique and thus allows it to walk off of local minima and into the global area.

This process of accepting the good steps and accepting detrimental steps based on the P value is continued until GSA finds the global area. When GSA finds the global area the P values will go to zero because C(s) will be close to C(o) . and the detrimental steps will be rejected. This allows GSA to find and converge on the global area.

In order to find C(r) it is necessary to find a random set of parameters r. These random parameters are found by the equation:

$$r = s + \Delta r \times y$$

where s is the set of the previous parameters used to compute the cost function C(s), Δr is a step size chosen by the user, and y is an array of normalized random numbers selected from a normal distribution with mean zero and standard deviation one.

The step size is the main discussion of this paper. The problem is that if the step size is too large, GSA will often miss the global optimum by simply jumping over it. Many times it is necessary to perform GSA over and over, reducing the step size each time in order to find the global optimum. This procedure may be too time consuming for all practical purposes. GSA with a variable step size should solve this problem.

This paper is based on a variable step size that is calculated by use of an acceptance ratio. If more than 90% of the steps in a block of GSA steps are accepted it means that GSA is far from the global and therefore the step size should be increased. If less than 20% of the steps in a block of GSA steps are accepted it means that GSA is close to the global and therefore the step size should be decreased. The incorporation of variable step size with GSA will be discussed in detail later.

Experimental

All computer programs were written in Fortran 77. The computers used were all 100% IBM compatible.

Simulated data was constructed by Gaussian curves with a range of 1–100. The peaks have bandwidths of 5 and amplitudes of either 10 or 5.

The responses of real data were taken from a UV-Vi s spectrophotometer.

For GSA coupled with the variable step size, ß was chosen by doing 100 iterations and adjusting ß until the average of the P values

FIGURE 8.5 (*Continued*)

was approximately 0.4. If the optimal cost function was not known an estimate was used. If GSAVS found a cost function below the estimated cost function a lower value was inserted. There is no need to readjust ß after the new minimum is selected, since ß is automatically adjusted after each step size change (this will be discussed further in the results and discussion section of this paper.)

Results and Discussion

In order to use the acceptance ratio variable step size with GSA, three major changes had to be made to GSA. They are general changes that work for every case tried in this paper. They were used in every analysis, unless otherwise specified.

First, the acceptance ratio was calculated at the right amount of steps in order to make GSAVS work. The calculation had to involve enough GSA iterations, so that the changes needed to be made to the step size could be determined. The calculation also had to be made after a small enough number of GSA iterations so that the procedure would not take an unnecessary amount of time. After much trial and error, it was discovered that the block of GSA iterations before an acceptance ratio calculation should be 20. An iteration block of 20 seemed to give enough information so that GSAVS could make an acceptable adjustment of the step size by use of the calculation of the acceptance ratio without wasting time with unnecessary iterations.

Second, the initial ß had to be adjusted so that GSAVS would work. If the suggested ß of GSA (such that P was between 0.5 and 0.9) (5) was used, the step size would always increase after the first acceptance ratio calculation. The reason for this was that ß was adjusted so that the P values were between 0.5 and 0.9, which caused about 80% of the first set of detrimental steps to be accepted. This would almost guarantee that the acceptance ratio would be greater than 90% and therefore increase the step size. Since an increase in the step size is not always necessary on the first acceptance ratio, something had to be done with the ß value. It was discovered that the best thing to do was to adjust ß so that the P values were between 0.2 and 0.7. This would reduce the chance of a forced increase in step size, while still allowing GSAVS to walk off of a local optimum.

The third and final change that was made to GSA was an automatic adjustment of ß. This change was necessary because when the step size changed it would sometimes considerably change the difference between C(r) and C(s) . This difference is a major factor in the calculation of the P value. Since the P value plays a very important role in the *GSA* technique, something needed to be done to preserve the consistency of the P value. Because the inconsistency in the P value was caused by the difference of C(r) and C(s) the following equation was derived:

$$\text{New } ß = \text{FDIF}/\text{DIF} \times \text{Old } ß$$

where *FDIF is* the average of the difference in C(r) and C(s) in the first three iterations. *DIF is* the current difference in C(r) and C(s), 01dß is the first ß value that is given by the user (this is the ß that is first used so that the first P values are between 0.2 and 0.7), and New ß is the ß

FIGURE 8.5 (*Continued*)

used after the step size change. New ß is calculated each time the step size is changed.

First, *GSAVS* was used to optimize a simple math equation. This optimization was compared with an optimization using *GSA*. The math equation optimized was:

$$R = X^2 + 2Y^2 - 0.3 \cos(3\pi x) - 0.4 \cos(4\pi Y) + 0.7$$

where X and Y are the parameters and *R* is the cost function. In this case *R* is minimized and the global optimum is *R* – *0.0* and (X, Y) – (0,0). The results are presented in Table 1.

OPTIMIZATION TECHNIQUE	INITIAL STEP SIZE	INITIAL X, Y	FINAL X, Y	FINAL R	NUMBER OF STEPS
GSAVS	0.15	0.85 0.85	0.000 0.000	0.00	2204
GSA	0.15	0.85 0.85	0.058 0.033	0.08	105

TABLE 1. **GSA and GSAVS Optimization of a Mathematical Function**

For this minimization, ß was adjusted so that the P values were between 0.5 and 0.9 for the case of *GSA* alone and 0.2 and 0.7 for *GSAVS*. Notice that the exact global was found using *GSAVS* and only the global area was found with *GSA* alone.

Figure 1 is a 3-0 plot of the response surface of this math function. Notice that at the starting points 0.85 and 0.85 the response is not on a local minimum. The optimization was started on a local minimum and also in the global area to see if *GSAVS* would converge to the global. The results are presented in Table 2.

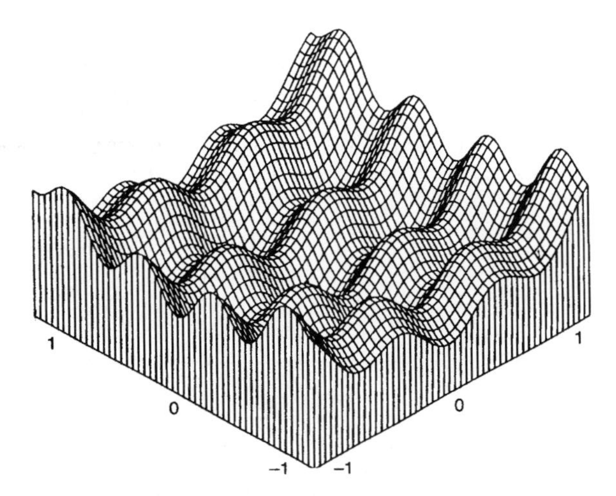

FIGURE 1.

FIGURE 8.5 ***(Continued)***

OPTIMIZATION TECHNIQUE	INITIAL STEP SIZE	INITIAL X, Y	FINAL X, Y	FINAL R	NUMBER OF STEPS
GSAVS	0.15	−0.5 −0.5	9.3E-7 4.7E-7	1.3E-12	2857
GSAVS	0.15	0.003 0.003	3.7E-8 4.0E-7	3.2E-13	4757

TABLE 2. GSAVS Optimization of a Math Function Starting on a Local Minimum

Table 2 shows that GSAVS is able to converge to the global from a local minimum and from the global area.

The final step size in these optimizations was decreased to the precision of the machine. During the last few iterations of GSAVS C(s) – C(r), which means that the step size is so small that it doesn't move anywhere when finding the random set of parameters.

Finally, GSAVS was used to perform a chemical analysis. The cost function was calculated by the following equation:

$$I \text{ rs} - cKI$$

where rs is the vector of instrumental response for the unknown sample, c is the vector of the unknown concentrations (the parameters of the optimization), and K is the K-matrix derived from Beer's law. K is calculated by dividing the concentrations of the pure component spectra into their corresponding instrumental responses. The bars around the equation indicate that the vector norm is taken.

First an optimization was performed using simulated UV-VIS data. The Guassian curves of a simulated two component system are represented in Figure 2. The analysis was performed using select wavelengths and all wavelengths. The results are presented in Table 3.

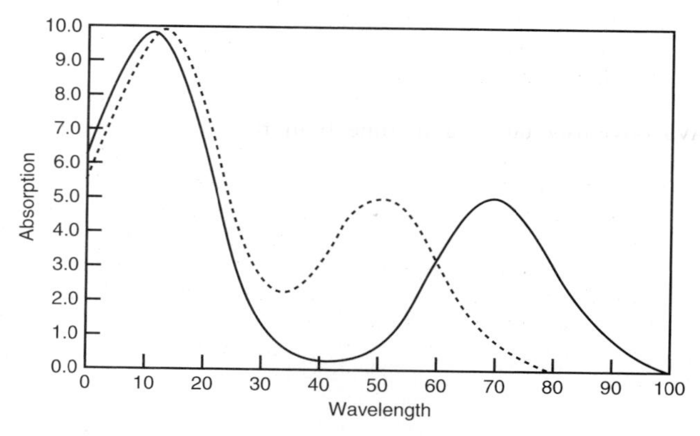

FIGURE 2.

FIGURE 8.5 ***(Continued)***

λ'S	INITIAL CONC.	FINAL CONC.	RELATIVE ERRORS*
ALL	0.001	0.5	0.0%
	0.001	0.5	0.0%
10.70	0.001	0.5	0.0%
	0.001	0.5	0.0%
1–30	0.01	0.5	0.0%
	0.01	0.5	0.0%
35–100	0.01	0.5	0.0%
	0.01	0.5	0.0%

*Relative errors were calculated by dividing the difference of the predicted concentration and the true concentration by the true concentration.

TABLE 3. **GSAVS Optimization of Simulated Chemical Predictions without Noise**

Notice that it does not make a difference if there is spectral overlap or not. GSAVS is able to find the global optimum. Because of this, the chemical analyses from this point on were performed using all wavelengths.

Noise was added to these data and the optimization of the chemical analysis was performed. Three and five percent relative noise was used because it should illustrate the worst case scenario. The results were compared to the classical least-squares approach (also known as the K-matrix approach. These results are presented in Table 4.

Note that the GSAVS gives better estimates at higher degrees of noise. The two components have different degrees of errors in all cases. One component consistently has a higher relative error than the other component due to spectral overlap. Note that GSAVS is not affected by the spectral overlap as much as the K-matrix approach seems to be (1.5% vs. 4.1% relative error). This suggests that GSAVS can be an effective tool for many different situations.

This study seems useless from a practical point of view because GSAVS obviously takes more time than the K-matrix approach. It seems pointless to do this analysis using GSAVS because of this. Actually, this is a precursor to other research ideas, such as analyzing for a single component in a mixture without knowledge of what the other components are. Obviously, the K-matrix approach could not solve this problem. The study presented in Table 4 was performed in order to test the validity of GSAVS for such problems as chemical analysis. It was assumed that if GSAVS did not work on this problem, it wouldn't be able to solve more complicated problems such as the one component analysis.

FIGURE 8.5 (*Continued*)

ANALYSIS	RSD	ESTIMATED CONC.	REL. ERROR
GSAVS	0.03	0.0997	0.3%
		0.5011	0.22%
GSAVS	0.05	0.0985	1.5%
		0.5013	0.26%
K-matrix	0.03	0.1010	1.0%
		0.4989	0.22%
K-matrix	0.05	0.1041	4.1%
		0.4965	0.7%

TABLE 4. **Optimization and CLS Results for Simulated Chemical Predictions with 3% and 5% Relative Noise**

In summary, GSA is an effective tool to use for global optimizations. Unfortunately, GSA does not always find the exact global optimum. Using the acceptance ratio to adjust step sizes accordingly allows GSAVS to converge to the exact global optimum. According to certain experts in the field of Chemometrics (namely Dr. Kalivas), global conditions are not always necessary or feasible; but it is good to know that if global conditions are needed, GSAVS exists to find them.

References

(1) Edgar, T.F.; Himmelblau, D.M. Optimization of Chemical Processing: McGraw-Hill: New York, 1998.

(2) Bayne, Charles K.; Rubin Ira B. Practical Experimental Designs and Optimization Methods for Chemists; VCH: Deerfield Beach, FL, 1986.

(3) Box, George, E.P.; Draper, Norman R. Empirical Model-Building and Response Surfaces; Wiley: New York, 1987.

(4) Kalivas, John H.; Roberts, Nancy; Sutter, Jon M. Anal. Chem. 1989, 61, 2024.

(5) Bohachevsky, Ihor O.; Johnson, Mark E.; Stein, Myron L. Technometrics 1986, 28, 209.

(6) van Laarhoven, P.J.M.; Aarts, E.H.L. Simulated Annealing: Theory and Applications; D. Reidel Publishing: Dordrecht, 1988.

(7) Vanderbilt, David; Louie, Streven G.J. Comput. Phys. 1984, 36, 259.

(8) Johnson, Mark E., Ed. Am. J. Math. Manag. Sci. 1988, 8 (3 & 4).

FIGURE 8.5 ***(Continued)***

In his executive summary, you will note the attention to details of paragraph structure, for example, which he adjusted appropriately for his audience. You will also note that he understands the loss of coherence that he risks by adhering to the convention of not using topic sentences for paragraphs in the experimental section. The other sections of his article are very well signaled by topic sentence and his abstract is conclusive. As a student, he is in no position to challenge the conventions of publishing in chemistry; what is important is that he understands the place of conventions in his field and that he has the flexibility to write both for chemistry professionals and his English professor, to whom he addressed his executive summary.

Finally, this sample clarifies our earlier point about the need for scientists to use symbolic language. At certain points, this student had to start writing in formulas because they expressed most clearly and accurately what he had to say. He simply adhered to the conventions of including formulas in a text; note his spacing and standard discussion of the elements in the formulas. This symbolic writing is absolutely appropriate to his audience, specialists in chemistry. It would be a big mistake, however, for him to write in formulas to a group of lay readers.

EXERCISES

1. Analyze the sample article for its effectiveness. You don't have to be a chemistry major to do some basic rhetorical analysis. You can compare the author's stated purpose and audience as described in the executive summary to the level of language, thoroughness of coverage, and degree of explicitness in the article.
2. Analyze the difference in publishing conventions among *Scientific American, Science,* and the *Journal of the American Medical Association* (JAMA). How would you describe their audiences, their purposes, their structures, and their formal conventions?

Solving Problems Through Policy Statements, Manuals, and Procedures

DISTINGUISHING AMONG POLICY STATEMENTS, MANUALS, AND PROCEDURES

There are real distinctions among policy statements, manuals, and procedures, as document types. These distinctions, however, are inconsistently applied within the professional world. The following discussions point out the distinctions among the three types of documents, considering them in a hierarchy by level of detail, beginning with the most general, the policy statement.

Policy Statements

Policy statements are extremely general rules. As rules, they have very general application. "Work shall be done safely" and "Decisions shall not be affected by race, creed, national origin, or gender" are examples of such general rules. Policy statements do not account for their own implications; if a policy is not well implemented, one cannot find the way to implement it within the language of the policy itself. The Ten Commandments provide a good example of some clear policy statements, but what happens to those who break them is a matter of speculation. And exactly how we are to honor our parents is not prescribed, as it would be in a procedure on the subject.

PREVIEW OF THE CHAPTER

Manuals

Manuals are descriptive documents. They present more specific information than policy statements do about narrower topics. Thus, manuals describing the duties of a certain position are common. Manuals may also describe the function of a group as a whole or of a piece of equipment. For example, the manual that comes with a personal computer describes how to put the equipment together and how to maintain it. Since initial start-up is a one-time activity, we don't consider this set of instructions in the manual to be procedural.

An important function of a manual is to document, that is, to record officially the status quo of the topic in question. Such documents as federal government facility descriptions and site plans fall within this category. These documents officially record information of a specific, descriptive sort, for example, a detailed description of all the buildings on a particular government site. Such a document will be used by planners and assessors.

The important point about manuals is that the descriptive material they contain does not provide the necessary level of detail to affect employee actions in specific situations. For example, a manual will contain a general description of the duties of an administrative assistant, but not a daily prescription of those duties. How to carry out these duties will depend on various circumstances that a manual doesn't intend to account for. So there is a great deal of flexibility in the use of manual descriptions. Unfortunately, most people think of procedures and manuals interchangeably.

Procedures

Procedures provide step-by-step instructions for performing a specific activity repeatedly. If written well, a procedure will ensure that the activity proceeds uniformly no matter who performs it. Thus, banks have set up procedures for using their cash machines. If people try to reverse the order of the procedures, they will not get any money. The procedure is absolute; and it must be so, because it has serious financial implications.

Besides ensuring uniformity, procedures should be regarded as a logical test of the activity concerned. Generally, however, the format of procedures tends to dominate their logical basis. People are so intimidated by procedures, knowing how seriously professionals take them, that they are tempted to regard them as above criticism, above analysis. This awe, though, is finally harmful and encourages the perpetuation of mindless activity.

Because procedures are so serious, sometimes with life-or-death consequences, they should be written sparingly. If readers are to focus intensely on procedures, writers need to ensure that the topic of the procedure deserves such attention. Thus, a procedure for miners to contact each other underground certainly seems necessary. But a procedure for miners to shower after their shifts is unnecessary, unless there is a safety consideration to such showering, for example, the need for miners in uranium mines to rinse off radioactive particles. Procedures on trivial subjects seem like opportunities for those with a bit of power to assert themselves to the many.

	POLICY STATEMENTS	MANUALS	PROCEDURES
Level of generality	High	Moderate	Low
Common forms	Rules Principles	Information Description	Directions
Purpose	Assert image	Provide guidelines	Specify actions
Functions	Identity	Reference	Performance

TABLE 9.1 **Distinguishing among Policy Statements, Manuals, and Procedures**

A good question to ask in order to decide whether a procedure is called for is this: does this topic have any safety, financial, or quality (especially, ethical) ramifications? If not, then the topic does not need to be turned into a procedure. If so, then it should definitely be written up as a procedure. And every procedure reader should be a troubleshooter, the role that Rick chose for himself after reading the classroom disruption procedure. Table 9.1 summarizes the distinctions among policy statements, manuals, and procedures.

CASE STUDY

RUTH WARNER'S OFFICE

Public School District 16 ▪ Administration Headquarters ▪ Elko, Nevada

Ruth Warner began to realize just how big this problem of the new reader was going to be for the sixteenth district. Recently the Elko School Board had approved a new reader for sixth-graders on civics, or embryonic political science. Our Global Family *was the latest book for sixth-graders and had the virtue of incorporating the end of the Cold War in its last chapter. As the elementary curriculum coordinator, Ruth had been especially pleased with this text chapter.*

Not everyone in Elko shared Ruth's enthusiasm, however. To her horror, she found that church groups and civic organizations were rising up in indignant protest over the text choice. They expressed their dissatisfaction in various ways, almost all of them uncivilly. Ruth received reports of their activities every day. One group sent a mass mailing to a sixth-grade teacher who was using the book. Another group went directly to the district superintendent, bypassing the classroom teacher, department head, principal, and several layers of administration in between.

Some individuals transferred their children to other schools. Others directed petitions to neighbors and students.

Ruth knew that the superintendent could be swayed by such protesters, despite their lack of reasonableness and their mob-rule tactics. Ruth herself was outraged by these displays. She decided to meet with Jack Kelsey, the superintendent, to remind him of state-level support for the procedures that the school board followed. Then she affirmed this support at the state level. With such clear-cut support from important sources, Ruth decided to seize the advantage. She would write a policy statement on what the school administration regarded as appropriate channels of protest. Ruth was determined to cut off the chaos of ad hoc protests. The protesters' energy, if channeled well, could be constructive; as it was currently being acted out, she saw it as destructive and irresponsible, as well as a very poor example to all children in the school district.

Although it would be difficult to obtain agreement on a policy statement, because bureaucrats are loathe to take clear stands on policy, Ruth would do what she could. These protesters who were giving her such a headache were actually doing her a favor as well; they were giving her the opportunity to point out the need for a clear policy. She doubted she'd get much argument on that point.

RHETORICAL PROBLEM SOLVING FOR POLICY STATEMENTS, MANUALS, AND PROCEDURES

Although policy statements, manuals, and procedures require the same rhetorical problem-solving process as reports do, their major concerns are different. Because the discourse of these documents is essentially directive, persuading the audience is not often a major part of the rhetorical problem, as it is in discursive reports. The writer does not have to present persuasive reasons for the reader to follow medical protocols, for example, or have to justify the logic of the company organization chart in the new employees' manual. Thus, rhetorical problem solving for these documents takes on a different cast.

Instead of worrying about persuasion, the writer must be concerned with usability, verification, and logically sound reasoning. The clarity and accessibility of the documents are crucial. In fact, manuals and procedures are the documents that most clearly adhere to the old-fashioned notion of hard-core technical writing. They seem to dwell in the land of objective truth, objective data, and unambiguous writing, where a word means one thing and one thing only.

This old-school impression of technical writing, that it deals just with the facts, ma'am, is basically true of manuals and procedures, but with one qualification. The level of detail in manuals and procedures will determine the level of truth that the document communicates. If, for example, Linda Rogers, a building supervisor, is writing a procedure on conducting fire hazard inspections, she will have a choice of the level of detail to go into. She might make the procedure essentially administrative, focusing on the paperwork to be completed, the reports to be filed, the areas to be inspected, on the assumption that she need not mandate to a trained fire inspector how to look for hazards. If she takes this high road, though, which is built on the weak foundation of assumption, she is letting herself in for all sorts of liability. If the procedure she writes is too vague and general, it will contain logical gaps that require the reader to improvise, thus rendering the inspection almost ad hoc instead of standardized. If, on the other hand, she is detailed about the steps in the procedure, she will find that she is back to the impressionistic world of human disagreement over the proper method, the goal, and the proper responsible party. Manuals and procedures assign responsibility; they are therefore often very difficult to write well and to approve.

Yet, despite the vagaries of their composition, manuals and procedures are extremely important documents with very serious ramifications. Badly written procedures can slow down response in places like nuclear power plants. Poorly written manuals can sit on the shelf, forever unused, whereas if they had been written well, they would be reliable descriptions of the way business is done in an organization. If a large corporation has a poorly written comptroller's manual, for example, then the corporation will receive numerous variations on the annual report to headquarters. The professional world takes manuals and procedures very seriously; they are far more complex than the notion of a procedure as a "recipe."

Policy statements are more obviously persuasive than manuals and procedures because they are rules. But policy statements are absolutes, not arguments. They are more abstract, more general, and are intended to reach a more varied audience than are manuals and procedures. They may be intended for internal audiences, external audiences, or both. Also, they may be drafted differently for different audiences, a situation that can lead to an ethical problem.

The challenge of writing good policy statements is to turn them into meaningful prose; they tend to be written as motherhood statements, which assert uncontestable and extremely general values. For example, this policy statement appears in one form or another in several American industries: "Workers will work in accordance with safe work practices." As a policy, this statement is hardly trendsetting. It indicates that safety is a concern of the employer and that federal agencies that oversee workplace safety, such as the

Occupational Safety and Health Administration (OSHA), have a presence of some sort at the facility. OSHA might conduct regular inspections; the personnel office might be contacted for quarterly accident reports. What this policy statement really means is that the company that posts it obeys federal laws about occupational safety. Need any company post a policy that says, in effect, that it respects federal law?

Ruth Warner's challenge will be to draft a policy statement that the district's administrators will approve and that will still have teeth in it in order to deal with the book opponents. Policy statements are difficult to write because they are a company's internal laws, yet won't necessarily have any legal weight. Implementing a policy often requires extensive legal review.

For the purposes of this chapter, the policy statement is a helpful concept because it makes the function and intent of manuals and procedures clearer. The point to remember is that policy statements are general rules, not concrete directions for people to follow to accomplish specific activities. Therefore, Rick Allen, whose case appears later in this chapter, needs a procedure, not a policy statement, to direct him in his time of chaos. The policy statement is the document that requires thoughtful planning; the procedure should specify appropriate actions in specific circumstances.

GUIDELINES FOR WRITING AN ENFORCEABLE POLICY STATEMENT

It is not easy to write an enforceable policy statement, because so many legalistic decisions go into it and because so many people must take a stand on its content. It's very difficult to get people to agree about such statements, regarding either their content or their tone. Generally, though, the following guidelines are helpful:

1. State the policy as simply as possible. Too many qualifications will both obscure the content and water down the force of the policy statement.
2. State only policies that can be enforced. It damages an organization's credibility to state a policy that it has no power or will to enforce.
3. Maintain the identity of the policy statement by avoiding general descriptions and procedural detail.

The enforceability of a policy statement is a vital consideration. "We suggest that students attend 80 percent of their classes" is unenforceable, whereas "five or more absences result in an automatic F for the class" is specific and clear enough to be enforceable, as long as it is also legally possible. A more moderate statement of the absence policy might be "five or more absences constitute grounds for failing the course." Thus, the instructor would have some flexibility to judge individual circumstances. Most instructors would want to make a distinction between students who had missed class because of illness and those who had missed class because of a bowling date.

The difficulty in writing a policy statement, then, is not so much in drafting the statement, except for the preceding three rules, as it is in getting agreement on the statement.

SAMPLE POLICY STATEMENT

Ruth Warner needs a policy statement to bring some order to the chaos of the Elko textbook rebellion. She needs a policy that will establish a channel communicating discontent with School Board and District decisions. In this way those making decisions won't waste their time with ad hoc attacks and complaints. Fortunately for Ruth, she has the autonomy to draft the policy herself and then get it approved by the School Board and the District Superintendent, only four people.

After analyzing the problem, Ruth knows what teeth she's going to put in the policy before she even knows what policy to advocate. The teeth will be that any route for complaints other than the one described by School Board policy will lead nowhere: the School Board will hear only those complaints that come via the approved route. Thus, the consequence of ignoring the policy will be that one's complaint will not get official consideration. Although most people don't like to encounter this sort of bureaucratic response, large organizations must have orderly channels of communication.

Since Ruth intends to be so strict about this policy and since she knows she has an enforceable consequence, she decides to make the policy statement itself as simple as possible. She could certainly write her policy to stymie any protest at all, but she doesn't really want to stifle protest; she wants to organize it. She decides on this simple wording: "Objections to School Board policy must be made in writing to the District Superintendent."

Although the Superintendent has no objection to being the conduit for complaints about School Board policy, he does comment that Ruth will be lucky if her audience—the people of Elko—will know what kind of issue the School Board would rule on. He suggests that she somehow explain what topics would be covered by her policy statement. To solve this problem, Ruth decides to add a description of the scope of School Board decisions: "Such issues as scheduling, discipline, text selection, administrative hiring, and standardized testing come under School Board purview." "What are the citizens of Elko, Nevada, going to know from 'purview'?" asks the Superintendent; so Ruth makes a change in the policy wording: "The School Board rules on scheduling. Discipline, text selection, administrative hiring (as opposed to teacher and staff hiring), and testing." This revision allowed Ruth to sharpen her own thinking and exclude the vague language in her first draft of the explanation: "such issues as," "under . . . purview." The fact is, those are the issues the School Board addresses; why imply that they are some of the many others, when there are no others?

Having addressed the problems of audience issues and clarity, Ruth then knows it's time to draft the consequence section of the policy statement. Again, she wants to write it clearly, but not necessarily in tones that would threaten an outsider audience. She needs a fine line between firmness and intimidation. She tries and rejects the colloquial—"If you don't submit your complaint in writing, we won't read it"—because it sounds too threatening, because it sets up an antagonistic we-you relationship with the audience, and because she's afraid all those negatives might confuse the readers. Instead, she chooses a more neutral version: "Only those complaints submitted in writing to the Superintendent will be addressed."

The new policy statement, then, is clear, specific, and understandable by an outsider audience:

> Objections to School Board policy must be submitted in writing to the District Superintendent. The School Board rules on scheduling, discipline, text election, administrative hiring (as opposed to teacher and staff hiring), and standardized testing. Only those complaints submitted in writing to the Superintendent will be addressed.

This is an enforceable policy, written in explanatory tones; it should encourage cooperation, not discourage participation.

CASE STUDY

BOB HANLEY'S OFFICE

St. Luke's Hospital ▪ Denver, Colorado

Bob Hanley knew that the fierce Rocky Mountain heat was actually a plague visited on him by a punitive desert god. He loosened his tie in his small, boxlike office, kept barely cool by its few vents from the central air-conditioning system. Someone had played with the thermostat again, and the temperature was 80 degrees. If you turned the thermostat down too abruptly, or directly, or harshly—Bob didn't know exactly what the crime was, but he knew the punishment well—the gods put the system on vacation. Trying to work in an 80-degree office in a building with windows that wouldn't open was all but impossible. Twenty years ago, when St. Luke's was a university instead of a hospital, the students who had been protesting the Vietnam War from his office wouldn't have let something like hermetically sealed windows stop them from cooling off the place.

Despite these less than ideal atmospheric and theological conditions, he had to summon up the patience that he didn't have to train his new administrative assistant. Bob usually enjoyed his job as a hospital administrator in charge of personnel, but he didn't have a moment to spare. Training a new assistant consumed much of his time, and often his finally trained assistants then took better positions elsewhere. This almost constant process meant that he had to absorb the time crunch, taking home briefcases full of paperwork each night for several weeks.

Bob decided to talk to Doris Lippincot, his counterpart at Denver General, a woman with twenty-five years of experience in the business. One of the old school, Doris had proved an invaluable ally, since she knew many of the people Bob worked with. But she also enjoyed watching this college graduate flounder in his new position. Bob had earned a degree in hospital administration when it was still a fairly new degree. He then spent three years learning the ropes, as if from scratch, when he worked with her at Denver General. "I would think you might have solved your training problem by now, Bob," Doris brayed. "You've only been through this about once a year for the past five years." Doris was really enjoying his misery.

It was hard, though, for Doris to dangle Bob along too much in such miserable heat. Instead, she took pity on him. "Why don't you commission a manual for your administrative assistant position," she suggested. The suggestion was so simple and so reasonable that Bob could have jumped out the window for not thinking of it himself. He would contact the technical writing group several floors beneath him, beneath the whole hospital for that matter, and see whether the writers could bail him out in time for the arrival of the latest in his long string of assistants.

GUIDELINES FOR PRODUCING A FUNCTIONAL MANUAL

Besides adhering to the principle that the content of manuals should be descriptive, not procedural or policy-oriented, the biggest job in putting together a manual is document design. There are many aspects to document design, but they all converge in a single purpose: to make the manual functional. Document design is an issue with all types of professional documents, and we devote Chapter 12 to it, but we introduce it here in connection with manuals because the task of making a manual readable is a challenge.

Formatting

One important aspect of document design that is usually considered important for the wrong reasons is formatting. Formatting is important because it sets up a consistent structure for the manual, a structure that readers can predict. Formatting is not so important simply for aesthetics or image. A system of headers, footers, and other internal identifiers must be consistent, in order to signal the reader appropriately. Manuals contain the largest accumulations of prose in a work environment, yet must somehow make their information accessible. Formatting consistently makes this access possible, because it sets up expectations in the reader. When these expectations are fulfilled, the reading goes faster, with greater comprehension. This business of predictability is very important to encourage in readers, since many studies has correlated the ability to predict text with a high level of comprehension.

Readability

Besides having consistent formatting, the manual should be readable. The writer's style certainly affects readability, and some style concerns are covered in the Appendix. But readability is also a matter of form. Here we refer to *form* as distinct from *formatting,* which is primarily a concern in word processing. Form concerns paragraph length, sentence type, and all other matters of arranging the words on the page. Generally, the more successful manuals do not rely on large, unbroken chunks of written material. To encourage the harried and bored reader, the successful manuals provide breaks, such as bulleted lists, tables, charts, figures, and all the variety of technical illustration available. Since manual material is essentially descriptive, the illustrations and graphics are appropriate for carrying the message to the reader.

Organization

Although this may seem so obvious that it need not be said, a manual must also be organized logically. Given sufficient time, most departments would put together organized manuals. But manuals are so often delivered in a rush to satisfy internal demand for documentation that their disorganization is often shocking. The usual logical organizing schemes are functional, covering tasks in order of their importance, or are chronological, describing activities from the beginning to the end of the working day. The bare minimum of organizing schemes would be alphabetizing topics by chapter. Another important organizer for manuals is the index. People need some way to get to the information they need quickly and comprehensively. The best indexes offer thorough cross-referencing, providing synonymous terms at various levels of technicality or formality.

Professional and technical manuals are often revised, since all aspects of the workplace are constantly changing. To accommodate almost constant revision, the manual author needs some provision for indicating which revision is reflected in the manual, perhaps by indicating this information on each page. The author should also consider numbering each chapter separately.

Thus, each chapter would have a numeric starter that renders the chapter independent of the rest of the manual. For example, the pages of Chapter 3 would be numbered 3-1, 3-2, and so on. With separate chapter numbering, when a section or an entire chapter is revised, only the chapter involved will have to be renumbered. With continuous pagination, however, any revision of sections or chapters would require the renumbering of the entire manual.

Appendixes and Attachments

Appendixes are sometimes called attachments, but the two are distinct. Appendixes are usually substantial self-contained sections of text; attachments are usually short supplements, like copies of forms and other paperwork. Appendixes and attachments are similar, however, in that both are supplementary to the main text; both present pertinent but secondary information that would be intrusive if put into the main text of the manual.

SAMPLE OUTLINE FOR A MANUAL

Bob Hanley's manual for his administrative assistant will succeed if he follows the principles of consistent formatting, attention to readability, and clear organization. Although we don't need to include the text of a whole manual here, we can certainly outline topics for it (see Figure 9.1). Even in this outline, we can see attention to signaling the reader through repetition of key words and phrases, such as "applicants," "overview," and "current," and good summary-level chapter titles.

Bob has a good plan for his manual. His arrangement of topics is logical, his chapter headings are descriptive and logically organized (according to the job search process), and he includes an index.

The four appendixes in Bob's manual contain information important enough to go in the manual, but not important enough to go in the main text. Appendixes 1 and 2, current organization charts for the whole hospital and for the personnel department, would require constant updating, so it would be foolish for Bob to include these in the body of the manual. Appendixes 3 and 4, legal and illegal interview questions and pertinent Colorado legislation, contain related information, some of which elaborates on points made in the manual but which is of secondary importance to the text. St. Luke's is more concerned to get federal legislation straight than it is to get Colorado legislation straight, because the federal rulings are more complex than are the state-level rulings.

Bob's manual as planned is thorough, logically organized, and informative. It should be enough to teach his new administrative assistants the essence of their job.

A HANDBOOK FOR THE ADMINISTRATIVE ASSISTANT IN THE PERSONNEL DEPARTMENT

Chapter 1:	An Overview of St. Luke's Hospital
Chapter 2:	General Description of the Personnel Office
Chapter 3:	Overview of Administrative Assistant Duties
Chapter 4:	Writing Advertisements
Chapter 5:	Screening Applicants
Chapter 6:	Acknowledging Applicants
Chapter 7:	Setting Up Interviews
Chapter 8:	Interview Duties
Chapter 9:	Coordinating with Hiring Departments
Chapter 10:	Handling Unsolicited Applications
Chapter 11:	Orienting New Employees
Chapter 12:	Handling Terminated Employees
Chapter 13:	Reporting Commitments
Chapter 14:	Monitoring Employment Legislation
Appendix 1:	Current Organization Chart: St. Luke's Hospital
Appendix 2:	Current Organization Chart: Personnel Department
Appendix 3:	Legal and Illegal Interview Questions
Appendix 4:	Pertinent Colorado Legislation Index

FIGURE 9.1 **Sample Outline for a Manual**

CASE STUDY

PROFESSOR RICK ALLEN'S OFFICE

Room 1595 ▪ College of Veterinary Medicine
St. Luke's University ▪ Denver, Colorado

The semester had no sooner begun than Rick Allen wished he had stayed at his cabin in the Unitas. His immensely popular course, Ethics in Animal Research, had become a staging area for protests from the increasingly vocal and militant animal rights group, Animals Like Us. Following St. Luke's open door policy for all courses taught at the university, his course in research ethics was always assigned to a large amphitheater classroom to accommodate not only students who had registered for his course, but also members of the public who wanted to attend his lectures. There was no way of controlling who attended his lectures, which literally left the door open to public protesters to the use of animals in medical research. But Rick wasn't sure how much longer he could control the situation. He was getting sick of being shouted down in his own classroom. Today some of the quieter students left in the middle of the class, apparently intimidated by the dramatics and perhaps disgusted by the lack of order in the ivy-covered halls of St. Luke's.

What worried him the most were the students who seemed in control through a sort of self-possession that Rick had never had and really couldn't imagine having, since he wasn't inclined to any absolutist positions about anything. He thought he had organized his life to function in a similarly nonabsolutist environment, a university, but was now wishing he had at least a little strong advice about how to handle the difficult situation in his class.

Rick felt that if he or the university mishandled the protesters, then he and the university might be liable. The St. Luke's faculty was considered in loco parentis, and so could not comfortably take an adversarial position with the protesting students. Rick felt that he personally could not take such a position because it violated his sense of the humanistic education. But he also realized that someone might get hurt if he didn't cool down the situation.

After checking with the administration, Rick found that a certain document did exist to respond to his dilemma—"Policies and Procedures in Case of Classroom Disruption." When Rick read the document, he had to laugh. The procedure, which was actually written at the time of the Gulf War as a revision of a Vietnam era procedure on protests, was a mishmash of badly ordered information, few firm directions, and plenty of waffling language, qualifications that made it almost impossible to understand exactly what was expected of professors who found themselves confronted with disrupted classes. In order to clarify the procedure for himself and to press administrators to take a firm stance on the issue, Rick decided to critique and rewrite the procedure, with the hope that exposing the vagueness and waffling would force the original authors to take a clearer stand. Rick had his work cut out for him, as we can see from reading the following document:

POLICIES AND PROCEDURES IN CASE OF CLASSROOM DISRUPTION

The following provides the faculty member or teaching assistant facing willful classroom disruption with guidelines that he or she may wish to follow in attempting to control the situation. These guidelines have been formulated to provide the instructor, the peaceable students in the class, and the institution maximum protection under the university Code of Conduct and the legal structure of the courts. If followed, they would also permit a relatively uniform across-campus response to disruptive situations, with consequent obvious advantages.

The suggested procedure in case of intentional class disruption, whether by students enrolled in the class or by others, is as follows:

1. The instructor should attempt to persuade the disrupters to desist and leave the classroom. At this point, threats of criminal, civil, or disciplinary sanctions are not appropriate.
2. If these efforts are unsuccessful, the instructor may wish to dismiss the class. If he/she does not and the situation escalates at any time to where life or property is endangered, the instructor should then dismiss the class without delay.
3. If the class remains and the disruption continues, the instructor, or a messenger, if feasible, should telephone the university police department and describe the difficulty. The police department will immediately contact the Dean of Student Affairs or his/her representative, who will proceed to the classroom at once. At the same time, a police officer, ordinarily in plain dress, will be dispatched to assist. While awaiting aid, the instructor should continue attempting the deescalation of the situation. On arrival of professional assistance, the instructor should relinquish authority to these persons.
4. If the instructor is unable to follow the procedure in Item 3 and wishes to retain his or her class and continue efforts to restore order, he/she should inform those interfering with the class that further disruptive conduct will subject them to university disciplinary proceedings and criminal or civil charges. If he/she is unsuccessful, the class should be dismissed.

Whenever disruption occurs, the instructor should:

1. Attempt to describe and identify the disrupters.
2. Secure detailed written statements from as many students as possible, describing the disruption and the role of the participants.
3. Prepare a report for the Vice President of Academic Affairs, embodying the data in items 1 and 2 above.

Only with clear statements of witnesses identifying the offenders and describing their wrongful conduct can the university investigate the disruptive event and support disciplinary action and civil charges.

Rick was confused; what was he obliged to do and what did he really have a choice to do? This procedure gave him so many options that he finally couldn't use it as written. It would have to be revised.

GUIDELINES FOR WRITING A LOGICAL PROCEDURE

In beginning to think about a revision of the procedures for dealing with classroom disruptions, Rick discovered three main guideline principles for writing procedures:

1. Include only procedural discourse in the procedure; eliminate policy and manual material.
2. Organize the procedure by process instead of by participant.
3. Review all procedures for their logical soundness.

One of the most important functions of a procedure is to get the procedural information, the steps for performing the activity, separated from policy statements and manual descriptions. A procedure should be a very pragmatically written document. It is not the place for philosophizing and speculating. Neither is it the place for justifying (the job of policy statements) or describing (the job of manuals). If a procedure mixes the three levels of information, then the procedure operator may have difficulty distinguishing the essential actions required by the procedure. Even if the user can distinguish the actions from the policy, the time required to do so is still costly.

Procedures are often organized according to who is performing the action. If the procedure is performed by only one person, this organization is not a problem. But if more than one person has a part to play in the procedure, then instruction based on a participant's role is a big mistake. It is like trying to write a procedure for a *pas de deux* by first describing all the prima's steps and then describing all her partner's steps. Such a procedure would give no sense of the *pas de deux* as a whole; it wouldn't integrate the activities of the two people who together make a whole out of a series of steps.

The justification for writing procedures by performer is that people need to know what their part is in a procedure. While it is true that people need to know their responsibility in a procedure, they also need to know how their part coordinates with other people's parts; that way, they can act more cooperatively and thus enhance the procedure's successful completion. While the procedure performer doesn't need to know everything that everybody else does, the sequence of activities must be clear.

The sequence of steps within an activity must be logically sound and should be the main emphasis of a procedure. Even if some people tend to be interested only in their own part, procedure writers should minimize this possibility. One of the most notable characteristics of a professional is a general concern for quality instead of a myopic concern with one's personal duties. Professionals are not expected to complain, "That's not my job," if they receive a request. Instead, they must see to it that the person who does the job gets the message.

Essential Characteristics of Procedures

Procedure writers must keep in mind the following five essential characteristics of procedures:

1. Accuracy
2. Feasibility
3. Specificity
4. Honesty
5. Thoroughness

Testers can walk through draft procedures to determine their accuracy, but feasibility is often a matter of logical analysis, because sometimes feasibility of "solutions" in procedures can't be tested under normal conditions. For example, the procedure can state that classroom instructors should get a message to the police department if their class is disrupted, but can make no guarantee in advance that such an action will be feasible for the instructor. What if the disrupters are blocking the door or access to a telephone? To provide a feasible way for the instructor to reach the police, the procedure writer might have to propose another solution, which would require the approval of all those responsible for the procedure and which might even prompt new policy decisions.

The specificity of procedures is a matter of constant evaluation. Whatever level of specificity an author chooses, one rule of thumb is constant: a procedure should limit itself to one action only. One noteworthy aspect of the procedure on classroom disruption is that several steps are embedded within each numbered step in the procedure. For example, step 3 contains at least five individual steps, depending on what decisions the administration would make about vagueness. As Rick rewrites this procedure, he'll need to separate individual steps. Discrete steps not only enhance procedure readability, but also allow author and reviewer to get a clear sense of the procedure's logic.

The honesty with which a procedure is written is an ethical consideration that has serious ramifications. If, for example, Jason Wells, a midlevel executive, must write procedures, which are to be reviewed by outside reviewers, to document how his group does business, then he must be sure to report how business is actually done. If he tries to create more appealing procedures than are actually used, he will be lying, of course, and will also probably get caught by the Quality Control group in his own organization, which will audit his group by the extent to which his activities are procedurally correct. If his own organization doesn't catch him, then the outside reviewers will, which would be all the more humiliating as far as his organization is concerned. Many the job search group has had its work come to nothing when the affirmative action auditor notes that decisions were not made in accordance with affirmative action procedures. It does no good, then, to write procedures that bear little resemblance to the reality of the work environment, but authors are often tempted to do so when external auditors are reviewing them.

Thoroughness is contingent on a procedure's specificity. Thus, a badly written procedure might contain three badly written steps that amount to something like this: (1) Begin project in accordance with rules learned in training; (2) Complete project by deadline; and (3) Draft a report for the General Manager on the success of the project. While easy to write, this procedure is absurdly general and therefore very difficult to perform thoroughly. The link between level of specificity and thoroughness, then, is clear.

There is another aspect of thoroughness that is often overlooked in procedure writing: alternative action in decision areas. In the two schools of procedure-writing theory, one school says that those who carry out procedures are thinking human beings and the other school says that they are "knuckle draggers." Those who believe the former will leave the decision making to the user. That's what Rick's university administrators did. Those who believe the latter will try to write procedures that allow no decision making by the users. There is a middle ground, however, between these optimistic and cynical schools: the prudent school. This school advocates the idea that users and operators are capable of judgment, but should not be left without guidance. Thus, the users should not have to invent their own alternatives to decision areas because the author will determine which actions are appropriate based on possible outcomes.

Model for Procedure Writing. The work Dr. Daniel Plung and associates did with procedures at the Waste Isolation Pilot Plant Project in Carlsbad, New Mexico, provides a fine model of prudently written procedures, both operations procedures and administrative procedures. The procedures are prudently written because they give sufficient guidance to people making decisions, without oversimplifying the procedure to eliminate all decision areas. Dr. Plung assumes that people who carry out procedural work are neither knuckle-draggers nor free spirits.

He bases his graphic procedure format on the flowcharts in computer science. Plung's system uses different symbols to indicate different logical operations required of the user—acting, warning, deciding, cross-referencing, and qualifying. The rectangle, circle, and diamond constitute the main action line in a procedure, which runs down the middle of a page.

Sample Action Step. Note that the person performing the action is identified at the beginning. Note, too, that only one action is contained in the step.

Technician tests the
water sample

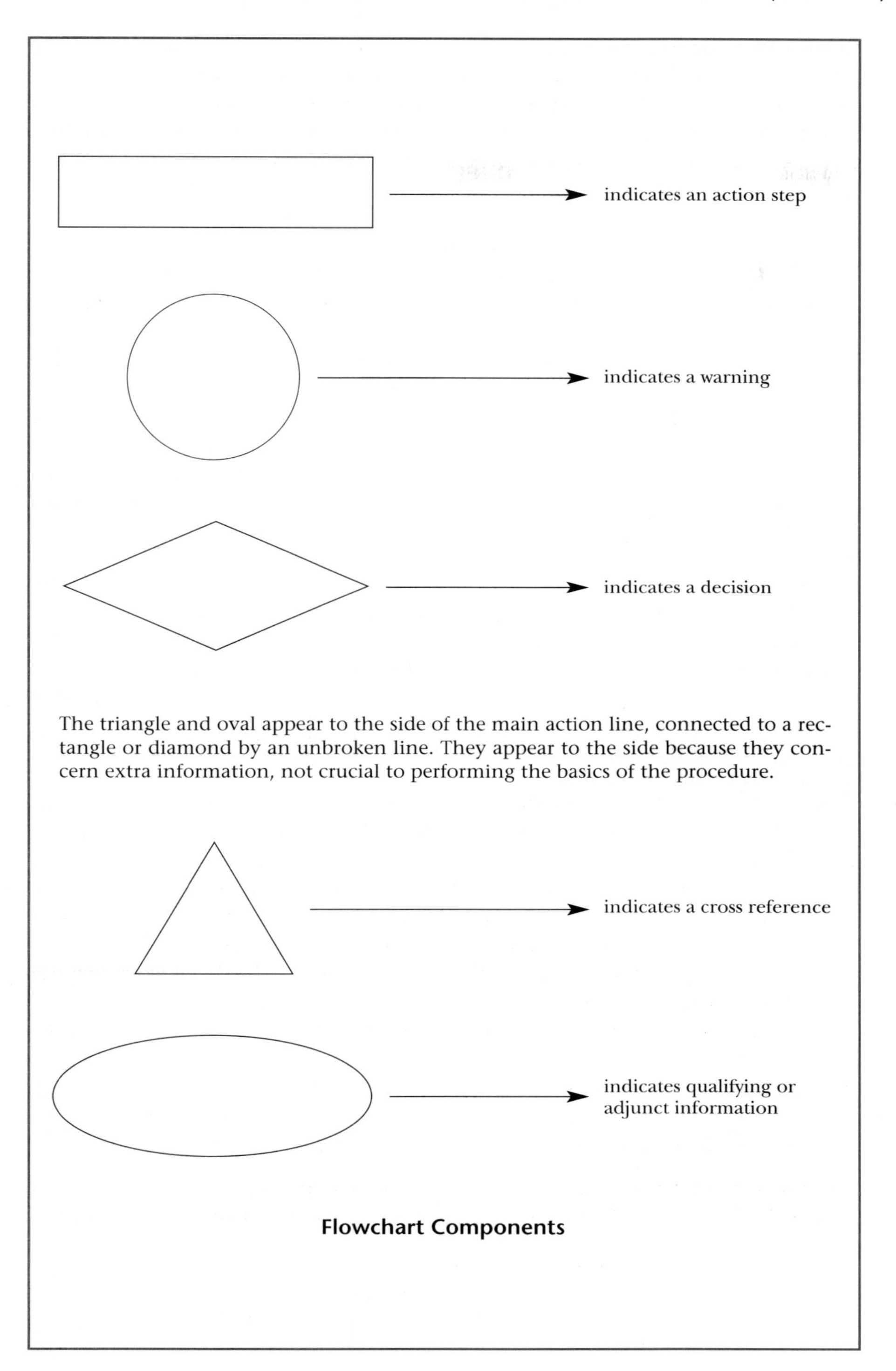

The triangle and oval appear to the side of the main action line, connected to a rectangle or diamond by an unbroken line. They appear to the side because they concern extra information, not crucial to performing the basics of the procedure.

Flowchart Components

Sample Warning. A warning will sit right on top of an action step, so the person who should heed the warning need not be identified in the warning itself. The warning is worded so that the most important information appears first in the statement. Many warnings in other forms obscure the crucial information in qualifications and prepositional phrases.

Sample Decision Step. Note that the decision step is written as a question, because more than one outcome is possible. People tend to assume that decision areas in administrative procedures are pro forma, but for the procedure to be well written, all reasonable consequences should be accounted for. Just because a general manager tends to sign off on any report that crosses his desk does not mean that his replacement will rubber-stamp work so routinely. The principles of the procedure must override the personalities of employees who currently hold the positions referred to.

In this example the decision step presents only two possibilities, yes and no. The procedure author draws these consequences out in alternate sections of the procedure, immediately following the diamond. Other lines and arrows indicate where the particular path of consequences rejoins the main action line of the procedure.

Sample Cross-Reference. Cross-references serve a couple of important functions in a procedure. First, as in the one above, the cross-reference will set up a loop of actions, where a loop is needed. For example, until a certain approval is obtained, the user may need to be in a cycle of reviews; until the appropriate signature has been obtained, there should be no progress in the procedure. Another function of the cross-reference is to direct the user to related procedures and appended information at the back of the current procedure. Cross-references help the user make major moves through a complex procedure that has many alternative paths to account for.

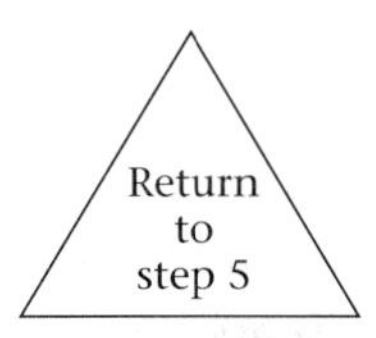

Sample Note. The main principle governing the use of notes is that they concern only secondary information. Notes don't affect the activities on the main action line; rather, they contain only adjunct information and qualifications. Procedure writers need to keep notes at a minimum because they can distract from the procedure's essential actions. If the writer finds that a procedure seems to require a greater number of notes, chances are that the author needs to write some of the descriptive information in a manual or to cross-reference the information in an appendix at the back of the procedure.

Other copies should go to the General Manager and Vice General Manager if the loss exceeds 5k

SAMPLE ADMINISTRATIVE PROCEDURE

When put together, these graphics form a clear path of actions that follows the logic of process. Figure 9.2 shows a sample administrative procedure presented in a flowchart per Dr. Plung's model. In addition to the flowchart, there are four categories on the first page of the procedure that orient the reader to the procedure's limits (scope statement), and any peculiar definitions, references, and general statements that pertain to the procedure as a whole. The general statements are usually some kind of rules pertinent specifically to the procedure; they are not rules general enough to qualify as policy statements.

For people accustomed to standard procedures, this procedural format can appear daunting at first. Since each step truly is only one step and since each decision area is called out as a real decision instead of a rubber stamp area, the procedure appears quite lengthy. The fact is, though, that it is the explicitness of the procedure that makes it seem longer than the standard procedure. Rick Allen will find that when he rewrites his procedure for classroom disruption, he will have many more steps than the four that are currently making up the procedure.

Note, though, that this explicit format makes some realities much clearer than they would be in the old embedded style. First, we can see exactly how bureaucratically rigid this process of getting an administrative leave is. One misstep, and a person's request would be nullified. And since the process takes so long, it would be difficult to summon the stamina to start the application all over again.

A scope statement at the beginning of the procedure does not concern essential action, but explains to whom the procedure applies, what circumstances it covers, and who falls outside its provenance.

Any terms that might be ambiguous or that need to be understood in a very narrow sense or that are too technical for the procedure audience should be defined before the procedural steps themselves are described.

In this section, list all references used to write the procedure.

PROCEDURE 15: APPLYING FOR ADMINISTRATIVE LEAVE

I. Scope: This procedure describes the process for obtaining administrative leave. It applies to professors and administrators. Others wishing administrative leave, particularly hourly workers, should follow procedure 25.

II. Definitions:

Administrative leave: a leave of absence taken to accommodate consulting work or off-campus research.

Administrator: all employees from category A-6 on.

Professional leave: a leave of absence taken to accommodate a sabbatical.

Professor: permanent faculty on 9-month appointment, tenured or untenured.

III. References:

St. Luke's University Handbook, 1996

IV. General:

A. Leave proposals are considered twice a year, in December and in April.

B. The decision process takes at least three months, under the best of circumstances. It is therefore wise to begin it early.

C. Any applicants who do not follow this procedure will automatically be dropped from consideration.

V. Procedure:

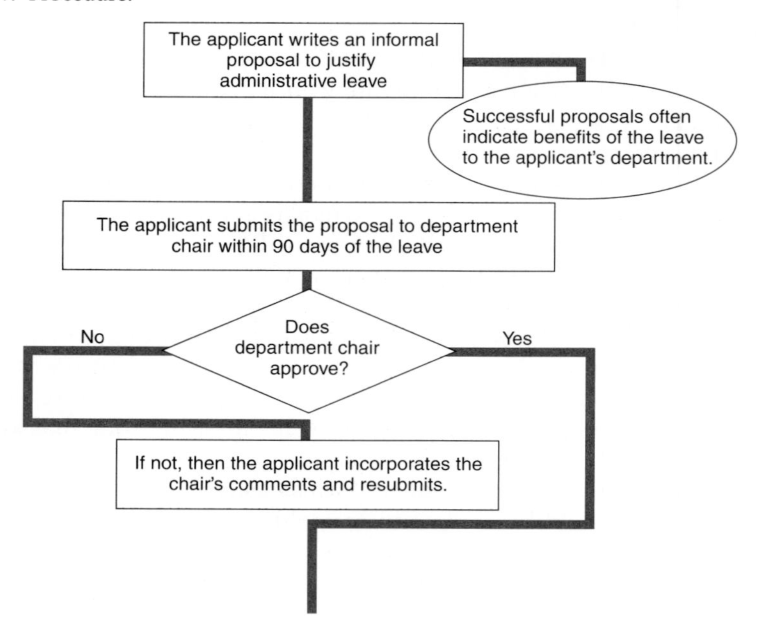

FIGURE 9.2 **Sample Administrative Procedure**

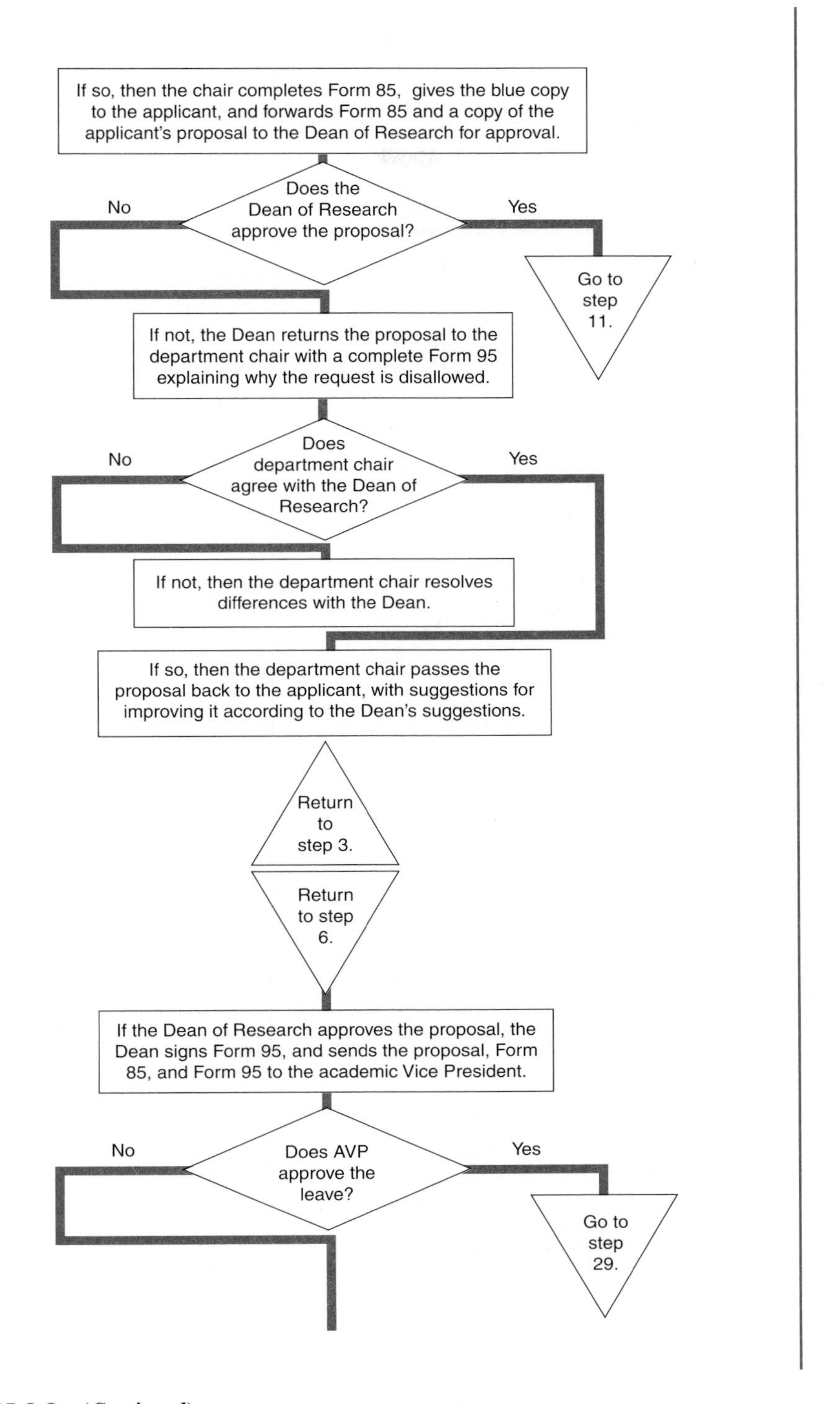
If so, then the chair completes Form 85, gives the blue copy to the applicant, and forwards Form 85 and a copy of the applicant's proposal to the Dean of Research for approval.
Does the Dean of Research approve the proposal?
No
Yes
Go to step 11.
If not, the Dean returns the proposal to the department chair with a complete Form 95 explaining why the request is disallowed.
Does department chair agree with the Dean of Research?
No
Yes
If not, then the department chair resolves differences with the Dean.
If so, then the department chair passes the proposal back to the applicant, with suggestions for improving it according to the Dean's suggestions.
Return to step 3.
Return to step 6.
If the Dean of Research approves the proposal, the Dean signs Form 95, and sends the proposal, Form 85, and Form 95 to the academic Vice President.
Does AVP approve the leave?
No
Yes
Go to step 29.

FIGURE 9.2 (*Continued*)

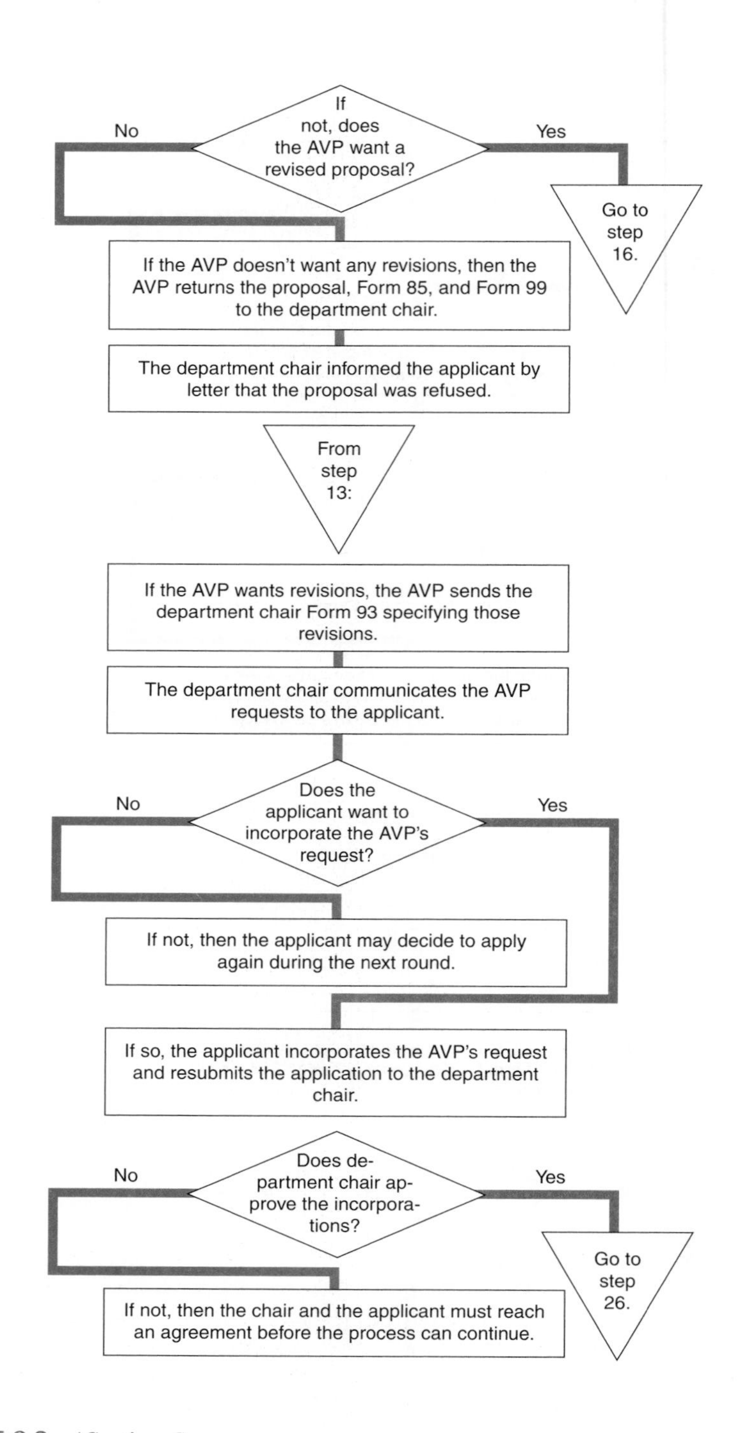

FIGURE 9.2 (*Continued*)

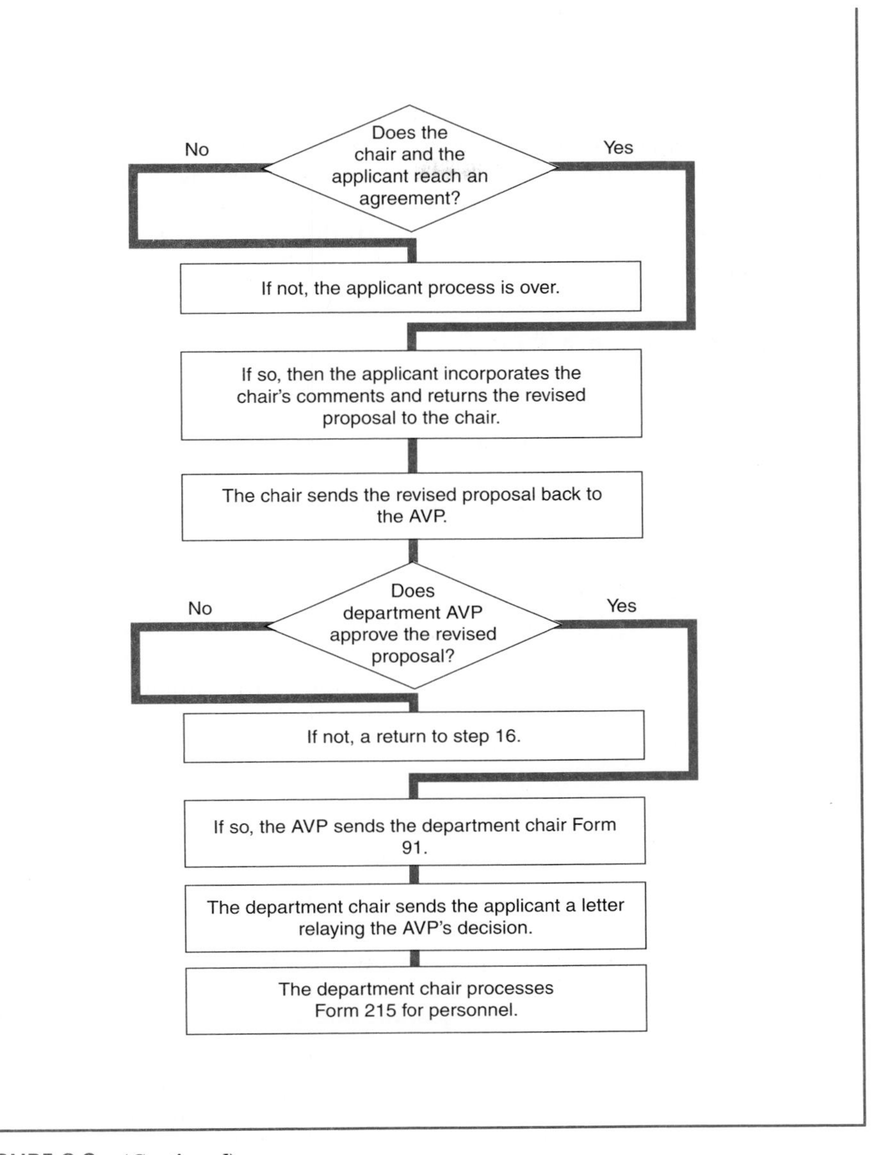

FIGURE 9.2 (*Continued*)

Another truth that becomes apparent is that this procedure puts a great deal of power in the hands of the department chair. Although this may or may not be good news for the applicant, it is a fact that any procedure reviewers should consider. Administrators might question how much they want department chairs in charge of administrative leave. Does that option open the way for any possible abuses? That's a question that reviewers would have to face.

The point is that the format of the procedure makes such biases clear for all reviewers. If any of the alternatives in the decision areas had not been specified, this error would appear graphically as an arm that attached to nothing else in the procedure, a path that dead-ended. As this procedure stands, it is not perfect, but it does try to account specifically for all the consequences of the decision areas. Whether reviewers would agree with these consequences or not is a question they would have to ask themselves during the procedure review process. But at least with this format, they could tell what the consequences were.

SAMPLE OPERATIONS PROCEDURE

In general, operations procedures tend to have fewer decision areas than administrative procedures do. This is the tendency partly because the procedures will not be so specific that they provide lessons in standard technical operations for the user. They will focus instead on a sequence of actions that constitutes an operational whole. Thus, for example, a technician would not find a procedure on how to change the oil in the company car, because this is a standard activity that has little degree of specialization attached to it. On the other hand, the technician would find procedures on how to work a specialized piece of equipment, like a one-of-a-kind nuclear waste forklift, or procedures on how to respond to a nuclear spill on the premises. All the principles that the technician had learned about cleaning up would be qualified by site-specific rules and parameters. The procedure in Figure 9.3 is a simple operations procedure that conveys the flavor of such procedures.

Although it may seem that this procedure has been written for knuckle-draggers, there are two good reasons for its specificity. First, there is a concern for quality control. The Loadem Company wants to be able to guarantee professional loading for its customers, who might decide to find another shipping company if they get poor results. The second reason is that Loadem's policy to penalize packers for damaged goods could result in constant disputes about who was responsible for damage. To ensure that the packer is not responsible, Loadem requires a signed inspection at the end of the process. This inspection protects the packers and furnishes Loadem with the documentation necessary to recover any damage fee from the truckers.

It is inspection step 16, however, that provides a loophole for the packer. The implicit decision required by an inspector has been flattened out in this procedure, as if the inspector will always find that the pallet has been properly packed. In order to make it clear that this inspection is not a rubber stamp, the procedure should be revised to make step 16 a decision instead of an action.

Although other operations procedures may be far more technical, in that they contain more technical information, their structure will be an elaboration of this sample operations procedure.

PROCEDURE 2: PACKING THE VEGETABLE SHIPPING PALLETS

I. Scope: This procedure covers the 10-square-foot pallets used at LOADEM for shipping all vegetables except tomatoes. All forklift operators are required to work in accordance with this procedure.

II. Definitions:

Vegetable shipping pallets: refers to the SURPAK pallets that are 10 feet square, made from wood, not metal.

III. References:

OSHA Guideline 35

IV. General:

A. All damage to the vegetables incurred by improper packing will be charged against the packer's account.

V. Procedure:

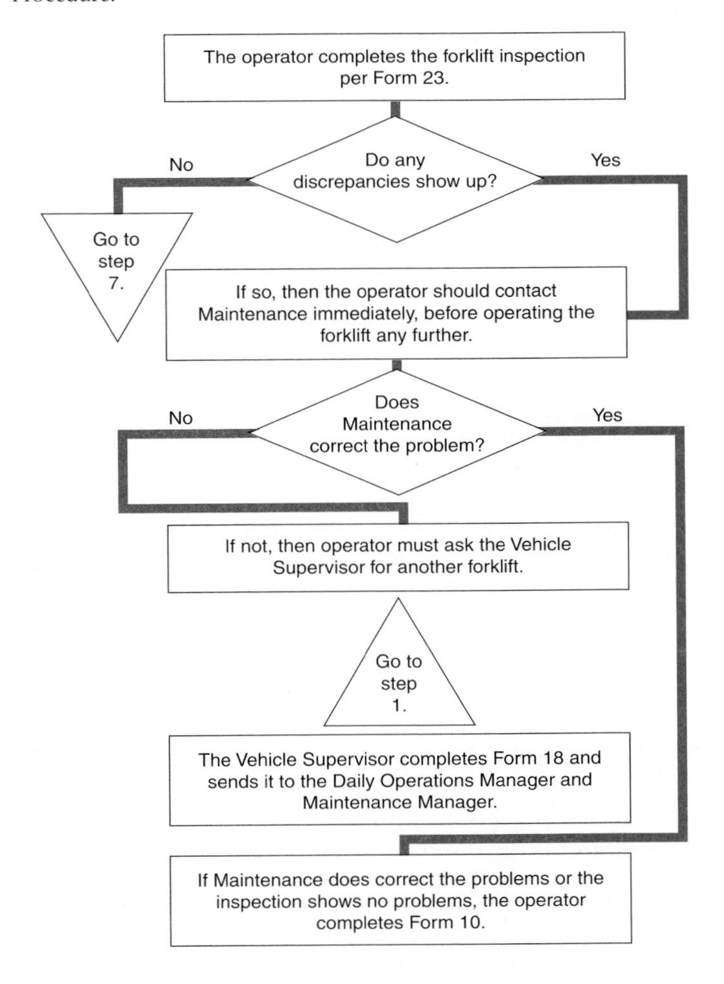

FIGURE 9.3 **Sample operations procedure**

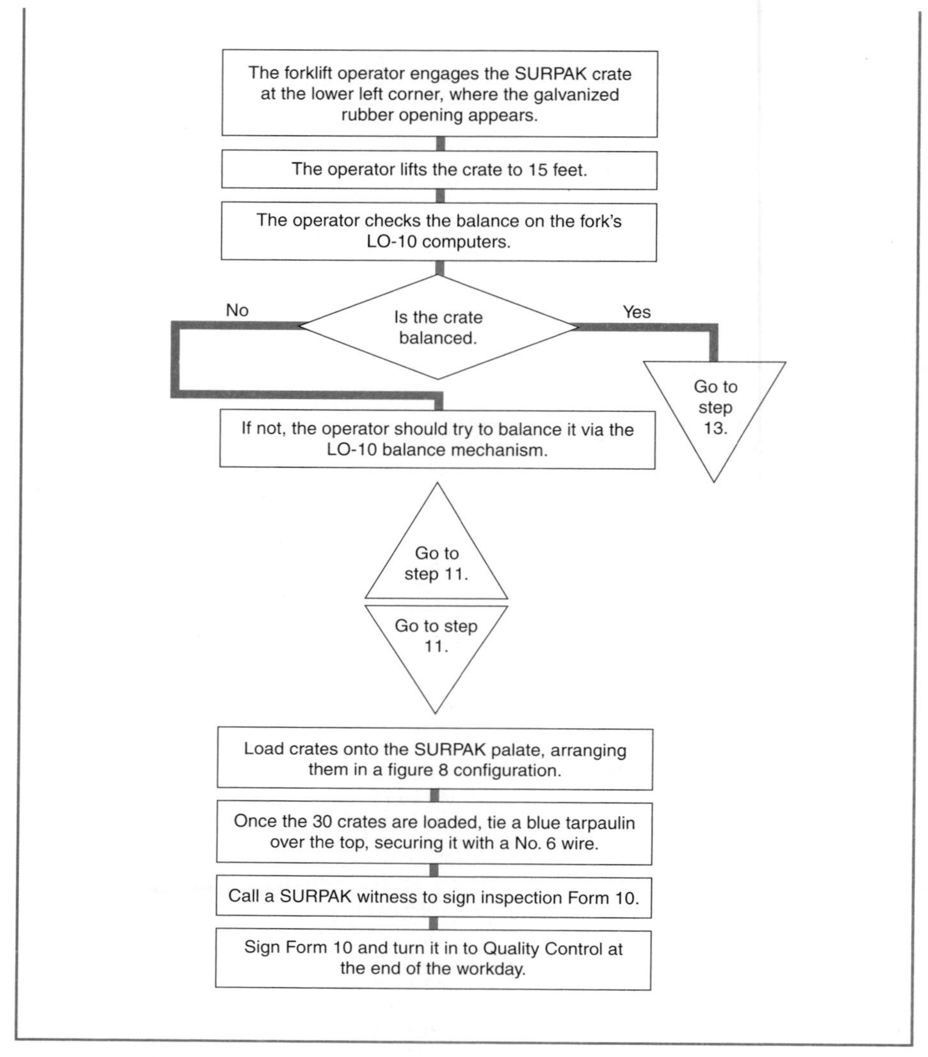

FIGURE 9.3 ***(Continued)***

ANALYZING AN ADMINISTRATIVE PROCEDURE

Professor Allen is so disgusted with the administration's poor attempt to commit itself to a procedure that he decides to try to motivate a revision. It is a task worth doing: not only are he and the university facing liability from several directions—parents, involved students, bystander students, staff, professors—but also as a matter of principle the university should have a coherent policy in place on such a potentially dangerous issue. Rick knows that it will take a long time to get agreement on a procedure about such a controversial issue, but a good place to start would be to make explicit all the embedded decision areas in the existing procedure. Figure 9.4 presents Rick's revision and analysis of the classroom disruption procedure.

PROCEDURE FOR HANDLING CLASSROOM DISRUPTION

Note to Administrators: I have converted our current procedure on classroom disruption into a flowchart to point out some problems with it. You will see below how many of the decisions are thrown in the lap of the instructor, with few guidelines for the instructor's protection. My comments appear throughout the procedure in brackets.

I. Scope: This procedure applies to the entire teaching staff at St. Luke's University, tenured and nontenured, full- and part-time. It applies to any disruption in a classroom that could endanger students or property [and/or faculty?]. Disruption that is not physically [or psychically?] threatening, i.e., uncooperative attitudes, disagreement with the professor and long debate about the disagreement, does not fall under this procedure. Such benign disruption is covered under Procedure TBD (to be determined) [I assume there must be another such procedure].

II. Definitions:

Disruption: behavior that is physically threatening to students or property. [We don't want to forget faculty here either or psychic disturbances.] It impedes an instructor's ability to conduct class.

III. References:

St. Luke's University Code of Conduct

IV. General: (None)

V. Procedure:

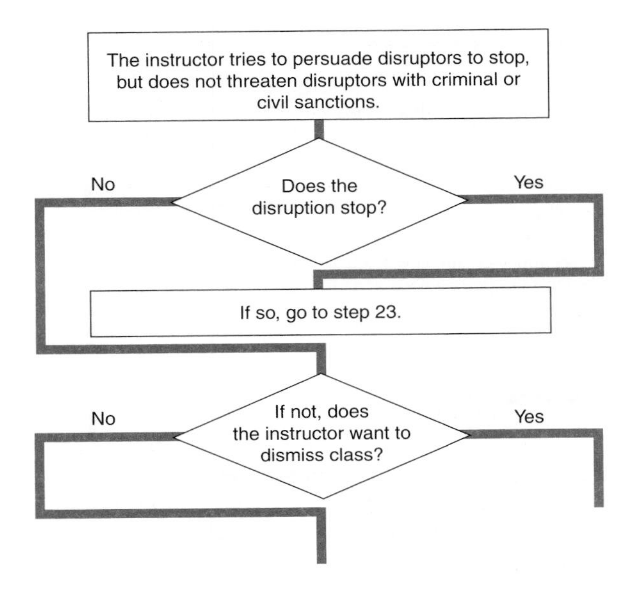

FIGURE 9.4 **Rick Allen's analysis of the classroom disruption procedure**

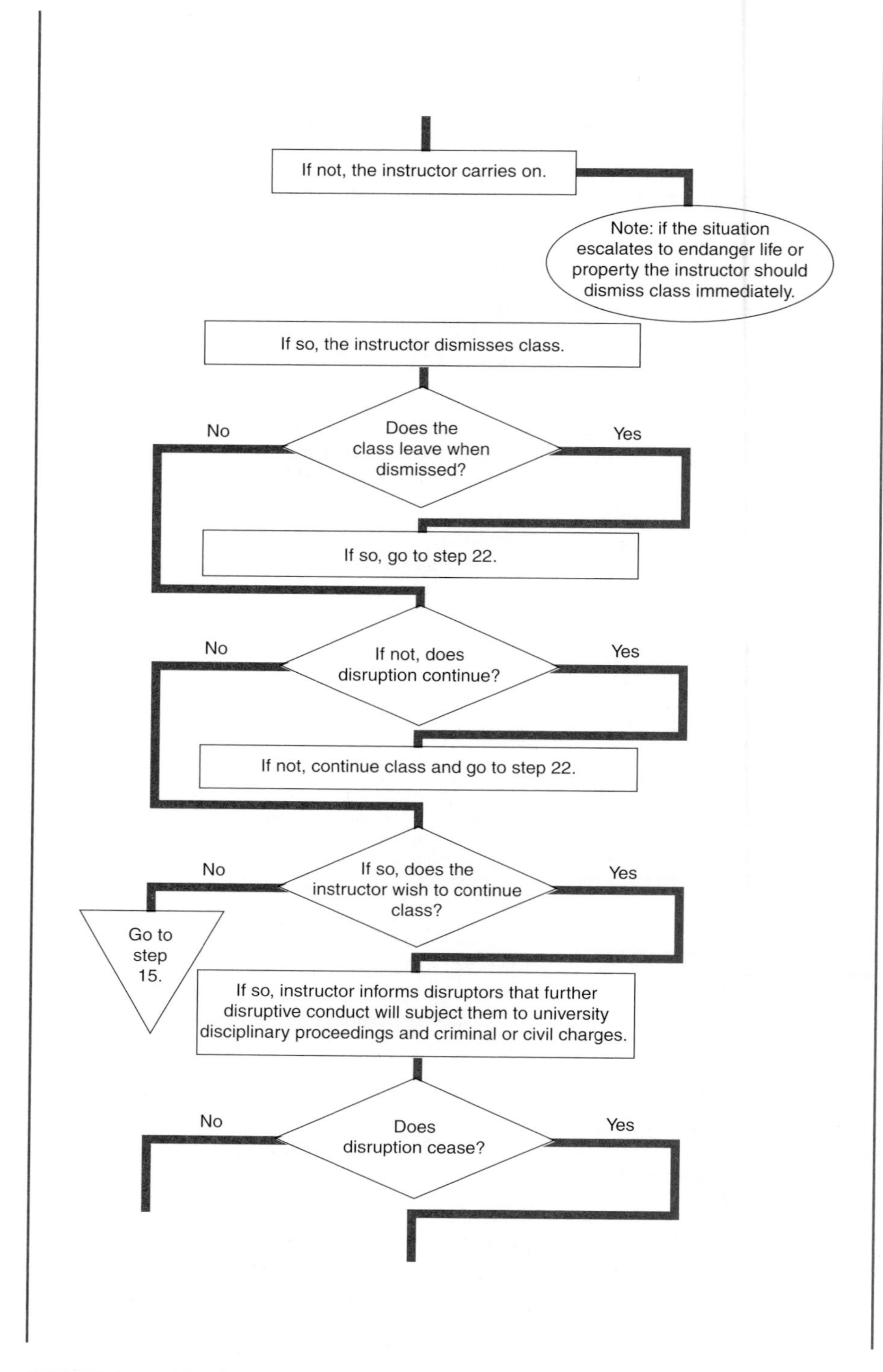

FIGURE 9.4 (*Continued*)

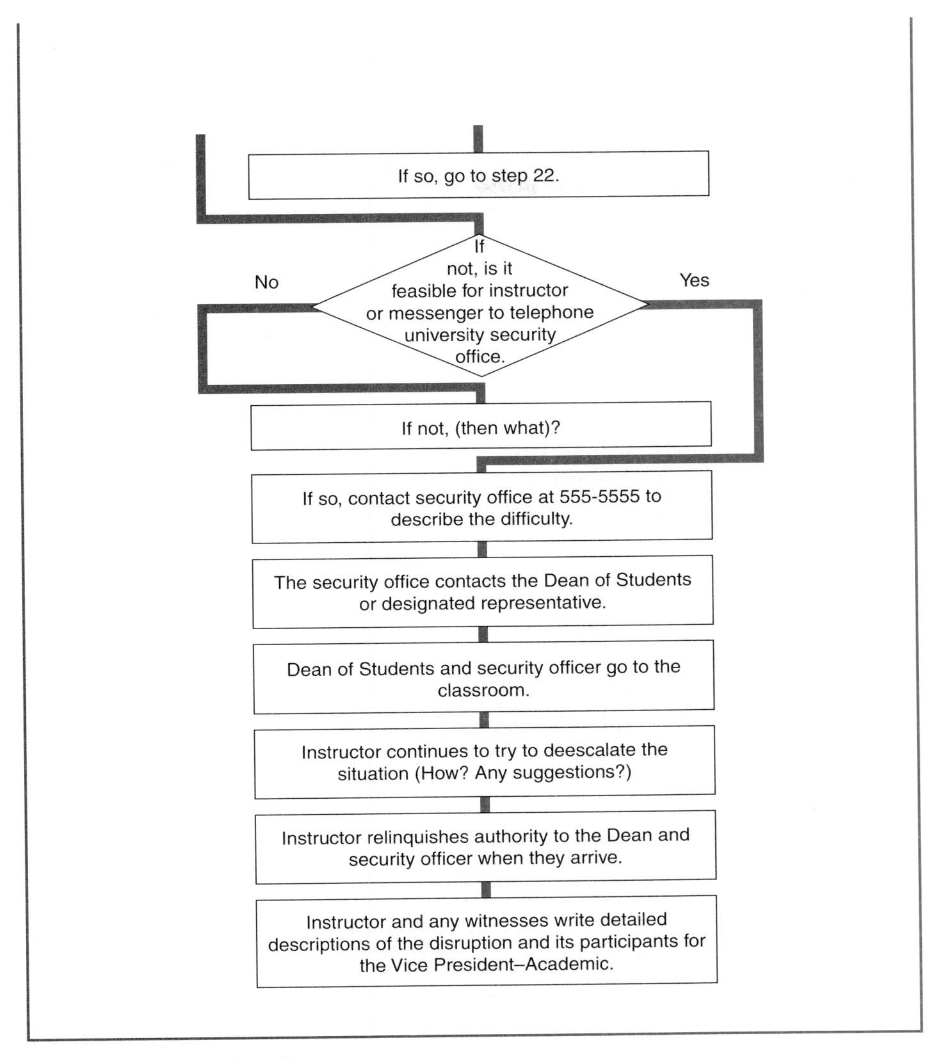

FIGURE 9.4 (*Continued*)

When flowcharted, the procedure looks as unwieldy and uninstructive as Rick suspected it was. In fact, the flowchart shows how weak the procedure really is. If a professor's simple logical analysis could get at the weaknesses in the procedure, a skilled trial lawyer could have a wonderful time with it. Even assuming that the procedure were on the right track, we can see that there are holes in it at steps 16 and 20; that is, the procedure reaches a dead end, with absolutely no guidance for the instructor who has come to that pass.

One of the most amazing points about Rick's analysis is seeing how many steps actually were embedded in a procedure that listed only four procedural steps, with another three added on the end as an afterthought. Rick's twenty-two-step analysis is as lean as he can make it without violating any of the decision areas embedded in the procedure's logic. On the other hand, the three steps added on the end of the procedure actually constitute only one step in Rick's analysis of the procedure.

Converting a procedure into a flowchart is an invaluable tool for detecting logical problems both in procedures that one is writing and in procedures that one has been assigned to review. You may recall that the professional profiles in Chapter 1 listed procedure review as a major part of a professional's job. Using flowcharts gives the reviewer a method and a very graphic representation of both procedural chaos and procedural quality.

CONCLUSIONS

This chapter distinguishes among three different types of directive discourse: policy statements, manuals, and procedures. Policy statements are very general rules; the primary concern is to write them so that they say something meaningful. Manuals provide descriptive and basic information; they are not as general as policies or as specific as procedures. They present guidelines but not to the point of describing a single process step-by-step, the way a procedure does. The major concern with writing manuals is designing them appropriately for their audience and organizing them logically. Procedures present step-by-step directions for performing a certain activity in a uniform manner. This uniformity is necessary for any activity that has important quality, financial, or safety implications. Procedures are essential when a process must be performed with no deviation.

POSTSCRIPT

Ruth Warner's policy statement was partially successful. Those who bothered to read it tended to abide by it, but the textbook issue was so hot that Ruth and the others in the school district's administration found they had to initiate some meetings and public forums to put themselves at the center of the dispute, all the better to put the problem into some kind of controlled deliberation.

Bob Hanley's manual was a great success. His new administrative assistant, Joe, spent a day with it while he was still training, and then used it as a reference until he was skilled in all aspects of the job. Joe realized, in fact, that he should update the manual to add another duty to the job description: maintain the Administrative Assistant's Handbook. He did maintain it faithfully; and when Joe decided to leave to become a law student, Bob turned the training over to Jeannine. The switch from Joe to Jeannine was the smoothest staff transition that Bob ever had.

Rick's procedure efforts were a qualified success. He certainly made it clear to St. Luke's University administration that the current procedure on classroom disruption was inadequate. But because the higher-ups could never agree on policies that would clarify actions in the procedure, the old procedure remained on the books. Rick got another teaching job, because his insistence about having a reasonable procedure in place got on some administrative nerves. The next year one student was hurt in a disrupted class at St. Luke's and successfully sued the university. Part of the suit's success derived from the fact that the university had no coherent procedure in place to deal with classroom disruption. Very soon thereafter, St. Luke's University became St. Luke's Hospital.

EXERCISES

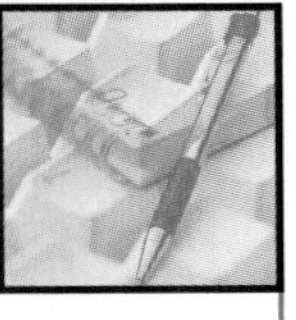

1. Using Rick's procedure analysis as a base, revise the classroom disruption procedure. Write a flowcharted procedure that will resolve the open issues in the first draft and will streamline the procedure as a whole. Exchange revised procedures and critique a partner's revision.
2. Obtain a procedure sample from a business or an organization in your community. Critique that procedure orally for your class according to the principles discussed in this chapter.
3. Analyze some of the policies you may find in your school's bulletin. Do they fulfill this text's criteria for policies?

ASSIGNMENT

1. Convert the accompanying sample draft procedure into a logical flowcharted procedure that adheres to the principles discussed in this chapter. You should note that this draft procedure is not organized by the logic of its actions, but by another criterion: responsible person.

REPORT AND INVESTIGATION OF OCCUPATIONAL INJURY OR ILLNESS

I. POLICY

It is the policy of this contractor to provide medical treatment to employees injured while working at a corporate facility, to take all necessary actions which will eliminate work-related hazards, and to develop methods to achieve these goals.

II. SCOPE

This procedure specifies the appropriate actions to be taken, following an occupational injury or illness to a contractor employee.

III. DEFINITIONS

Occupational injury or illness: Any injury or illness that occurs or is alleged to have occurred in the course or as a direct result of an employee's work assignment

Investigation report: A study and subsequent report of pertinent information necessary to determine the accident cause and corrective actions necessary to prevent a recurrence

Serious injury: Any occupational injury that precludes the injured from returning to his or her regular work assignment and/or identified as a reportable injury for inclusion on OSHA Form No. 200

Recordable injury: Any occupational injury resulting in one or more of the following:

1. A full shift's absence from work
2. Any amputation
3. Any fracture
4. Any medical procedure exceeding first aid measures, e.g., suturing
5. Any loss of consciousness

Travel on company business: Any authorized travel or temporary assignment, outside the main headquarters

IV. GENERAL

The procedure describes actions needed to provide adequate medical treatment for injuries received by employees while in the course of their regular work assignment; establishes investigation and reporting systems necessary to compile information for use in the prevention of accidents and to supply information to appropriate state and federal agencies as required; and establishes a recording system of all job-related injuries.

V. RESPONSIBILITIES

1. Injured Employee
   A. If physically possible, inform supervisor immediately of a job-related injury or illness.
   B. Provide supervisor with pertinent information regarding accident.
   C. If traveling on company business, advise cognizant manager of accident or injury as soon as it is practical.
   D. Work according to safe working practices established in training.
2. Other Employees
   A. If injured employee is physically unable to report accident, advise cognizant manager of injury immediately.
   B. If able, provide assistance to injured employee.
   C. If a witness to the accident, provide supervisor with pertinent information regarding same.

3. Cognizant Manager of Injured Employee
   A. Advise Project Manager, Manager of Safety, and Manager of Personnel of accident or injury immediately.
   B. Assist Personnel Department in securing medical attention if required.
   C. Complete Record of Occupational Injury or Illness form and forward copies of same to Manager of Personnel and Manager of Safety.
   D. Conduct investigation of accident within 24 hours, forwarding copies of investigation report to Manager of Personnel, Manager of Safety, and Project Manager.
   E. Identify and implement corrective actions required to prevent recurrence of accident.

4. Personnel Manager
   A. Maintain emergency medical procedure for Project office.
   B. Maintain pertinent emergency notification and health information for all project participants.
   C. Secure emergency medical care for job injuries as required.
   D. Distribute pertinent accident or injury information to appropriate company organizations.

5. Manager of Safety
   A. Assist in accident investigation.
   B. Review and approve corrective action required to prevent recurrence of accident.
   C. Inspect area to ensure corrective actions identified have been implemented.
   D. Maintain OSHA Form No. 200 and make appropriate entries to same.

Solving Problems Through Letters and Memoranda

CASE STUDY

SUPER BOWL SUNDAY 3:20 P.M. JAY DALTON'S CUBICLE

The Write Solution, Inc. ▪ Santa Fe, New Mexico

The technical writers at The Write Solution, Inc., (WS) were really disgusted this time. They often had to work on weekends to finish last-minute jobs, but having to work this weekend was pushing the abuse too far. A major client, Texas Oilcats, annually contracted its stockholders' report with WS. The job was a big one: about 150 pages of text and as many photographs. It essentially paid WS's January expenses and brought a small bonus if produced correctly the first time, without revisions. But, as welcome as this client was to WS, Texas Oilcats was always difficult to work with. The salt was rubbed in the old wounds this time, because the staff was working on Super Bowl Sunday.

The staff—writers, graphic artists, and word processors—were well trained and conscientious, but because they were salaried, they worked overtime without compensation. It's difficult

PREVIEW OF THE CHAPTER

to consider the perks of salaried work when a person is missing the Super Bowl and in danger of missing it every year because the client is late. This abuse of staff was causing morale problems, and WS couldn't afford to lose anyone.

Besides not adhering to WS's schedule, the Oilcats had also argued about the bonus. For example, the Oilcats often didn't make a distinction between a second time through production to correct WS errors and a second time through because someone at the Oilcats changed a passage, reordered the table of contents, or withdrew a section from the galleys. Thus, WS often found itself losing bonus money because changes were needed to correct its "errors."

As Jay Dalton saw it, the fault was not in the stars but in WS. Although he couldn't deny that Texas Oilcats was a difficult client, he felt it was up to WS to protect itself from financial liability and its staff from abuse. It was clear to Jay that the Oilcats depended on WS; after five years of producing the report, WS could produce it more efficiently than any other technical writing contractor. Texas Oilcats was consistently happy with the final product, and it took less of the Oilcats's time to produce because WS was so experienced. That experience gave WS some leverage, in Jay's opinion.

How to start wielding that leverage? Jay had some ideas about this problem. He had thought about it every January for the past four years, as soon as it was evident that the Oilcats would be difficult to control. He also had hopes that a successful resolution of the Oilcats problems would demonstrate his own executive ability; that's how people got promoted to management at WS. In such a small company—only thirty employees—there weren't many managerial slots. Jay would do his best to get one. He would start by writing a few memos.

MEMOS AND LETTERS AS PROBLEM SOLVERS

As noted in Chapter 8, professional writing is classified into three categories: scientific, technical and professional, and administrative. The focus in this chapter is on the administrative category, which may be the most prevalent form of professional writing but which is most easily treated simplistically or even dismissed as essentially formulaic. Although the structural conventions of this kind of documentation are quite rigid and thus somewhat formulaic, careful writers must otherwise be as thoroughly critical of their administrative writing as they are of their technical and scientific writing. Good administrative writing is the grease that makes day-to-day business in the working world go smoothly.

Memos and letters, the primary examples of administrative writing, are two of the most frequently used means of solving problems in the professional world. Even though a banker may never write a feasibility study and an engineer may never write a lab report, they will both write letters and memos. These documents cut across all disciplinary boundaries.

Documenting with Memos and Letters

We start our discussion of administrative writing with the question "Why not simply pick up the phone or drop an e-mail message?" This is the fundamental question and one that all professionals should remember to ask as they do their day-to-day business. The fact is that much business does get conducted by phone, or in recent years, increasingly by e-mail. But even those telephone or e-mail contacts will often be followed up with a piece of paper, a memo or letter, that confirms the result of the discussion. Why would people double their labor in that way? The primary reason is the need to document, to keep a securer record of the way events transpire. This record may never be pulled out of the file, but if it is needed later to solve a dispute over the contents of the oral or e-mail agreement, it is, indeed, precious and utterly essential.

Professionals learn with experience what should be documented and what should not. Here are some guidelines to recognize the situations requiring documentation:

1. Oral agreements, especially with people outside your organization;
2. All personnel issues, whether you are the manager or the managed, including time and money agreements, all travel expenses, health-related issues, injustices, and potentially troubling situations;
3. Significant personal work accomplishments, which you should keep in a file for your use in revising your résumé and for support in your annual employment review;
4. All internal work agreements designating particular responsibilities, especially with people from other departments, and important when interdepartmental projects have absolute deadlines; and
5. All important requests that you make and those that are made of you.

This sampling gives some idea of the occasions in working life that require documentation. Although some of these, like significant work accomplishments, could be listed informally at the back of a calendar, they are all invaluable for tracking the work that professionals do and the working realities that they experience.

Distinguishing between Memos and Letters

Professionals use both memos and letters to solve problems and to document information. The primary difference between these two types of administrative writing is the audience each addresses. Memos are written to people inside the organization; letters are written to people outside the organization. This difference in audience is the basis for other distinctions between memos and letters.

Memos, because they are sent to insiders, are not as formal as letters. The style, format, and tone of a memo are less formal than those of the letter. However, the informal style is less important than the fact that a memo, like a letter, documents; that is, it records an opinion, an action, a plan, or a train of events. Professionals often record information for themselves in memos addressed to their own file simply to keep track of important issues.

Letters solve problems with outsiders—customers, clients, contractors, and so on. Because letters address an external audience, they are more formal in presentation, format, and, in particular, word choice than are memos. Chances are that an external audience will not share the same insider language appropriate within an organization. Since the external audience is, in a sense, uninitated, at least at the outset of correspondence on a particular subject, more focus is required on background and introductory information in a letter than in a memo. The letter writer is responsible for keeping the reader oriented to time, purpose, and history. Additionally, the tone in a letter may be more restrained, less open than in a memo, although the same level of sincerity is certainly possible and desirable. But we should also remember that greater restraint does not call for stiltedness or stuffiness.

In certain situations that call for a greater level of formality or special recognition, letters rather than memos are sent to people inside an organization. Examples of in-house letters include congratulatory letters, recommendation letters, letters of promotion, disciplinary letters, and unfortunately at times, letters of resignation or dismissal.

FORMAT OF A LETTER AND A MEMO

This chapter briefly reviews standard letter and memo format, but it is more important to understand the problem-solving function of memos and letters. There are two reasons for not dwelling on format. First, once at work in the professional world, writers will either turn the formatting job over to a secretary or look up company recommendations on letter formatting. Often companies set their own conventions for correspondence to achieve some uniformity and a company look. They almost always set memo formats, ordering stationary with that format imprinted on it. Second, the best-formatted letter in the world cannot overcome the structural and rhetorical gaffes that come from inadequate problem solving.

No one is going to ignore a well-written letter if the date appears on the left side instead of on the right. The truth is that many professionals often don't notice such offenses, whereas English teachers are notorious for noticing them. The workaday professional is most interested in the technical and rhetorical sophistication of the documentation: How well does it solve its technical problem? How well does it communicate its purpose to its audience, within a certain context and the writer's ethical constraints?

Standard Letter Format

In lieu of company formatting guidelines, the standard block letter format (see Figure 10.1) is useful. The essence of this format is that everything begins at the left margin.

Standard Memo Format

The standard memo format (see Figure 10.2) works well for communication between an employee and her or his immediate superior, for example. A memo may be necessary to ask for assignment extensions or to explain a missed deadline or for whatever administrative purposes an employer expects written communication. The writer centers the whole memo on a full-length page. Note that the memo dispenses with the salutation, addresses, complimentary close, and signature. The subject line needs a descriptive phrase; this line orients the reader to the memo's subject and sometimes to its purpose.

These standard formats should provide consistent-looking letters and memos for the student, job seeker, and private citizen, but employees of professional organizations may find these formats superseded by others particular to their organization. What will not be superseded, however, is the rhetorical discussion that follows in the rest of this chapter.

CONVENTIONS OF ADMINISTRATIVE WRITING

Because professionals must read so many memos and letters from so many sources—some estimates indicate as many as 140 pages per day in recent years—they have devised conventions to ensure the most efficient reading possible. Efficiency is the keyword here because no one reads these letters and memos for fun; professionals read and write these documents to do their work. The strength of administrative writing conventions is that they address the reader's concerns, not the writer's concerns. Thus, these conventions make a text very reader-friendly.

RYAN ABERNATHY
2840 Andrews Center - Littlefield, TN 44365

June 15, 2001

Ms. Ramona Weathers
Director of Marketing
ALY, Inc.
12 Century Square
Moscow, GA 30303

Dear Ms. Weathers:

I want to thank you for the interview you so graciously participated in last week. Your answers were most helpful and should go a long way to dispel some of the myths the public has about marketing.

Regarding your question about the number of marketing graduates coming from Moscow University each year, I did some research and found that the totals have been coming down about 10 percent each year. In 1999, MU graduated 512 majors; in 2000, 460: and in 2001, 416.

If you need anything further, please let me know. I'll send you a transcript of our interview next week and a copy of the draft article in about six weeks.

Sincerely,

Ryan Abernathy

Ryan Abernathy

Date
(Skip 3 lines)
Sender's return address
(Skip 1 line)
Recipient
Recipient's title, company, and address
(Skip 1 line) Salutation, starting with Dear and ending with colon
(Skip 1 line)
(Skip 1 line)
(Skip 1 line)
(Skip 1 line)
Complimentary close
(Skip 4 lines to printed name) Sender's signature
Sender's name, printed

FIGURE 10.1 **Standard block letter format**

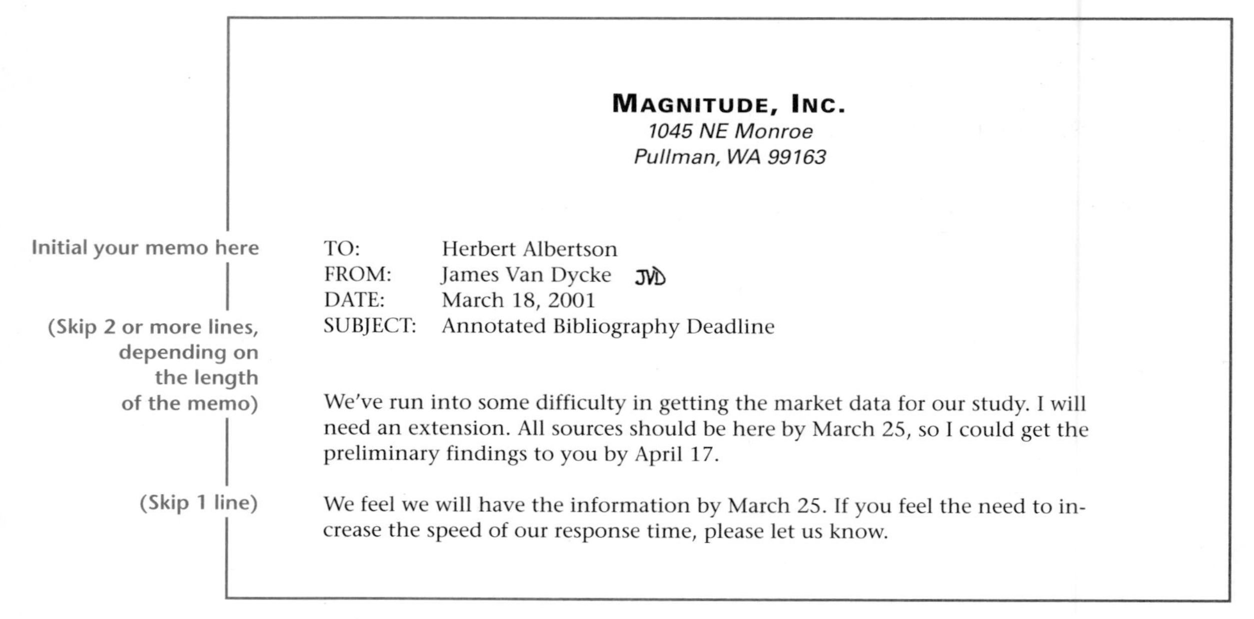

MAGNITUDE, INC.
1045 NE Monroe
Pullman, WA 99163

TO: Herbert Albertson
FROM: James Van Dycke JVD
DATE: March 18, 2001
SUBJECT: Annotated Bibliography Deadline

We've run into some difficulty in getting the market data for our study. I will need an extension. All sources should be here by March 25, so I could get the preliminary findings to you by April 17.

We feel we will have the information by March 25. If you feel the need to increase the speed of our response time, please let us know.

FIGURE 10.2 **Standard memo format**

Figure 10.3 presents a structural and rhetorical analysis of the memo. In other words, the memo is described as a professional would read it, with the information that the professional would focus on in italics. The same analysis applies to letters. For the most efficient, effective administrative writing, the writer must honor these conventional expectations of the reader. In the same way, scientists honor the conventional expectations of their readers when they structure their scientific articles by a standard order of standard topics.

Structure and Content

For both letters and memos, the basic structure is absolute: an introduction, a body, and a conclusion. Writers must always incorporate this structure into their administrative writing. However, this requirement doesn't mean that every memo and letter will have three paragraphs in it, one devoted to each part of the structure. Sometimes the whole structure will be taken care of in one paragraph, sometimes within a multipage document.

The fundamental administrative writing structure is elaborated in Figure 10.3. In this case, a memo illustrates the structure's parts, but a letter would work out exactly the same, except for the different letterhead formatting (see Figures 10.1 and 10.2). Note how well this structure addresses the rhetorical issues of purpose, audience, and context. The fourth issue, ethical stance, calls for the writer's full attention. There is nothing in this structure that automatically incorporates attention to ethical stance.

A few points about the model memo are worthy of attention. First, note that requests for action are pulled out of the body of the memo and put in

MEMORANDUM (with accompanying memo head)

TO: Direct addressee
FROM: Signator (Remember, this is not necessarily the author of the document.)
DATE: As close to the date signed as possible
SUBJECT: Keywords to describe the problem to be solved

The first paragraph provides an introduction, consisting of the following elements:

1. *A purpose statement,* which tells why the author is writing the memo, for example, to request help, to provide information, to request a meeting.
2. *Background information,* as needed, which should be kept brief.
3. If the memo is intended to provide an *answer* to some question, the answer should be summarized here. Memos are not detective stories to be read for pleasure; the sooner the author can make the main point, the better. It is stylistically and practically offensive to play coy in a memo.
4. If this memo is in response to another, provide *reference to other correspondence.* The references usually take the form of letter numbers or correspondence dates.

The middle paragraph or paragraphs form the *body* of the memo. Here the author will develop *pertinent details, explanations, or arguments* required to expand on the point of the memo. The author may also provide *alternatives for solutions to problems* on which the recipient must make a decision. The author should not rely on the recipient to devise solutions for which he or she has done no thinking, especially if the recipient is a higher-ranking person on the organizational chart.

The final paragraph provides the *conclusion.* Here the author may request specifically any desired action, including the dates by which the action should be completed. The author should comment in closing on his or her *willingness to cooperate.* This should be done with a formal statement such as "If I can help with resources, please let me know."

gg/EC/MO [Typist's initials, author's initials, and ghost writer's initials, if necessary.]

Enclosures [Number and titles of any enclosures are listed here.]

[Memo number may be noted below the enclosures or, if there are none, below the initials.]

FIGURE 10.3 **Structural and rhetorical analysis of administrative writing in a memo format**

the conclusion. Doing this isolates the action so that the reader does not skip over requests that are embedded in the details of the problem in the body. The main question on any reader's mind upon receiving administrative wrting is, What do I have to do? And if *what* is the first concern of the reader, *when* is the second concern, so the author must always specify the date by which an action is to be accomplished.

It's also worth repeating that any answer that the administrative writing has to deliver should appear immediately in the introduction. The explanations and rationale can appear in the body, but the reader should know as soon as possible what the answer is. So, for example, if a manager asks for three worthy candidates from the fifty who applied for a position, the search committee should list the three names in the first paragraph of the memo that makes the recommendations. Writers may resist delivering bad news so directly, but if it is embedded in the body of the writing, the reader may miss it. If it's personal bad news, the reader will only resent being strung along before the boom is lowered.

Length

Sometimes technical and professional writing students are given rather arbitrary length specifications. They are told not to write long sentences, long paragraphs, or long pages. Although such limitations may be necessary when writing newspaper articles, they are not necessary in administrative writing. Conventions of length, such as limiting memos and letters to one page, are simply superstition, put forth as if there were some quantitative formula for strong professional and technical writing. The fact is that clear, explicit prose is so easy to read that it doesn't matter whether a memo runs over a page, or a job application runs over a page, or a résumé runs over a page. The professional reader will not be counting pages but will be looking for substance; and if the writer writes substantively and clearly, then deviations from so-called standard length will not arouse hostility. To assume that readers will be offended by a so-called nonstandard length sells the professional readers short. They are not mindless cogs in a wheel but rather are thoughtful people trying to do their work well. In fact, they are apt to get a bit hostile if a one-page formula letter is so badly written that it conveys nothing and thus requires the reader to waste time using other means to clarify the message.

READING AND WRITING PROFESSIONALLY, AT TWO LEVELS

Whenever professionals read memos and letters, they read them on more than one level. The literal level, what the document actually says, certainly needs to be clear to the recipient; but the implicit level, what is not said very directly, must also be clear. Seasoned professionals read on both levels after picking up the company way of doing business. Entry-level professionals can certainly put these principles to good use as well.

The Literal Level

The reader must be able to understand the literal level of a letter or memo. As obvious as this injunction sounds, it is often easier said than done. For example, a quick reading of the memo shown in Figure 10.4 leaves the reader with the impression that the memo has no purpose; the memo simply appears to record a lot of information that is apparently useless.

Keeping in mind the conventions described in Figure 10.3, we can surmise how Wilcox will read this memo from Walton. He will glance at the subject line for orientation, skip to the last paragraph for his action and the date by which he needs to complete that action, go back to read the first paragraph, all the while skimming the body as needed. After reading this memo, James Wilcox has just one question: why is Russ Walton telling me this? Because there is no clear statement of purpose in the memo, Wilcox has no way of understanding the writer's intention. And if the purpose of the memo

JetStream, Inc.

Memorandum

TO: James Wilcox, Legal Counsel
FROM: Russ Walton, Accounting
DATE: January 15, 2000
SUBJECT: Annual Report

Our client for the past several years has been Mixum Paint and Supply. We often do several jobs for Mixum with the understanding that this work is to be paid on a commission basis.

Last year Mixum had Accounting prepare a report to summarize its actions with various subsidiaries. This report was a rush job and as such did not partake of the same status as the usual contingency jobs would.

In our billing for the special report we have received determinations that our basis for payment will not be met as we had thought it would. Subsequent talks with Mixum confirmed that there had been an error that Accounting must avoid in the future.

I wanted to hear from you about this.

cc: Joanna Belmont

FIGURE 10.4 **Memo with no apparent purpose**

isn't clear, the reader will have problems puzzling out the literal level of meaning.

If the writer violates the conventions or isn't sure of the purpose of the memo and so is unable to make purpose and action clear, then the reader will become frustrated and either file the memo in the trash or telephone the message to the author. A phone call takes time that need not be spent if the memo is written correctly in the first place. And the fact that Walton has even written about his trouble with Mixum tips off Wilcox that he will, indeed, have to call to straighten out the Walton request. The issue must be important; otherwise, Wilcox would never have put his request in writing.

An in-depth look at the memo reveals its weaknesses. First, the subject is very vague. All Wilcox knows is that the memo is about some annual report, but whose report this is and what year the report was produced are mysteries. The subject line of a memo should accurately describe the content of the memo by key words. The best way to think of this function of the subject line is to imagine having to code a document in an index; in this example, the code "Annual Report" gets the reader only to genre, or type of document, referred to. It does not indicate what Wilcox should focus his attention on. The to, from, date, and subject lines are not pro forma parts of a memo. They have a specific function and form a substantive part of the document.

The memo's first paragraph is puzzling. It is in the first paragraph that the reader expects to see a purpose statement, one that would clearly elaborate on the information in the subject line. This memo, however, contains no clear purpose statement. What has the writer accomplished in the first paragraph? From what Wilcox knows about the subject, he can tell that the writer is providing some background information. Although it is true that background information does go in the introduction, this information is secondary to a purpose statement. The conventions of memo writing require that all secondary information be kept to a minimum. The aim is to provide as much essential information as necessary, in as concise a form as possible.

The subsequent paragraphs focus on a particular problem that the Accounting Department has had with Mixum. The reader sees that there has been a misunderstanding and that Mixum seems to be holding out on some kind of payment that Accounting feels is its due. But this whole scenario has a kind of shady feel to it, because, for one thing, the language is so abstract; "as such did not partake of the same status as the usual contingency jobs would" is a good example of this vagueness. Is a contingency job the same thing as a commission job? Using substitute terms for the same idea can be very confusing to a careful reader. The first part of the phrase is simply wordy and badly written; the author is having trouble writing something directly and concretely.

"We have received determinations that our basis for payment will not be met as we had thought it would" provides another example of the problem in this memo. Here Wilcox wonders who made the "determinations," whether Accounting will be paid at all or whether the client intends to dribble out payments, and whether any direct meeting has occurred about this issue between Accounting and its client. The use of "determinations" has two regrettable results: it lets in all kinds of possibilities that the reader cannot interpret for lack of information, and it attempts to dress up what is only vagueness in the end with a falsely elegant, legal sounding term.

The final paragraph, where the reader expects to see future actions spelled out, is as vague as the rest of the memo. In the first place, the construction "wanted" suggests that Walton had wanted Wilcox's help at one time, but now has given up, and so just wanted to pass along this note of disappointment. Even if Wilcox overlooks the problem of tense in this sentence, however, he is still left with the fact that Walton requests no specific action. If the memo is simply informational, then it should say so in language like this: "I wanted you to know the background of this situation before Mixum's lawyers call you next week." Then Walton might have closed with an offer of help. A thorough reading of the memo brings the reader no closer to any certainty about what the writer is talking about than a careless reading does. The reason this memo is so dysfunctional is that all the conventions of memo writing are broken by someone who lacks the ability or experience to break with conventions confidently and convincingly.

Analysis of the Revised Memo. If Wilcox were to "translate" Walton's memo, he would find that Walton has made a very simple request in a very complicated way; that is, his request is bogged down in ambiguity and vagueness. Essentially, Walton wants Wilcox to review the payment problem to see whether Accounting has any legal leverage with Mixum. Figure 10.5 presents Walton's message within the conventions of memo writing.

The revised memo follows the conventions of administrative writing and is a few paragraphs longer than the incomprehensible memo. Yet the added length affects the quality of the memo positively because the memo does all the work it is intended to do. Therefore, no follow-up action is required of the reader. Wilcox has no need to make phone calls or visits to Walton to clarify the translated version of the memo. So, even though the second memo is a bit longer, it is far more efficient than the first.

The revised memo follows the conventions of memo writing so well that Wilcox can read efficiently the way professionals do, from subject line to last paragraph, and back up to the first paragraph. The second paragraph concisely describes a situation about which Wilcox has no information at all, so the description occupies two paragraphs, one to describe the problem and one to describe Walton's attempted solution.

In addition to fulfilling all the conventions, Walton's revised memo also addresses the reader well, anticipating accurately what Wilcox may and may not know about the situation. For all Walton knows, Wilcox may think that Mixum is one of JetStream's own subsidiaries instead of a client. So it is appropriate for Walton to identify Mixum, even though the information may seem obvious. This sort of thoroughness saves extra time later on. It allows the reader to get the literal message as efficiently as possible.

The most important virtue of Walton's revised memo is that it is conclusive. He thought through what he wanted to say; in fact, he drafted what he wanted to say before sending off his ideas. Walton's first memo looks like a rough draft because it is so inconclusive, uncertain, unorganized, and vague. That's the way draft writing looks. As long as the writer doesn't actually send

JetStream, Inc.

Memorandum

TO: James Wilcox, Legal Counsel
FROM: Russ Walton, Accounting
DATE: January 15, 2001
SUBJECT: Request for Legal Review

I need your help on a matter with one of our clients, Mixum Paint and Supply. The Mixum file needs a legal review because of a mix-up Accounting had with the company last month.

Briefly, we did some extra work for Mixum on the assumption that this work would be paid for in the same manner as the other commission (or per piece) work. Mixum, however, claims that this extra work is covered by the regular fee schedule, some 25 percent less than the commission schedule.

I've talked to Mixum's administrative manager, Dot Taylor, but she sees no way for the work to be paid for as I think it should. She referred me to Mixum's attorney, Ralph Wainright, if I have any further questions.

Will you look at this file by January 25th to see whether we have any legal leverage to claim the work as commission work? I need to know your opinion so that I can institute a suit, if necessary.

If you need further information, please let me know.

fm/enclosures: Mixum file
December file

FIGURE 10.5 **Revision of the purposeless memo**

the draft version, it doesn't matter what it looks like. It may well take a draft for the author to figure out what needs to be said and to discover what the memo's purpose really is. Especially when a person wants to write a memo in response to a touchy problem, it is useful to draft the memo first so that it can be transformed into readable text for its audience.

Remember, the strength of administrative writing conventions is that they address the reader's concerns, not the writer's concerns. Clarity is absolutely necessary for the reader to understand the literal level of the message.

Common Errors in Administrative Writing at the Literal Level. Although Walton's original memo is certainly badly written, it is not unusual. It instructively embodies the four most common errors committed by administrative writers:

1. Neglecting purpose statements
2. Burying the main point in the body
3. Including background information not pertinent to the reader's purposes (a result of poor audience analysis)
4. Calling for only vague actions

The Implicit Level

We mentioned earlier that professionals read on two levels at once, the literal level and the implicit level. This section takes up the implications of administrative writing, which tend to be political. Everyone in an organization, from the mailroom attendant up, reads administrative writing at the implicit level. Authors should be able to read their own writing at this level to avoid inadvertently putting themselves at a disadvantage.

There is a lot to read in a memo on a literal level, and the conventions of memo writing help standardize the process. But the implicit level—which is largely political—may yield far more related information than the reader could ever get from the literal level.

Elements to Consider at the Implicit Level. There are several elements that contribute to one's understanding at the implicit level.

Writer's Tone. The most obvious element contributing to implications in memo writing, tone conveys the importance of a subject for the writer. The tone tells whether an assignment seems urgent *to the writer.* This information is important, not because the reader must also believe the situation is urgent but because the reader can be alerted to others' likely responses. The emotions associated with the topic and the value attached to the topic constitute the tone of a document.

Identity of the Author and Indication of a Ghostwriter. The From line gives the signator of the memo. But the person who signs the memo is not necessarily the author. A highly placed executive, in fact, will rarely be the author of a memo or letter, choosing to delegate such work to administrative assistants, deputy managers, or secretaries. Therefore, if the document lists a highly placed executive as the author, the canny reader immediately looks for clues as to the true author of the memo. One place that information can be found is at the bottom of the memo, where the typist, who may not necessarily be the author's secretary, lists typist's initials, signator's initials, and ghostwriter's (or actual author's) initials. This information is important because it tells the reader whose policy, or problem, or position is currently in favor with the signator. Furthermore, such information tells the reader who needs to be persuaded or dissuaded about any issue that might be causing trouble.

For example, Mark Thibadeaux receives a memo from Assistant Comptroller Sharon Weist, telling him that the company is changing its policy on copy machine use. At the bottom of the memo Mark sees this: jm/SW/JH. This tells him that Janet Monroe typed the memo, which is reasonable since she is Sharon's secretary. It also tells him that JH, Jeff Holz, is the true author of the memo, which implies that Holz is the one responsible for this policy change. Because Mark doesn't agree with the change, he can start trying to reverse it by speaking to Jeff. Sharon will be more likely to reverse her decision if Jeff encourages her to do so.

Recipients of Copies. The copy notation will appear in different places in a letter or memo depending on a company's habit, but it will usually appear directly below the heading or at the bottom of the document, as it does in the first version of Walton's memo.

The people listed in this section indicate who is also involved in the situation under discussion. In effect, the copy notation provides further context about the topic. Some of this context is simply organizational, as when an author will send a copy to a subordinate who is in charge of a project under discussion but who has no actual administrative authority in the matter.

More than organizational information, though, the copy notation provides political information, answering such questions as the following: Who are the "players" in this situation? Who are the major powers behind this policy? Who is being groomed by the author? Who is being subordinated or snubbed in the situation, by virtue of being excluded from the copy notation? Who are the insiders, and who are the outsiders?

The copy notation also can be a persuasive device in a letter or memo. If, for example, the author feels that the reader will be hesitant or reluctant or unenthusiastic about taking a requested action, the author may send a copy to the reader's supervisor or the author's own supervisor to apply some extra pressure. For example, if Olivia Draper wants to revise the marketing approach her department has for a new product that's not doing well, she might get more immediate attention if she sends copies of her memo to her boss to other key members of the managerial/marketing team. Intelligent use of copies eliminates the need for any kind of desperate tone because several readers are in on the problem that needs to be solved. The rationale is that concerted attention will take care of problems more efficiently than sequential contacts, one letter or memo at a time.

But here's a caveat. We need to point out here that some caution is necessary in this business of sending copies to multiple readers. If Olivia is not careful to propose her revision in a reasonable, thoughtful way, she runs the risk of angering her boss, who could think that she is going above his head before he has had a chance to respond to her ideas. It is crucially important that she make her primary reader, her boss in this case, know that his help and decisions are important in getting the revision done. If she is careful in creating the right tone, her boss will be favorably disposed to considering her ideas, and perhaps even more important, she will not run the risk of turning off the others who have received copies. After all, they may be very favorably inclined toward supporting her boss, and she certainly doesn't want to end up doing nothing but shooting herself in the foot.

Another prudent use of sending copies is to alert higher-ups to actions the author is taking to resolve certain potentially messy problems. If, for example, James Blankley has discovered a design flaw in the plans for a nuclear waste forklift, he will want to address this problem in written communication because it has potentially damaging consequences. Because of safety concems and legal liability questions, James will write a memo to the head of the design group, thus documenting the problem, and he will send copies to his own boss and other higher-ups (like the Manager of Safety) who should be aware of the problem, that is, *cognizant* of the problem, to use the workplace language. The copy notation should eliminate any hesitation in the design group about correcting the problem. Although overuse of copies can be annoying, especially if used mainly for self-promotion, their reasonable and prudent use can promote better business practices.

Even though reading for implications may seem Machiavellian in an academic discussion, it is a necessity for sophisticated action in the corporate world. All experienced employees read for what is implied as well as what is directly stated. Professionals need to incorporate sophisticated reading strategies into their everyday lives to be more effective in their work.

ANALYZING THE ADMINISTRATIVE WRITING PROCESS

Jay Dalton wasn't an experienced technical writer for nothing; he knew how to solve problems with memos and letters. The process he went through can be outlined as follows:

1. Isolate and articulate the specific problems that need to be solved.
2. If more than one problem needs to be solved and the problems are not related, then prioritize the problems and take them on in separate letters or memos.
3. If the audience is internal, write a memo.
4. If the audience is external, write a letter.
5. If the problem is a personnel problem within an organization, write a letter. A letter carries more official weight than the informal memo does.
6. Follow the conventions of administrative writing (see Figure 10.3) for structuring the documentation.
7. Clarify the literal purpose by writing a draft of the documentation. The writing of a draft may bring up the principle behind the specific problem; in other words, drafting is actually part of the thinking process in writing and is not an indicator that the author is inefficient or inexperienced. Especially if the issue is one the author feels strongly about, the documentation should be drafted and revised. If the author writes a memo or letter about something that really makes him or her angry, writing a draft will allow the author to vent emotions that, if sent along in the final document, might alienate the reader. There's nothing wrong with indicating one's displeasure, but the drafting process allows the writer to decide how much to tone down the language, not necessarily to save the reader's feelings but rather to protect the writer.

8. Once the literal purpose is clear, state it in the first paragraph of the document. Be sure to make the document conclusive. Therefore, if the document is intended to relay an opinion, state it up front. Don't lead up to it dramatically in the final paragraph.
9. Review the draft for its implications. Check through all the elements that contribute to an implied reading of the document.
10. If the memo is intended to propose a solution to a certain problem, then present the reader with alternatives to choose from.
11. If the memo is intended to transmit a policy or solution to subordinates, then present a rationale as well. Remember that subordinates, even if they don't hold professional titles, deserve the courtesy of basic explanations.
12. Proofread the document after it is typed or word processed. Remember that the signator is responsible for any errors in the document, whether or not the signator is responsible for the actual word processing.

Isolating the Main Problem

As item 1 in the previous list states, writers should isolate and articulate, at least to themselves, the specific problems that need to be solved. There may be a way to solve more than one problem at a time as long as their overall presentation is uncomplicated. In fact, it may be most efficient to solve related problems at the same time. But the writer needs to be sure that the relationships are real and that the presentation is clear. For example, Jay had to decide here whether the problem with Texas Oilcats was an isolated problem or whether it was the most disturbing manifestation of an underlying problem.

Identifying Underlying Problems

The distinction between an isolated problem and one that underlies other problems is an important one to consider. If a problem is mislabeled "isolated" and treated as such, yet is actually only a manifestation of a deeper problem, then any solution for that isolated problem will be ultimately inadequate. The following example demonstrates this issue:

> John Smith is disturbed about the difficult attitude that he has noticed recently in his subordinate, Kathy Marsh. Whereas Kathy used to be supportive and cooperative, she is now sarcastic and truculent. She seems to be sniping instead of communicating with him, and frankly, this bad attitude of hers makes him want to avoid her. As he sits down with his problem, he brainstorms options for dealing with it:
>
> **OPTION 1:** He can ignore the problem, do nothing, and wait for Kathy to change back to her old supportive self.
>
> **OPTION 2:** He can call her into his office to discuss this problem and to get more information.
>
> **OPTION 3:** He can write a letter to her file, citing her difficult behavior as cause for a lower-than-average raise.
>
> **OPTION 4:** He can write a memo to all employees about the importance of professional behavior in the workplace.

Of the four options, his most reasonable choice is number 2. If he selects this option, he will discover that Kathy is in the middle of a divorce and is suffering from upheaval in her life. John will also realize that he has no reason to take her attitude personally; Kathy is just running on a short fuse because of the stress of her divorce. Assuming that they have a satisfactory conference, there will probably be no need to document this problem at all, so no memo or letter will be necessary.

If John chooses the first option, to wait, he may also get a satisfactory result. After several months, as Kathy gets adjusted, she perhaps will calm down, and things will return to normal. But avoiding discussion of the problem will increase the time necessary for its solution and may build up a wall that will be difficult to overcome once Kathy regains her good temper; in the meantime, John will have suffered, and he may find it difficult to be open with Kathy once she decides to cooperate again. Kathy needs to know that John is feeling some discomfort from her difficult behavior.

The third option would be disproportionate, given the long good relationship John and Kathy have had before the sniping began. If John writes a letter to her file, he will be relieved a bit because he has taken some action, but he will also have to understand that this documentation could hurt Kathy in the long run. If John leaves the letter in her file even after she recovers from her difficulties, she will have a permanent blot on her record even though her difficulties were transient. In that light, John's documentation would seem a bit harsh and unforgiving. He could avoid any kind of permanent damage by removing the document from her file as soon as he felt better. But, if he intends to make the letter a permanent part of her file, no matter what, then he is ethically bound to let her read the letter and respond to it. If company policy does not require such fairness and John does not initiate it, then he should ask himself some questions about whether he would want his manager to treat him so punitively if the situation were recast with him in the role of sniper.

Although the fourth option may seem ludicrous, it is popular with those who want to avoid confronting a single subordinate. It is a sort of "scattershot" approach to send out a general lecture in response to a specific problem that concerns a specific employee. It is also an approach that managers use all too often in business. The problem with taking this approach is that it can cause more instead of less trouble. The intended subordinate may overlook the connection, and the other subordinates may get offended by such a general lecture that they can see no cause for. The behavior is not unlike that of the preacher who scolds the assembled congregation for not coming to church. Not much percentage in that.

In getting back to Jay Dalton's situation, we find that he wants to take the most efficient approach possible to the Oilcats problem. After thinking about it for some time, he realizes that this problem is, indeed, an isolated incident. There is no other account for which he and the other employees have to do such unreasonable deadline work. He will address the problem as a single problem, then, instead of pulling in other "related" problems as long as he's going to be proposing solutions. It's important to take problems on one at a time unless the specific problem is actually a symptom of a larger one.

Applying the Principles of Rhetorical Problem Solving

Having decided the nature of the problem, Jay considers it in light of the rhetorical problem-solving process that has always been useful for him. This process is presented in detail in Chapter 3.

Jay determines his audience first. He decides to address more than one audience at a time in this memo, because several of the people he reports to are concerned with the problem. His immediate manager, Jim Rothschild, fields all the work with the Oilcats, but the main contact between WS and Texas Oilcats is actually Jim's manager, Tom Fields. Jay decides that he wants Tom to be aware that the Oilcats is a problem account and that he has a workable solution to the problem; however, he respects the organizational channels in WS and doesn't want to jump over his own manager's head with all his bright ideas. Therefore, he decides to make Jim his primary reader but also to send a copy of his memo to Tom so that he can be aware of the situation.

Jay's notion of the purpose of his memo is already clear. He wants to propose that a chronic problem be solved. Since that is his purpose, he will organize his memo accordingly. An intelligent strategy for a problem-and-solution memo is to describe the problem concisely but to focus mainly on the solution to the problem, by providing alternative solutions as the audience might require them.

The context for this memo is fairly complicated. Recall that context often concerns the political and emotional aspects of a situation. So, for example, a touchy issue, one that the recipient doesn't want to hear about, would require a different strategy from that of an issue that seems to be a simple matter of fact. Because personnel problems are particularly touchy issues, they should be approached with great sensitivity. A new operations policy, on the other hand, rarely evokes a strong emotional reaction. In the Texas Oilcats case, the problem is the struggle between the employees' interests and the company's interest, which often coincide, but not always. To satisfy the company's best interest—to please a difficult and demanding client—the employees must work unreasonable hours for a few weeks during the year. Jay decides he must be careful to show that he understands the company's interest in this problem but that it is also in the company's interest to treat staff respectfully.

Finally, the ethical considerations in this problem are substantial. Jay must first determine what his ethical stance is in this situation and then what actions will be in response to that stance. His position is that in a small company like this one, management should treat employees as partners. The management, in fact, does treat WS employees that way in a number of policies; for example, WS employees may choose profit-sharing programs and protected savings programs. Nevertheless, Jay decides that since the employee abuse over Texas Oilcats has been going on so long, he is obliged to voice his concern. He also believes that there are several alternatives to consider, that the situation is not a simple, clear-cut matter. He will be able to live with himself better knowing that he has defended his principles and has tried to be constructive.

SAMPLE PROFESSIONAL MEMO WRITTEN TO SOLVE A PROBLEM

After putting his problem through the rhetorical problem-solving process, Jay is much clearer about the direction he will take in his memo. All this prewriting analysis will save him several drafts of the final copy. Figure 10.6 presents the final copy of Jay's memo.

THE WRITE SOLUTION

Memorandum

January 31, 2001

Subject: Proposed Solution to the Texas Oilcats Deadline Crunches
From: Jay Dalton JD
Telephone: 509-236-2198
To: Jim Rothschild
cc: Tom Fields

We seem to have done such a good job accommodating Texas Oilcats that the company has no need to change its pattern of doing all its business with us at the last minute. This pattern has caused a lot of trouble within our own ranks. I think the solution is to contract help for these crunches, making sure that the Oilcats bear the cost.

Since the Oilcats are such an important client, we are in a difficult position. All our other clients who tried to ignore our deadlines were either educated or refused. We learned that turning away a few difficult clients was worth it, both for financial and morale reasons. But the Oilcats account pays our January bills and then some. If we turn this account down, we will take a major loss: we have tried education, with no success. If we try education again in some kind of formal, serious way, we might well be wasting our time and we might also come across as if we are asking the Oilcats to solve our own problems. At this point, the Oilcats have no problem—we should remember this.

On the other hand, if we solve this problem on our own, we will not have to engage the Oilcats in a time-consuming and unguaranteed education process, yet we will still respect our own people. My idea is that we fill in the labor gap with contract editors and printers.

(continued)

FIGURE 10.6 **Jay Dalton's memo**

My estimates show that the cost of contract labor for the overtime hours we normally work for the Oilcats would be about 10 percent of the usual cost. We should tack this 10 percent onto the usual fee: the Oilcats will gladly bear the extra cost because the quality not only will remain consistent but also will probably be enhanced. Our own people, who will not be burned out on the job, will be sharper during the process, thus eliminating some of the costly alterations (and further chance for error) that the process usually demands.

The gain in morale from this move will make the step worthwhile. We have made a point of hiring reliable, consistent people. We have treated them respectfully—except for this job. If we don't take some action on this problem, we risk losing our people to places that pay overtime. When employees work overtime, they want to get paid for it. Our system makes no provision for overtime pay for salaried employees.

What we need, simply put, are more hands on the job. If we add a few hands, we will have no further problem with the Oilcats: we will retain our own people, who know the Oilcats account inside and out: and we will have upheld our own policy of putting our own people ahead of any unreasonable demands of our clients.

Please let me hear what you think of this proposal by February 15. I have to meet with the Oilcats then to set up next year's schedule and budget.

FIGURE 10.6 (*Continued*)

A few things about Jay's memo deserve comment. First, WS's memo heading style is different from the heading style that JetStream uses. The point is simply that different companies use different memo styles. It is unnecessary, therefore, to memorize heading formats as long as the style contains essential parts—*date, subject, to* and *from* lines, and copy notation. The advantage of WS's style is that the reader sees the *to* line and copy notation together, all the better to analyze audience and purpose of the memo. Neither this style nor any other, however, will account for any blind copies, which are copies sent to people not listed on the memo itself. Jay's memo is a straightforward proposal and problem analysis; his sending a copy to Jim's manager, Tom, is a message to Jim that he wants serious consideration of this proposal. It is also a message to Tom that he is a constructive problem solver with administrative capacity.

One thing that bothers Jay about his memo is that it exceeds one page. However, Jay has done the reasonable thing: said what he had to say as concisely as possible. If a memo must exceed one page, it should be clear and well-written so that the reader will not feel put out. Jay's memo adheres to these guidelines.

One of Jay's more successful accomplishments in this memo is making it extremely explicit, which can be seen in three places: his subject line, thesis, and topic sentences. The subject line tells the reader that Jay is proposing a solution to the Oilcats problem. It doesn't state exactly what that solution is, but it cues the reader, eliminating the need for a lot of identification in the first paragraph. Jay knows that his audience knows as much (Jim) or almost as much (Tom) about the Oilcats account as he does, so he is relieved of the need to provide a lot of background or introductory information. In essence, the introduction to Jay's memo appears in the subject line.

Despite the fact that the memo exceeds a page, it is not an onerous job to read it because the reader knows immediately where Jay intends to go in the memo. The last sentence of the first paragraph contains his thesis, his main point, his conclusion, his reason for writing, his results—all these terms are a way of naming the most important information that the writer has to relay. This placement does the reader the courtesy of getting to the point. Once the reader knows the point, she or he can switch gears to examine the logic of Jay's argument.

This order—a signaled solution, followed by arguments in favor of it—is the opposite of the order seen in the hard sell. This other rhetoric often reviews logical support before presenting the solution or altemative that is being argued for. For example, television advertising often begins like this: "Do you want to feel like a king? Do you want to get rid of the drudgery in your life? Do you want to have the nicest yard in your neighborhood? Then you'll love Edge-O-Matic!" The "reasons" are presented before the solution, all to build suspense, excitement, and interest in the product. In professional writing all such suspense-building should appear in the public relations and public affairs writing departments, the most frankly persuasive arms in any organization. The workaday departments rely on a much more explicit, straightforward presentation. Jay's audience is already interested in his solutions; it is Jim's job to be interested in Jay's solutions, and just to ensure that added bit of attention, Jay sent a copy of his memo to Tom. Jay's thesis statement clearly states his conclusion and, as a bonus, explains who is going to pay for the solution, always a major concern when someone proposes to bring in more people to help with a project.

The rest of his memo reads smoothly because of his excellent topic sentences. The test for topic sentences is simple: read only the first sentence in every paragraph to discern any order and sense in the memo. Following is a paragraph consisting of only the topic sentences of Jay's memo:

> We seem to have done such a good job accommodating Texas Oilcats that the company has no need to change its pattern of doing all its business with us at the last minute. Since the Oilcats are such an important client, we are in a difficult position. On the other hand, if we solve this problem on our own, we will not have to engage the Oilcats in a time-consuming and unguaranteed education process, yet we will still respect our own people. My estimates show that the cost of contract labor for the overtime hours we normally work for the Oilcats is about 10 percent of the usual cost. The gain in morale from this move will make the step worthwhile. What we need, simply put, are more hands on the job. Please let me hear what you think of this proposal by February 15.

These topic sentences give a very accurate précis of the whole memo. All the major topics—cost, benefits, problem analysis—are clearly signaled. And each paragraph devotes itself to the function determined in the topic sentence—introductory, developmental, concluding. Jay's memo will require significantly less follow-up time for his reader than would be the case with Mr. Walton's original memo. Jim Rothschild will not have to make follow-up calls or visits.

The principles of solving problems in memos and letters are constant, although their application is myriad, given the infinite variety we find in human nature.

POSTSCRIPT

Jim liked Jay's suggestions very much. The Oilcats grumbled about the extra cost but didn't want to lose the "experts" on its report, so it agreed to the change and then rushed its efforts even less. The extra help eased employee tensions considerably. The quality of the document was enhanced. The WS employees put away their résumés for a while. Jay's good efforts brought him a raise and a promotion to line manager. Jay slept very well for a few months, until the next crisis presented itself.

TRANSMITTAL LETTERS AS PROBLEM SOLVERS

Of the many types of letters that companies write, one type tends to be overlooked—the transmittal letter. Customer-oriented letters are likely to get handled by the public affairs or public relations branches in companies, but the other professionals, including scientists, engineers, and technicians, will have to write transmittal letters. The transmittal letter is a sort of preface to the document it sends. It may summarize the document's organization and structure; it may explain, in the case of a completion report, what expectations found in the proposal did not work out in the actual research. These discrepancies should be explained at a summary level. It is worth considering the problems that transmittal letters address.

Functions of a Transmittal Letter

The three major functions of the transmittal letter are

1. To reach a wide audience for a regularly produced document
2. To record production and delivery of an important document
3. To orient the main addressees to any changes in purpose, format, or audience that may have been decided on from issue to issue

In fact, any change that might be seen as a discrepancy from previous conventions and expectations needs to be signaled in the transmittal letter. Even if the recipients initiated the changes, it is important to remind them in the transmittal letter so that they can adjust their expectations for the way they should read the document.

People are so accustomed to reading standard reports in a certain way that it throws them if the report doesn't look familiar in a particular issuance. And, given professionals' huge reading requirements, familiarity is essential. For that reason, the transmittal letter must clue them in to the fact that something about the familiar document has changed. Readers would be shocked, for example, if *The Wall Street Journal* began to publish in a *National Enquirer* format. So, too, are professional and technical report readers shocked by any violation of their reading expectations.

Transmittal Letters as Documentation. A very important function of the transmittal letter is to record the fact that a report was produced and distributed on time, per previous agreements with the recipients. When companies are doing contract work, for example, reports are often contractually required and constitute a major source of income. Much of the work done for the federal government is this sort of contract work, most commonly associated with the Department of Energy and the Department of Defense. The contractors to these agencies may be paid on a fee system that requires delivery of status reports that document that certain contracted work was completed. It's very important that these reports are truthful, since they become the basis for paying the contractor. It's also important for the transmittal letter to be signed only after the sender is satisfied that the report is truthful, because the transmittal letter is a sort of certification for whatever it transmits. The signator should be able to stand by the information enclosed.

Copy Notation for a Transmittal Letter

It is not uncommon for transmittal letters of major quarterly reports, for example, to contain numerous names in the copy notation. Besides the main addressees, who might number ten, one might find a copy notation containing thirty-five names. In such cases it is unreasonable to consider all these entries as audience problems when drafting the report. Such a list is only functional, serving as a distribution list. It may have some political significance in its honorific additions or punitive subtractions, but intensive attention to the list as an audience issue is not necessary. Many people included in the copy notation shelve their copy of the report, keeping it only for reference. In fact, certain libraries and reading rooms often appear in copy notations for major reports.

The major concern for the author is to update the list of recipients. Although a secretary keeps the typed list current, the author is ultimately responsible for providing the names of recipients.

SAMPLE PROFESSIONAL TRANSMITTAL LETTERS

The two accompanying sample transmittal letters indicate some of the functions that have been discussed. The first letter (Figure 10.7) is from The Write Solution to the Texas Oilcats. It is the transmittal letter WS has been sending the Oilcats for the past five years, with updated names on the distribution list, which is lengthy since it lists the recipients of an annual report. Many people would have an interest in such a report.

The second letter (Figure 10.8) is from WS to another client, who requested some changes to the format of the report that WS prepared. That change is summarized for the client, so that she won't be stymied by a new format.

THE WRITE SOLUTION

7355 La Tertulia Dr, Santa Fe, NM 88000

January 31, 2000

Mr. Cecil Simons, Manager
Public Affairs
Texas Oilcats
1499 Live Oak Boulevard
Dallas, TX 54666

Dear Mr. Simons:

We are pleased to deliver Oilcats Internationale for your review and approval. It seems the Oilcats have had another good year. We hope the coming year holds the same success for you.

Please contact me if you have any comments or suggestions about the report.

Sincerely,

Tom Fields

Tom Fields, Manager
Production

sr/TF/JW

cc:
M. Cummings, WS
J. Cushing, WS
A. Campbell, WS
W. Jenkins, WS
B. Williston, WS
E. Gonzales
R. Winchell
M. Aguirre
S. Schatz
A. Montfort
R. Hitchcock

FIGURE 10.7 **Transmittal letter from the Write Solution to Texas Oilcats**

THE WRITE SOLUTION

7355 La Tertulia Dr, Santa Fe, NM 88000

January 31, 2000

Ms. Rachel Ramon, President
Southwest Technologies
P.O. Box 199
Albuquerque, NM 88102

Dear Ms. Ramon:

We are pleased to deliver the hydrology report for the second quarter, 2000. As usual, the production process was very smooth. The report's revised format seems to be a real improvement for the reader.

I would remind you of the four changes in this issue that we agreed on in our meeting last July 25th:

1. We changed the photography from black and white to full color.
2. We published in the new cover format worked out with your public affairs people last November.
3. We reordered the report to eliminate two small sections—fire safety laser applications—and to combine three other sections under the heading "Education." This new order appears below.
4. Finally, we revised the back cover logo per instructions from Southwest's public affairs department.

The new order of chapters is as follows:

I. Highlights of the Second Quarter
II. Laser Research
III. Defense Work
IV. Automotive
V. Education

This new order should enhance coherence in the report. If you have any questions or comments, please call me.

Sincerely,

Tom Fields

Tom Fields, Manager
Production

sr/TF/JW

cc: J. Dalton
R. Winchell
M. Aguirre
S. Schatz
A. Montfort
R. Hitchcock

FIGURE 10.8 **Sample professional transmittal letter**

WRITING FOR SOMEONE ELSE'S SIGNATURE

As we noted in Chapter 3, one way that problems get solved in the professional world is for subordinates to write memos and letters for their managers to sign, so that subordinates become ghostwriters, in effect. This responsibility can demonstrate how well-suited a subordinate may be to make the move from salaried professional to manager. Such a demonstration would require that the professional show evidence not only of writing competence but also of problem-solving competence.

Considering the Audience

The difficulty about ghostwriting a memo or letter is that even though there may be only one addressee—no copies, blind or noted—the writer must always address at least two readers, the addressee and the signator. This is a matter of addressing audience correctly, and professionals will find this job easier to do, the better they know the signator.

Analyzing the Rest of the Rhetorical Situation

The second thing to remember about writing for someone else's signature is that it is crucial to analyze the rhetorical aspects of the job, sometimes referred to as "scoping out the job," before anything gets written. This process requires that the writer must know exactly who is to be addressed and for what purpose, what the context is (political and otherwise), and what the signator's ethical stance is on the subject. In other words, authors must be sure to put the writing task through the rhetorical problem-solving process in the same way they would if they had initiated the task themselves.

Proofing the Document

The third important point about ghostwriting is that signators may well insist upon a standard of perfection from the ghostwriter that they would never expect of themselves. This added responsibility can be weighty, because the writer cannot count on the signator to find all the errors that need correcting at the draft stage. That situation might mean that there would be errors in the final stage and that there would have to be numerous revisions for minute offenses.

When any signator or manager demands perfection of a document, the writer has no choice except to work with a partner to proof the document. Remember that computer programs that check spelling will find obvious typographical errors, but not the homonyms that constitute improper usage (*there* for *their* or *they're*) or the errors that form real, but inappropriate, words (*pubic* for *public*, for example). The writer and a partner will have to undertake the painstaking process of going through the document page by page. It is intense work but will often reveal errors that the writer just can't find after having read several versions of the same page.

CONCLUSIONS

This chapter emphasizes the rigorous analysis that administrative writing requires. It is conventional in that professionals have expectations that the structure needs to fulfill. But it is a matter of individual problem solving as well, since each occasion for a memo or a letter requires the writer to ask, first of all, whether any documentation is really necessary, and second, the questions involved in the rhetorical problem-solving process. Length is less important than clarity and logic; format is less important than essential structure. Problem solving through memos and letters is the most common and potentially effective means that any professional has with which to work more effciently.

EXERCISES

1. Interview a professional or graduate student in your field to determine what kinds of situations require documentation. Explain to your interviewee what your understanding of documentation is and when it is appropriate or necessary, and then find out how and to what extent the interviewee's understanding and experience are the same.
2. Case Study

The Situation

You have just been promoted to management. To celebrate, you are taking a vacation before you start your new position, Manager of Daily Operations. You stop by your new office to check your mail before you drive to the airport. You are already in your Bermuda shorts; you cannot stay longer than one hour in your office, or you will miss your plane. When you get to your desk, you see that you have three problematic situations already waiting for you.

The Task

Write the documentation you feel is appropriate for the following situations. First, you must determine what about the situation requires documentation. Then, you need to analyze the rhetorical situation. Finally, you should remember to use copies effectively.

After you've written the documentation that you think appropriate for these questions, write a memo to your instructor providing a rationale for all your important decisions, both your business decisions and your compositional decisions, such as whom to address, whom to send copies to, and the like.

THE PROBLEMATIC SITUATIONS

A. In a move calculated to impress you, Sam Young, your administrative assistant, authorized two secretaries to track the comings and goings of all sixty-five employees in your building. The employees are now very angry to have received memos, signed by Sam with a copy to you, totaling lost work from late arrivals and early departures. The sixty-five then authored a memo to you, complaining that the company no longer trusts them and they don't appreciate being nickled and dimed to death.

B. The engineering group that reports to Daily Operations designed a special forklift for moving nuclear wastes. The prototype is ready, and sales are lined up. As you look over the design, however, you discover that the forklift has inherent instability that will make it lurch on the slightest grade (the prototype was tested in a flat area).

C. You read in the weekly accomplishments report that the scientific group reporting to Daily Operations is very close to perfecting the no-run nylon hosiery that the world has been waiting for. However, you know, but the scientific group does not, that the no-run nylon project is about to be scrapped. The scientific group never made clear the importance of the project or its nearly completed status to the budget people. The project will be dropped in a week if you do not step in. You are certainly not anxious to lose either the project or the cooperative stance that the scientific group has the reputation for.

ASSIGNMENT

1. With an eye to the organization chart provided here, and bearing in mind the process of analyzing the rhetorical situation, write the summary memo that will most clearly and honestly describe the following problem. Feel free to define Project X as you wish.

   Project X is behind schedule. The project is supposed to be completed in one month, but because of various problems, you are, in effect, three months behind schedule. As the lead on the project, you feel you must write an overview of the situation for your manager's signature. The general manager must know what is going on so that he can authorize steps to help you meet the deadline. In business, the bad news must be revealed quickly so that the consequences can be mitigated if at all possible. People don't simply let deadlines slide by without taking some kind of action, at the least informing responsible managers in enough time for them to try to change the outcome. Your overview should be addressed to the general manager, who has a long-standing professional relationship with the supplier (his cousin).

Contributers to the Problem

A. The supplier, C Corp, has consistently missed delivery dates, once delivering material three weeks after the contracted date.
B. The technician from the engineering group has been out of the office five times in the last month. This technician is the sole expert on the computer graphics necessary for your project.
C. Your own scientific group's work has been hampered by miscalculations on the part of Mr. Hamilton, who will retire in a year after thirty-five years of service to the company.

Complicating Factors

A. Your manager wants to push Hamilton into early retirement.
B. Hamilton taught you everything you know about the company.
C. Your company will lose 50 percent of its fee if the deadline is not met.

Organization Chart

General Manager
(Les Michaels)
Addressee

Engineering
Group Manager
(Karen Wolcott)

Computer Technician
(Sam Cruikshank)
Contributor

Scientific Group Manager
(Rudy Larsen)
Signator

Project Lead
(You)
Writer

(Mr. Hamilton)
Contributor

The Rhetorical Situation

A. You are working for a promotion to management. You have a problem, though, because as the lead, you are responsible for the success of the project but have no true administrative power.
B. Les Michaels is always interested in the bottom line.
C. According to the organization chart, you are on the same level as the uncooperative computer technician.
D. Your own manager, Rudy Larsen, will be loathe to inform Les that there is a problem unless he has also thought of a solution.

After you have written the overview that you think is appropriate given the constraints of this situation, exchange with a classmate, who, acting as your manager Rudy Larsen, will decide whether to sign the memo as written or to send the memo back for changes.

TERM PROJECT OPTION

Write a transmittal letter to your instructor to introduce your major project. Not only will you want the letter to appear in some recognizable format, but you will also want to adhere to the demands of the transmittal letter outlined in this chapter. Be sure to document any changes from your proposal that your instructor might be suprised by. For example, if you have decided to address your project to a different audience from the one you designated in your proposal, you'll need to account for that change.

Solving Problems in the Professional Job Search

CASE STUDY

EVENING RUSH HOUR

Interstate 205 ■ Atlanta, Georgia

John Loomis sat up straight in the driver's seat on his commute home as he felt the ambivalence of the past few months resolve into a cold certainty: he must begin another job search. This decision was the one he dreaded because it meant taking on a very time-consuming, painstaking "second job" to find a new one. His current job as staff geologist at GEOLAB more than filled his workweek. But he and GEOLAB had not worked out.

For the third time in as many years, John was passed over for promotion to a middle management position. More serious was his doubt about the direction that GEOLAB had taken, especially the contract work with industries responsible for pollution, work that he found more difficult to excuse. His unflattering findings about these industries' activities may well have been responsible for his static position. He was more than competent, but his competence had proved to be rather inconvenient for

PREVIEW OF THE CHAPTER

(continued)

GEOLAB's management. This latest deferment on his promotion sent him a clear message: be a team player, or be ignored. Although John was willing to cooperate with a company's often contradictory missions, he was unwilling to bury the results of his work. He knew that other companies would appreciate his scruples; he need not accept that his work and his values had nothing to do with each other.

His decision made, John began to realize that his new job search would be a far cry from his first. As a graduating geologist, John had put together a rather traditional cover letter and résumé for the university placement office, which put them on file for the recruiters to read. He then interviewed with six different companies, three of whom made him an offer. He chose GEOLAB because its offer was the most generous. The whole procedure that time was simple. He had only to follow the placement office's suggested résumé format and attach a rather perfunctory cover letter that said, in effect, "You'll see the basics of my education and experience on the enclosed résumé, and I'll be looking forward to talking to you about an entry-level position." He made the ten copies of these documents that the placement office required and then waited for the messages to materialize on his answering machine.

Now there were no recruiters lined up to interview him. His five years out of school seemed like a lifetime. It was hard to recall the brief points on the placement office handout, but John had an ace up his sleeve: he had experience with the hiring end of the job search. This thought encouraged him, so he turned up the air conditioning and began to think through the job search process.

FUNDAMENTALS OF THE JOB SEARCH

John Loomis's job search contains some sophisticated twists and turns that an entry-level search would not simply because he has some professional experience and has been out of school for five years. Yet, John is not so very advanced. For example, he does not have a graduate degree; some of this book's readers, on the other hand, will be going on to graduate or professional school before taking on a very serious job search. Also, as a traditional student, John may be younger than some readers who are nontraditional students. He may be less experienced than our readers are as well.

John's story is important because it depicts the deeper rhetorical considerations that we all must accommodate when engaged in a serious job search. His story provides a feel for the decisions that professionals make as they conduct their search. Those contemplating getting their first job out of college may not care much about the finer points of the search; they just want to do anything to get that first job and may assume that their college credentials will do the trick for them. Yet, unless some magical guarantee that downturns in the economy will never happen again, very few job seekers will be able to rely only on college credentials. Most will have to distinguish themselves from the pack of other job seekers, all with similar college credentials.

Appealing to Employers

The most important point we have to make in this chapter is that job seekers must persuade potential employers that they are the best candidates for the job; they cannot rely on their college credentials or their job experience to speak for them. So, as they look for a job, they must take an actively persuasive role in their job application materials: the cover letter and the résumé. If they do not take on this persuasive job, they will have a longer, more discouraging search. A persuasive cover letter and résumé, on the other hand, will give them the edge to proceed to the interview. Persuasive letters and résumés will gain leverage for the writer, because more people will be interested, and, thus, gain enhanced options.

In Chapter 3, we made the point that the quality of technical problem solving is only as good as the ability to convey it effectively. In Chapter 12, we say that document design can make a good document better, but it cannot make a bad document good. The same principle holds true for the job search: the applicant's credentials and experience are only as good as the presentation of them. Although an attractive-looking résumé can make good credentials look better, it cannot make bad credentials look good. In fact, an expensive résumé that says virtually nothing always looks absurd. On the other hand, the writer's ability to select and highlight those details that will appeal to a prospective employer will enhance a mediocre-looking background. Rhetorical skill will separate the job seeker from the mass.

Applying for jobs, then, is not simply a matter of formatting a cover letter and résumé according to some formula and then waiting for the employers to choose according to the objective criteria advertised. First, the writer is not the passive applicant in this process but has some control over how the materials will be received: rhetorical control. Careful writers can present their case more effectively than other applicants do. Second, the application process is not strictly objective. People, not machines, are being considered. And the application reviewers know all too well that mistakes in hiring are very difficult to overcome and can make people's lives miserable in the process. Think how much of their waking time people spend at work.

We will explore the rhetoric of the job application process, getting essential principles that transform a standard-looking résumé and cover letter into outstanding ones. John's case extends beyond simple formulas and into the harder questions that people need to be asking as they look for a job. We will also look at a student résumé, with an eye to making the best of the student's lack of experience and education.

Remember that John won't be doing anything that an entry-level applicant won't be doing, unless the applicant plans to work only through recruiters on campus. Even those who fall into this category should take John's search seriously because, like him, they may one day decide they want a new job.

Common Job Application Errors

John understood the sort of work that lay ahead of him, because at GEOLAB he had served on three reviewing committees for new hydrology managers. In the course of this review work, he must have read through five hundred application packages that included cover letters, résumé, recommendations, and GEOLAB forms. Being on the other side of the application package, as evaluator applicant, had been a shocking experience. John remembered some of the rejects:

- The "canned" letters
- The perfunctory letters
- The illiterate letters
- The uninformative résumés
- The résumé with suspicious gaps in time
- The inflated résumés

John was taken aback by the overall arrogance—or was it naïveté?—of those applicants who seemed to assume that their fervent desire to work at GEOLAB warranted special consideration. These applicants didn't understand that their desire was shared by all the other applicants, that the reviewers had duties other than reviewing applications, and that most applicants fulfilled the basic objective requirements for entering first-line management and, therefore, had to make a persuasive case for themselves in order to stand out. Figure 11.1 presents one of those rejected letters.

There is clearly more wrong with this letter than there is right. It starts with the addressee information and the salutation. Apparently Tom Jones Smith made little effort to find out who would be an appropriate recipient of his unsolicited letter to Magnitude, Inc., since no one, not even the personnel director, is named specifically in the addressee information. But that kind of information is not hard to find; most corporate websites list the names of at least the people who have positions that deal with the public most often: directors of customer services, public relations, and personnel. If that information is not available through a website, a simple call to the corporation will more than likely turn up the name of the personnel director or of somebody who deals with unsolicited letters of application.

Tom Jones Smith
1045 N.E. Monroe St.
Pullman, WA 99163

January 3, 2001

Magnitude, Inc.
1017 Broadway
Independence, IN 20109
Attn: Hiring Dept.

Dear Sir,

As I am ready to graduate this semester, I am eager to speak with you about suitable job openings.

I have known about the advantages of engineers and have prepared since high school for working in a big company like yours.

My résumé lists all my qualifications for employment.

I look forward to a call from you to set up an interview at your earliest convenience.

Sincerely,

Tom Jones Smith

Tom Jones Smith

FIGURE 11.1 Rejected application letter

Then there's the salutation. Tom is operating under the unfortunate and woefully outdated assumption that anyone who would read letters of application would be a man. If Tom is really stuck and simply can't find anyone in particular to address his letter to, then he needs to address his letter to the Personnel Department in general, which no doubt will be made up of both men and women. The only appropriate salutation in these circumstances is Ladies and Gentlemen, skipping the Dear and ending with a colon.

The body of Tom's letter is both remarkably vague and presumptuous. Tom doesn't bother to mention where he will be graduating from, or which degree in which field he will be receiving. It's anyone's guess what he means by "I have known about the advantages of engineers." Does he mean advantages that engineers have in employment? Or in the business world? Or maybe just in life? He also shows that he knows nothing about the company to which he's writing except that it's "big." Further, he presumes that his enclosed résumé will speak for itself in touting his qualifications. This is probably his biggest mistake. The cover letter must guide the reader to the most convincing, outstanding elements of the applicant's record. No matter how well constructed it is, the résumé itself cannot act alone as the applicant's promoter. That job is the applicant's first, and then with some careful thought and choosing, that of his or her recommenders.

Tom's last mistake is his presumption that an interview will be forthcoming: "I look forward to a call from you to set up an interview at your earliest convenience." Even if Tom's readers have bothered to read to this point in his letter, this one sentence would clinch the deal: the next stop for Tom's letter and résumé is the recycling bin. We would hope that it is just naïveté that makes Tom think that an interview would come in due course, but whether it's simply naïveté or a less attractive case of arrogance, Tom needs to learn that there are no givens or sure things in the business of applying for jobs. A good dose of humility is the antidote for Tom's final mistake.

The rejected applications bore an uncanny resemblance to John's own first cover letter and résumé. Perhaps his father's years of service at GEOLAB *had* played a part in his getting that entry-level position. Nor did many of the applicants John screened receive much help from their recommenders, many of whom responded either as if they hardly knew the applicant or as if this applicant were a trusty collie, loyal to the end.

It had been grim work to sort through those packages, some obviously put together at great expense. The reviewing committee eliminated half the applicants after a ten-second reading of the application materials. The rest were pared down almost as quickly, when a second reading revealed gaps in time or unanswered questions about such things as the number of job switches in an applicant's résumé.

The short-listed applications were more carefully considered and then ranked after some debate among committee members, for substantive accomplishments and recommendations. The quality of the application package itself mattered even more at this point. Professionals must write well; their judgments must be well supported, their style succinct, and their attention to detail meticulous. These characteristics would be evidenced in their cover letters by their ability to anticipate and counter any arguments against their "pitches." In fact, the cover letter itself was more of a pitch than John had ever realized when as a student with a sought-after major, he could pitch randomly. Those applicants at the top of the short list were called for interviews that would begin an evaluation cycle of a much more intense nature.

John well understood, after so many hours with other people's application packages, that his package, if thoughtful, distinctive, and substantive, would get his foot in the interviewer's door. He had no intention of being rejected because of his own poorly written documents.

THE JOB SEARCH PROCESS

John's new job search wouldn't be taking place under the favorable auspices that had surrounded his job search as a graduating senior in geology. He and his friends in engineering, hotel administration, and business had been fortunate then that the economics in their fields produced interviewers bidding for the best of their graduation class. His friends in the social sciences and humanities, on the other hand, had not been so warmly greeted for twenty years, and so they had to tackle the more complicated job search immediately out of college. Now John would have to concentrate as his friends with English and history majors had done from their first search.

John felt fortunate he still had an income to pay the bills while trying to square his values with his job description. As an older and a wiser searcher, he knew this process would take time. His friend Cheryl, an accountant, searched on her own for a year before finding a job in her home state that allowed her to make the jump into management. Although she had sent out seventy-five impeccable letters and résumés, her search was prolonged by applying to companies that had not advertised positions. John knew he could expect a shorter search time since there was still some movement in his own discipline.

With his decision made to take on another job search, John ate a hearty dinner for the first time in weeks, then settled down to work. He first drew up a list, breaking down the tasks ahead of him as follows:

1. Reactivate placement file.
2. Approach recommenders.
3. Research possible employers, and find position descriptions.
4. Update résumé.
5. Write cover letters.

John knew that the planning and detail work he put into the front part of his job search would shorten appreciably his overall search time. He cheered at the thought of the graceful resignation memo he would write GEOLAB. There would be no need to burn bridges; he might find himself working for a former GEOLAB employee one day. The professional world, he found, can be a rather close-knit community. Professionals switch companies as easily as mercenaries switch armies. The days of the one-company career are over.

Reactivate Placement File

John knew this job placement service was well worth the nominal fee of $25 at his alma mater. Not only could he use the service to house his recommendations, but he also could expect the office to forward to him any listings appropriate for his position and field. The confidential recommendations he had solicited were more appropriately handled through such a professional intermediary than by John himself. The résumés he had reviewed that listed references separately at the bottom of the page appeared amateurish and of dubious confidentiality. And even though he didn't expect to find many leads appropriate to his experience, he felt it worthwhile to review the listings that were available.

To update his file, John wrote to the placement office requesting a revision of his dossier—now six years old—and included a check for the fee. He then updated the placement file's summary-level forms on positions held and classes taken. Since he had taken postgraduate courses in Atlanta, he requested that transcripts of those classes be sent to the placement office in Gainesville.

John carefully revised his objectives statement. Now that he had clarified his objectives, he could improve on the tired, but often considered standard, entry-level objective he had written six years ago: "To find an entry-level geologist's position in industry or research." He changed this to "Manage a research and development group investigating pollution in aquatic environments." I may as well shoot for the job I really want, he reasoned.

His objective updated, John then perused the list of recommenders he had written on his college form. Of the seven he had listed originally, only two could comment with any authority on his recent work. One recommender was his major undergraduate geology professor, whom he saw at the national geology conferences. Another was one of his Atlanta-based graduate professors. The other five he had lost touch with. John decided to replace those five with three current recommendations among those who knew his work at GEOLAB. He would tell the placement office to destroy the old recommendations from his fraternity president and his supervisor at Ed's Eats, where he had worked summers.

John could cross the first step off his list. This step had required three letters—two to the placement office and one to his Atlanta school—and two hours of work. It was April 22, a week after his decision to begin his search.

Approach Recommenders

Although this second step on his list would not be time-consuming, John knew it could be tricky. He needed to find people who knew him well, were enthusiastic supporters, were reliable, and could write well. These apparently obvious qualities were not so easy to find combined in single individuals. His immediate manager, Sue Jackson, was definitely enthusiastic. She had supported him for promotion the three times he had bid. She also knew him well, having been his manager for three-and-a-half years. She wrote well. But her performance in the reliability category gave him pause. Though Sue, no doubt, would maintain confidentiality at GEOLAB, she might take a very long time to get this correspondence written. It had once taken her two months to write a recommendation for John's coworker Alice, whose job search was thereby prolonged. John decided to ask Sue after approaching some more efficient candidates; he was in a hurry.

Toby Graham and John had worked together on a big project last year. They had respected one another, worked well together, and collaborated on the project's final report. It was a bonus that Toby had come to GEOLAB from SeaScience, which had been sorry to lose him and which was high on John's list of possible employers. Toby gladly took on the task of recommending John.

Nathalie Schuster and Marv Roebling were similarly suitable. Nathalie was Sue's boss but had been John's immediate supervisor before Sue was hired. Marv was the company's comptroller and had consulted him often about his section's budget and schedule priorities. Marv's recommendation would lend evidence in support of John's bid to come into his new company as someone not only technically proficient but fiscally proficient as well.

Now that he had enlisted three extremely reliable recommenders, John decided to ask Sue. She was delighted to oblige and, predictably, responded within nine weeks of the placement office's request. Because his other recommenders had been prompt, John's search was never stalled by Sue's busy schedule. Neither was Sue put in the position of hearing from Nathalie—her boss—that John was looking around. Most important, John had a current recommendation from his immediate manager, and, although slow in coming, it would be useful for the next two years.

As of April 25, then, John was ready to start his job search in earnest.

Research Possible Employers, and Find Position Descriptions

John knew what he valued about his work: rigorous science, important projects, ecological commitment. Now he had to find the company that also valued those things. Although he had built up several impressions of other companies from his fellow geologists, he decided to research formally the ones he was most interested in, SeaScience and Aquatics, Inc. He was interested mainly in the types of projects these companies took on and in their clients.

To do this research, John obtained copies of those companies' annual reports, which gave general descriptions of past and current projects, as well as partial client lists. From this information, John could see that SeaScience was probably the company he should aim for; it had performed several analyses for the state of Alaska after the oil spill in Prince William Sound. Aquatics, Inc., was contracted with several of the same companies he already had done work for, so John decided to put that company on hold but not to eliminate it completely.

However, John didn't want to limit his search to one company, so he continued his research. He scanned the professional journals and the Internet for employment listings. He also contacted the employment centers of the professional societies he knew and respected. From these sources he obtained another dozen listings, although none for SeaScience. He then did the basic annual report and some more Internet work on the dozen listed companies, from which he could pare his list to eight. With nine contacts to be made, John was ready to get his résumé updated and his cover letters written.

Since John did this research while he arranged for recommenders, he was still only two weeks into his job search.

Update Résumé

With a shudder, John stared at his old résumé. He had updated it only cursorily in the past few years, usually to send to chairs of panels on which he had presented a few papers, to provide them with background information for their introductions. This old résumé was embarrassingly uninformative and a sure candidate for the rejects pile. In his anxiety to keep it to a page, which is what his placement office had advised him to do, he had what looked like just an outline for a complete résumé.

Basics of a Strong Résumé. Having been on the other side of the hiring process, reviewing others' applications rather than submitting his own, John understood the basics of a strong résumé.

Purpose. The purpose of a strong résumé is to give a concise picture of the applicant through well-selected and well-explained details. Because people read résumés for dates, dates are crucial. The dates should present a clear picture of the applicant's activities for the past several years. The applicant who hasn't been working for a few years should explain that gap on the résumé.

Audience. Because the applicant doesn't know who will be reading the résumé and cannot count on that person's familiarity with local detail, concise explanations of the local and regional details are required. For example, the reader won't necessarily know what Job's Daughters is. Remember to orient the reader. Remember, too, to highlight through boldface type, italic type, or capitalization those details that readers must notice.

Arrangement. There are several ways to organize résumés—for example, by dates (chronological résumé) and by categories of work performed (functional résumé). The chronological résumé (listing education, experience, and so on) is very common, especially for new college graduates, when education is the applicant's greatest single asset. But the functional résumé is gaining ascendancy, even for the new graduate, because it places greater emphasis and attention on what the applicant has done, even if it has been largely limited to an academic environment. The point is to choose one organizational pattern and stick with it consistently. That pattern sets up reader expectations that will be violated if the writer is inconsistent.

Sample College Résumé

While reading through John's old college résumé (see Figure 11.2), examine how well he seems to have considered purpose, audience, and arrangement. This résumé is not an impressive document. Of course, in 1995, John had no

RÉSUMÉ

JOHN LOOMIS
P.O. Box 3857
Gainesville, FL 32601

Objective:	To find an entry-level geologist's position in industry or research
PERSONAL:	Single Weight—180 lbs. Height—6' 2"
EDUCATION:	1995—B.S. in Geology, University of Florida, Gainesville, FL Senior Thesis: "Effects of the Glencoe Slide" Major Professor: Dr. Fred Thompson 1991—High School Diploma, Glencoe High School, Orlando, FL
EXPERIENCE:	1994—Summer: Ed's Eats, Orlando, FL, waiter 1993–1994—Sigma Delta Alpha Fraternity House, University of Florida, Gainesville, FL, cook 1991–1993—Fair Oaks Country Club, Gainesville, FL, caddy
AWARDS:	1994—First place, 300-m Freestyle, University of Florida Swim Team
REFERENCES:	Please contact The Placement Center Student Pavilion, Box 94 University of Florida Gainesville, FL 32706

FIGURE 11.2 **John Loomis's college résumé**

professional experience in geology, but he had made very little of the experience that he did have. Following are some of the mistakes John made, whether for lack of good advice from the Placement Office and others in the know, or in spite of it:

- He failed to annotate his senior thesis; a brief summary would have sufficed.
- He forgot to mention that he had financed his undergraduate education by himself through grants, loans, and scholarships, which was impressive information about an undergraduate that any prospective employer would appreciate knowing.
- He undercut his Most Valuable Team Member award by omitting the fact that it was awarded in part by his winning a first place in the individual medley that set a national record, officiated in an NCAA competition.
- His grade-point average (GPA) was missing but was provided on the placement office forms that the recruiters, at least some of them, had read. His was impressive, a 3.9, which undoubtedly accounted for the six interviews he did get. So, for that entry-level position, he had been fortunate to have two backups, the placement office forms and his father's association with GEOLAB.

Plan for Updating the College Résumé. For his current purposes, his résumé needed a massive overhaul.

Revising Content. John's first task was to revise content. He needed to add information pertaining to the past six years, provide information that was missing, and delete items that no longer were necessary or appropriate. He also decided that he would take a more functional approach to the arrangement of information, emphasizing his professional activities of the past six years rather than his education, which though still an asset, is no longer his primary asset.

Objective Statement. He would omit the objective statement. For one thing, he would make his objective clear and more convincing in his cover letter. This letter would establish a strong connection between his management objective and his qualifications. In other words, he could demonstrate that his objective was warranted. Objectives are just pipe dreams if you can't make a case supporting them.

For another thing, he rarely ever saw an objective statement on any résumé except an entry-level one, where it was usually so vague (as his was) that it said very little. He certainly didn't want his résumé to look amateurish.

Finally, he didn't want to limit the way he was categorized by his reviewers. Chances were that an objective statement specific enough to be substantive might give a reviewer grounds for dismissing him from consideration for a position he might not have foreseen as a possibility. He wanted to keep his options open and would never reject out of hand a combined professional and managerial position, should someone want to consider him for it.

Personal Information. In the personal information category, he would omit his height and weight information, a frivolous vanity he had indulged in the first time around; although his height was the same, his weight was not. This information was irrelevant.

Professional Activities. In the professional experience category, John needed a full description of his work over the past six years. He would add descriptive information to give an overall view of his duties, and a significant accomplishments category to highlight the substantive projects he performed alone as well as the ones he participated in with others. In this way, he could reasonably believe that this experience would stand out from the experiences of those who had listed only their experience, without explanation. John would also call out his current administrative experience, since he was clearly interested in moving into a management position.

Publications and Papers. John could now add a new category, related in substance to the experience category: publications and papers. He had one article in a respectable journal and two conference papers. This work helped establish his professionalism as a geologist, contributing to the academic work in his discipline in addition to the applied work he did for a living.

Education. In the education category, he would add his postgraduate courses and more fully describe his senior thesis. He would also add the name of his graduate advisor, a major figure in geology.

Awards. In the awards category besides better describing his swimming, he could add his GEOLAB award for best professional, another testament to the consistent high quality of his everyday work. Of course, he would put this award first to adhere to the reverse chronological order he used throughout the résumé. In that way, his readers would catch the most current information in each category with the least amount of reading.

References. In the references section, John might list the names and addresses of his recommenders. Application reviewers sometimes prefer to know from whom to expect letters of recommendation.

Attending to Readability. As for overall readability of the résumé, John knew that he should consider certain principles.

Length. Since he was no longer a fresh college graduate with little to report, John found that it would be very difficult to keep his résumé to the one-page limit he had formerly adhered to. Now he had really substantive professional accomplishments to his credit. He also had a bit of persuading to do about his managerial preparedness, and he intended to do this explicitly, but succinctly. He would throw caution to the wind and extend to a second page.

Although he was uneasy at first about this decision, he remembered that all the top candidates he had reviewed as a selection team member had two-page résumés. This extension to a second page was required because of the detail, summaries, and persuasive tasks their résumés had to accomplish.

White Space. It was interesting, in fact, that the two-page résumés John had read actually were easier to read than most of the one-page résumés his selection committee had received, for this reason: they used white space attractively and pragmatically, so that the whole résumé appeared balanced and proportioned. The white space also helped emphasize and subordinate information unobtrusively. He wouldn't have to rely solely on typography to accomplish these rhetorical tasks.

Typography. John would enjoy changing font and point size on his Macintosh PC. Palatino was particularly readable, with an edge of grace that might set his résumé apart from the turgid standard fonts. He flirted with the idea of a humorous font, like Western script, but resisted, realizing he was beginning to enjoy this job search far more than he should. At least he wasn't considering a small portrait of himself in the upper right-hand corner of the résumé, like the one that the selection committee had all enjoyed laughing at. It had broken the tedium of their search but was quickly discarded as unprofessional. Nor did it help the author's case that he was not particularly attractive. And, as one selection committee member had pointed out, such a portrait was a rather sly way of appealing to stereotypical considerations that the law had gone a long way to putting out of bounds.

John would also highlight and subordinate information with boldface type, italic type, and underlining. He would set up a consistent system that his reader could rely on to signal first-, second-, and third-level categories.

Sample Professional Résumé

After working for two evenings, John was satisfied with the results of his work. Combining all the considerations of rhetoric—his purposes and audience—with format considerations to promote readability, John produced the résumé shown in Figure 11.3, putting it together carefully, as if he were solving a puzzle. He was surprised by the difference six years had made in his résumé: the new one made the contrast rather dramatic.

John's résumé took a good week to fix up. He was now three weeks into his search, staring at May 5 on his calendar.

Write Cover Letters

Now came the most difficult part of John's problem-solving task: writing cover letters that effectively correlated the categories of his résumé and the demands of the positions he wanted to apply for. His updated résumé would give prospective employers a condensed picture of what he had done; the cover letter would explain how his past was a useful basis on which to build a future together.

Again, from his selection committee experience, John knew that his cover letter should embody certain principles:

1. Adapt, where legitimate, for audience needs.
2. Inform through summary.
3. Persuade through evidence.
4. Synthesize as necessary to correlate experience and requirements of the position.

RÉSUMÉ as of 5/1/01

JOHN LOOMIS
52 Tuxedo Park
Atlanta, GA 30300
jloomis@yahoo.com

PERSONAL Home Telephone: 404-563-0987
Office Telephone: 404-371-2112

PROFESSIONAL ACTIVITIES ***1995 TO PRESENT*** Senior Staff Geologist
GEOLAB
30 Peachtree Drive
Atlanta, GA 30300

Description:

Administer field studies for five of the ten largest clients, including NASA, DOE, and DOT
Administer budget for geology and hydrology groups, approximately 50 employees in all
Conduct all training for geology and hydrology field-workers

Significant Accomplishments: Achieved consistent EXCELLENT rating from DOE and DOT, which earned GEOLAB $5 million in bonuses over 1996–2000

Reduced geology budget 30% and hydrology budget 40% by eliminating duplication of labor from several procedures

Discovered role of gophers in undermining Chattahoochee River levees

Publications and Papers: "Effects of the Glencoe Slide." 35, Geology Journal, 1997, 18–35. "Further Observations on Glencoe." Paper delivered at the 35th Professional Geologists Association Conference, Pasadena, CA, October 1999. "Fault Mapping in Aquasystems," Paper delivered at the 12th Geology in America Conference, University of Minnesota, Minneapolis, MN, 1997.

FIGURE 11.3 **John Loomis's revised résumé**

John Loomis/Résumé
Page Two

Awards	1991—Best Professional; selected by GEOLAB International 1994—NCAA Gold Medal; 300-m Freestyle Champion; National record set (still unbroken) at the 75th NCAA competition, University of Alabama, Birmingham, AL
EDUCATION	***2000 TO PRESENT*** 15 hours (5 classes) toward an MS in Geology: Georgia State University, Atlanta, GA Current Research: "Mapping the Piedmont River Fault" Major Professor: Dr. Susan Richardson 1995 B.S. in Geology; University of Florida, Gainesville, FL 32706 GPA Overall—3.9; in major—4.0 Senior Thesis: "Effects of the Glencoe Slide" Research demonstrating the demise of Glencoe River beaver populations north of Orlando, as a result of the '79 slide Major Professor: Dr. Fred Thompson

REFERENCES			
	Toby Graham	Nathalie Schuster	Mary Roebling
	993 Zinn St.	4392 Gramsci Ave.	5321 Goldman St.
	Atlanta 30301	Atlanta, GA 30301	Atlanta, GA 30301
	404-371-9483	404-371-5943	404-371-4392

FIGURE 6.11 (*Continued*)

Adapt, Where Legitimate, for Audience Needs. This first point is qualified by the phrase *where legitimate* for both pragmatic and ethical reasons. The pragmatic reason is that it does no good to present false or exaggerated material in a cover letter because it will eventually be discovered and will have wasted the time of both the prospective employer and the job candidate. The interview, in fact, is in part designed to detect such falseness. So, for example, if John is not a real team player, then he should not try to project himself as such.

The ethical problem with "overadapting" for an audience is apparent, but worth stating, since many sources in society instruct people to do just that—overadapt. This advice resembles that which mothers from older generations used to give to their prepubescent daughters: pretend to be interested in everything he says. It is important to find a job, to be financially secure, to build a solid professional history. But there is no need to see this as the only important thing in life.

Society is not yet so unreasonable that every other facet of life must be subordinated to career. And if you proceed in the belief that all comes second to career, you will be in real danger of losing your integrity. You should be aware of your boundaries. Ask yourself, How far can I go to present myself as suitable for the position I want to apply for? If you have any sense of yourself, there will be clear limits to any answer to that question.

Inform through Summary. This second principle clarifies the difference between the cover letter and the résumé. While the cover letter should preview the important points in the résumé as they apply to the position description, it should not simply rehash all the details contained within the résumé. The résumé serves an outline function, arranging all the details necessary to document activity; the cover letter, however, deals in relationships between these documented activities and the positions applied for. These relationships have to be articulated and supported (the topic of the third principle) in a reasonable way. One of the worst things an applicant can do is to claim relationships where, in fact, none exist. Often such an application will be thrown out because of the hostility of the reader to the applicant's trying to "pull something" in the cover letter.

Persuade through Evidence The third point is also an important one. John had read too many cover letters (and had even written one himself) that were essentially perfunctory transmittal letters, referring the reader to the résumé. This strategy puts the onus on the reader to pull the applicant's case together. John remembered being simply too rushed and too tired to make a case for those who had not made their own. It had cut his work considerably when applicants had been professional enough to show him, very explicitly and directly, how their experience matched, exceeded, and approximated the position requirements.

Synthesize as Necessary to Correlate Experience and Requirements of the Position. Where holes were left that constituted gaps between the applicant's experience and the position description, the responsible applicants had addressed them, instead of trying to hide them. It was not uncommon for conscientious applicants to realize that they fulfilled all but a few of the requirements very well. The few mismatches were no reason not to apply for a job they wanted. Instead, the applicants suggested alternatives to the holes, strategies for overcoming them. Even when such arguments had to be discounted because the gaps were simply too big or the alternatives inadequate, John felt some respect for the honesty of these straightforward applicants.

He knew that not all his colleagues agreed with him, seeing a hole as a hole and perhaps admiring those who tried to cover up their inadequacies. But John was keeping the larger issue in mind: he might one day be working for one of these applicants. He preferred to think of working for one who would try to solve problems that existed to one who would try to cover them up.

Problems could probably be solved adequately once they were acknowledged. Refusal to acknowledge them never helped any company, he had learned. NASA's problems with the Hubble Space Telescope, designed for use in orbit around the earth, had demonstrated that principle clearly enough. John had read that the lens makers had ignored unfavorable test results because of their belief that a certain lens was superior. As a result, the space center's mission had failed to produce the photographs that it had sought to capture; moreover, this problem could not be corrected from Earth. Thus, the faulty equipment languished, awaiting a repair vehicle that would not reach it for some years to come.

Sample Position Description

John kept these principles in mind as he analyzed the position descriptions he had collected, including one from *Oceanique, Lmtd.* (Figure 11.4). The lack of description for SeaScience was a separate problem he would address later. His analysis of position descriptions was mainly to clarify for himself exactly what requirements he would need to address in his letter. Sometimes these requirements were specified; sometimes they were implied.

Although this position description was a bit daunting, since it emphasized an experienced manager, John was not put off. He had real administrative experience from three important projects, even though he didn't have a managerial title with GEOLAB. He would make this point in his cover letter.

From the sequence of professionals listed on the description, he suspected that this company might prefer an oceanographer; oceanographers might be majority in this group. But he could play a strong second as a geologist. In addition, he had done DOT contract work before, with smooth progress. His research background was strong although more terrestrial than

OCEANIQUE, LMTD.

4580 Falmouth Dr., Suite 925, Boston, MA 02200

POSITION TITLE
The Boston branch of OCEANIOUE, LMTD., requires a manager for its research division. Applicants should send current résumé and cover letter to Ms. Angela Rowan, Personnel Manager, Box 2155, Boston, MA 02200. Material should be postmarked by 6/1/01.

DUTIES
Supervise a 16-member team of oceanographers, geologists, hydrologists, and computer specialists concerned with physical description of New England fisheries. Administer multimillion-dollar budget for a US DOT project investigating the shipping routes in the North Atlantic for degradation.

REQUIREMENTS
Management experience, strong research background, B.S. in relevant physical or biological science required. Graduate education preferred. Salary negotiable.

FIGURE 11.4 *Sample position description*

aquatic, but he was serious about shifting emphases, and so would try to make that argument part of his case. With his proven administrative record, yet no previous managerial title, OCEANIQUE might see that his salary requirements could be less onerous than those of a Woods Hole manager, for example.

Sample Cover Letter for an Advertised Position

In response to this description, John wrote the letter shown in Figure 11.5, which highlights his attempts to synthesize his qualifications for his reader; that is, in several places John will spell out exactly how he is a good match for the position description. Throughout this letter, John tries to summarize and to persuade his reader. These communication skills take far more effort than simply referring to the revised résumé, good as it may be.

Because the difficult part of John's argument is to persuade his reader that he is qualified for a managerial position, despite the fact that he has never held one before, he devotes the bulk of the letter to that argument. Since research background is solid and well detailed in his résumé, he mentions it as further proof but does not detail it in the letter, because he wants to keep the letter concise. Only in very complicated situations would one want to write a cover letter that is longer than a page. Yet, it is important that the cover letter be complete enough to serve as a good introduction to the résumé.

John wrote seven other cover letters adapted to the position descriptions he had found in his research. These letters followed the principles he had used when writing to OCEANIQUE because they were all in response to specific position descriptions. John's letter to SeaScience would have to be a bit different, though, since he was not responding to a written description.

Submitting an Unsolicited Application

With the SeaScience application, John was trying to break into a closed system that, in fact, might not have any openings. He had to rely on his own knowledge about the company and on his research; that is, using his knowledge and research about the company, he had to argue that a mutually beneficial relationship could be established. John's first decision was that he would address his letter to Rick Stewart, Vice President in Charge of Research. John did this with some confidence because he had met Stewart at a conference and had discussed research interests with him. Stewart had seemed receptive and impressed by John's panel. Although normally he would not address a cover letter to an executive, John thought it was worth a try in this case because he had some previous dealings with this executive. For formal reasons, though, John would copy the letter to the manager of personnel at SeaScience. He didn't want Stewart to think that John expected him to do all the legwork for this contact with SeaScience; John could contact Stewart and still do his own legwork.

Since he was so interested in the company, John thought it would be worthwhile to mention his colleague Toby Graham, who was formerly employed by SeaScience. Doing that can be a risky strategy simply because the

May 5, 2001

52 Tuxedo Park
Atlanta, GA 30300

Ms. Angela Rowan, Personnel Manager
OCEANIOUE, LMTD.
Box 2155
Boston, MA 02200

Dear Ms. Rowan:

I am most interested in the Manager of Research position I saw advertised in March's issue of Geology Today. My previous experience administering projects for US DOT and my background in geology, hydrology, and aquatics are particularly suited to the requirements you specify in the position description.

I have been the lead on several important projects at GEOLAB, where I am now employed. I am experienced in fulfilling contractor expectations. For example, I have been the lead for the past three years on a US DOT project investigating the breakdown of Chattahoochee River levees. DOT appreciated the efficiency of my team's work, awarding us an EXCELLENT rating for all three years. My other contractor experience includes work for DOE on the stability of mine shafts in New Mexico and work for NASA, analyzing samples taken from the moon. Both of those projects received EXCELLENT ratings, as well.

My budget experience comes from lead positions on major projects. While I have yet to administer a multimillion-dollar budget, I have administered those in the $500k range. I assume your budget requires CPAF documentation, which I am practiced in. At GEOLAB, I have compiled and administered budgets on both the CPAF system and a standard corporate system.

It is the nature of the work you've advertised that attracts my attention—investigating possible environmental degradation in the North Atlantic shipping routes. I am looking for another position, not only to take on a management title, but also to take on the interesting work of environmental investigation. The GEOLAB projects don't have the same ecological emphasis as the one you've identified in the advertisement. My research background is hardly challenged by standard contract work. I am eager to put my postgraduate experience to work, especially in aquatic research. My previous work on aquatics has been well received.

My résumé, which I have enclosed for your review, lists my work and home phone numbers. I look forward to discussing this position in more detail, at your convenience.

Sincerely

John Loomis

John Loomis

FIGURE 11.5 John Loomis's cover letter in response to an advertised position description

mutual acquaintance might not be well regarded by the reader. But everything John knew about Toby Graham, that he was generally respected in his field, persuaded him that it would only help him to mention Graham's name.

For this letter, John followed these principles:

- Make the previous relationship with the addressee clear from the beginning of the letter.
- Mention a mutual acquaintance only if there is reason to believe the person is well regarded and only if the person is truly a mutual acquaintance. It is legitimate to use a contact as long as that person indeed can provide a means of access to whoever will be considering the applicant's request.
- Make a strong case for the mutually beneficial relationship that might be worked out, but make the case formally and respectfully, using the principles of persuasion and evidence that one would use to respond to a position description.

When applying this final principle, John must remember to balance his previous knowledge of the reader with his own status as an outsider. John may know Rick Stewart, but not as a coworker would, so he wouldn't want to overdo any informality.

Sample Cover Letter for an Unsolicited Application

John's letter to SeaScience appears in Figure 11.6. Two points are worth mentioning about this letter. First, John violated his own page-limit rule with a cover letter that ran one-and-a-half pages. This was a risky strategy, because he had no guarantee that Stewart would want to take the time to read such a long letter. All John could be sure of was that if he was going to send such a long letter, it would have to be well written.

The problem is not that people adhere to fixed page limits for any quantitative reason; the problem is that since so many people write poor letters, the chances that a letter will be poorly written, and thus defeat the author's purpose, increase with each paragraph. Thus, the poorer writers generally benefit from shorter letters. On the other hand, a good writer, who considers audience, who has a pleasant style, and who makes substantive points throughout the letter, can write a longer letter *when necessary*. In this case, a longer letter is justified by the fact that John is writing an unsolicited cover letter, so he has to do a bit more explaining than he would in a solicited letter. This letter length is also justified because John writes well. Even though he exceeds one page, he writes very efficiently, considering all that his letter must accomplish.

Another interesting point about this letter is that John ends it with the statement that he will call Stewart at a certain time. This strategy is also risky and not one to be used in standard letters. For example, if John were to use this tactic with OCEANIQUE, he might come off as pushy, trying to rush the search for his own convenience, as if he were the only applicant. He knows that he is not the only applicant, since he is responding to a nationally

May 7, 2001

52 Tuxedo Park
Atlanta, GA 30300
Dr. R. D. Stewart, Vice President
SeaScience
P.O. Box 85
St. Augustine, FL 30333

Dear Dr. Stewart:

I'm not sure that you will remember me without some explanation. I met you last year at the Geology in America conference in Fort Worth. You had attended the session I chaired on aquatic fault lines in the Atlantic, and we exchanged data a few weeks later. Toby Graham, formerly employed by SeaScience and now senior hydrologist at GEOLAB, told me he had quite a good experience at SeaScience and encouraged me to write you. Apparently he worked with you quite extensively in 1996 on the Cousteau project.

I am particularly interested in your company because of the environmental work you do. I very much admired SeaScience's study of the ozone hole over Antarctica and would like to devote my energies to such work. My experiences in administration and research might be attractive to Seascience, since I am accustomed to working on tight budgets and producing consistently verifiable research.

That said, I wanted to let you know that I am currently looking for another position and am most interested in anything that might be suitable at SeaScience. While the bulk of my experierce has been as lead of research projects, I am interested in taking on a formal managerial role in research.

My experience as the lead on several important projects at GEOLAB, where I am now employed, has taught me how to fulfill contractor expectations. For example, I have been the lead for the past three years on a US DOT project investigating the breakdown of Chattahoochee River levees. DOT appreciated the efficiency of my team's work, awarding us an EXCELLENT rating for all three years. My other contractor experience includes work for DOE on the stability of mine shafts in New Mexico and work for NASA, analyzing samples taken from the moon. Both of those projects received EXCELLENT ratings, as well.

My budget experience comes from lead positions on major projects. I have administered several in the $500k range. In Fort Worth you mentioned that your budget requires CPAF documentation, which I am practiced in. At GEOLAB, I compiled and administered budgets on both the CPAF system and a standard corporate system, so I can adjust to whatever budget documentation requirements SeaScience must meet.

FIGURE 11.6 **John Loomis's cover letter to inquire about a possible position**

R. D. Stweart
Page 2

I am sending a copy of this letter and résumé to Mr. Arnold in Sea-Science Personnel, just so that he can put the copy on file. But since you are the administrator of the Seascience Research Section, I wanted you to know about my job search. While I am applying for other positions advertised in the journals, I am particularly interested in coming to work for SeaScience, where I think we all would find a good match between mission and ability. I hope to call you the week of May 15 to discuss any offerings that might come up. In the meantime, I look forward to discussing my search further at your convenience.

Sincerely,

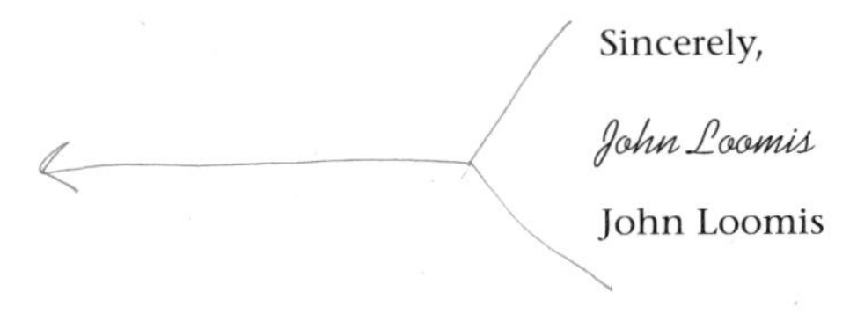

John Loomis

John Loomis

FIGURE 11.6 (***Continued***)

advertised search. But since John is the only applicant for a possible Sea-Science job, as far as he knows, this tactic is not unreasonable.

Furthermore, John's call not only will inquire about the status of the application but also will touch base with a colleague. He will inquire whether such a position is available, and he will also follow up on any possible difficulties with the way he has approached the company. In this case, then, the follow-up call is justified. It will accomplish business and will develop a contact that John already has.

With this letter to Stewart, John reached the end of his cover letter stage. Since he wrote each letter individually, sometimes adopting paragraphs from other letters but always approaching each letter as a fresh rhetorical problem, this stage was quite time-consuming. It is not surprising that by the time he was finished, it was the end of May, since John was still working full-time and living a life with his family.

POSTSCRIPT

On May 17, John called Rick Stewart at SeaScience. Since Stewart was out of town until May 25, John had no choice but to wait on his contact. He preferred to talk to Stewart before he talked to the personnel manager, who might toss John's application because he did not go through the orthodox channels. In the meantime, John heard from OCEANIQUE on May 20, where he was short-listed for the managerial position. Replies to his other seven letters were polite but unencouraging.

On May 25, John did talk to Stewart, but because Stewart called him. Stewart was pleased to get John's letter, enthusiastic about John's interest in the company, and encouraging about actually establishing a position for John. Since this position would be an additional one, John would have to be willing to wait for several weeks for the bureaucracy to go into gear. John was willing; he began his wait.

On June 10, OCEANIQUE called John for an interview. He went to Boston on June 15, returned to Atlanta on June 17, wrote a formal thank-you letter for the interview on June 18, and was offered a position on June 30. Although John was terribly pleased to have an offer so efficiently in hand, only ten weeks from the beginning of his search, he hesitated because of his desire to work at SeaScience. He called Stewart to apprise him of the situation, and Stewart asked him to hang on a little longer; the extra position was only halfway through the SeaScience system.

By July 15, OCEANIQUE began to press for an answer. John negotiated for some time, until August 1, and called Stewart again. This time Stewart had bad news. The extra position could not be approved until the next fiscal year, which began October 1. Stewart felt that John's chances would be excellent then but could not make any guarantees. John was stuck. He decided to push from another end.

On July 25, John informed his immediate supervisor, Sue, that he had received another offer. Sue was sorry to hear that he would be leaving but didn't seem to feel moved to make a counteroffer. John suspected that Nathalie Schuster, Sue's superior and his former supervisor, might be willing to counteroffer, so he wrote a memorandum explaining his offer, and copied the memo to Nathalie. As he predicted, Nathalie was indeed interested in a counteroffer, so she asked him for a week to consult with her superiors to determine how much and what position she could offer him. Within three days, she resumed with an offer for a 12 percent raise, but no change in position. John rejected this counteroffer since his position at OCEANIQUE would include a managerial position title and perks. In two more days, Nathalie came back with the 12 percent raise and a managerial title. She obviously wanted John to stay.

John then had to think about the reasons he had for wanting to leave GEOLAB in the first place. Besides his frustration over being passed over for management, he felt unfulfilled in his work, and he believed that his results were pushed under the rug because they were unflattering to the government contractors who had commissioned the studies. Was he willing to continue doing work for these contractors, knowing they were hostile to his results, after being stroked a bit by a salary increase and a managerial title?

At this point, John realized he would have to talk to Nathalie again to clarify this problem. Nathalie encouraged John to persist in his research, despite unflattering findings, and pledged to support his conclusions as much as she could. But she could not guarantee that he would receive the full support of upper management; in fact, he could expect to have to fight for his principles.

John appreciated Nathalie's honesty and began to perceive that perhaps he never would be able to have the kinds of guarantees he craved from his employer. His choices, then, were to stay where he knew the problems but had wrested some respect; to go to an unknown but receptive environment (at least it seemed receptive), to continue his search; or to wait for Sea-Science. His decision would not be an easy one; on the other hand, it was a decision he relished and a far cry from the sort of decision that he was making three months ago in the rush-hour traffic on I-285.

In the end, John decided to stay with GEOLAB, to try on this new hat in the company. He gracefully declined OCEANIQUE's offer by phone call and letter. He contacted Stewart with his decision, who was pleased that John had not counted out SeaScience. October was close; now that John had a managerial position at GEOLAB, it would be much easier to justify the new managerial slot for him. The job search, if it did nothing else, opened a number of new opportunities for John, making him realize how much he had to offer a company.

In the future, John would not see the search as a signal of the end of his career satisfaction, but rather as a chance to assert himself in his profession. Although he couldn't imagine that he would skip from company to company—he was not that ambitious—he could imagine taking on another job search when his values and work became too unrelated.

SHAPING UP A STUDENT RÉSUMÉ

In this section, we will work with an entry-level application by a student applying for an engineering job. In fact, the student is Joe Brown, whose professional profile appears in Chapter 1. From this profile, we know that Joe made it successfully into his engineering field. How much did his student résumé contribute to his success?

First Draft of the Résumé

The first draft of Joe's résumé (see Figure 11.7) doesn't look particularly promising. While reviewing it, note the points that could be strengthened to appeal to an engineering recruiter.

Purpose. Joe wants to convince prospective employers that he is well suited to an entry-level position. Because an entry-level position in engineering isn't a specialized calling, Joe needs to focus on the strengths in his academic preparation and work experience. Although the latter may not exactly relate to engineering, it will show a bit about his character as an employee.

Has Joe achieved his purpose with this résumé? He has made a good start but needs to do a better job of highlighting his strengths: significant work experience and academic diversity, branching off from engineering into computers and math.

Audience. Joe has appealed somewhat to his audience, which is recruiters for engineering firms. In other ways, he has missed them.

JOE BROWN
1015 Smithfield
Twin Falls, ID 83897
(208) 534-9786
joe_brown@falls.com

OBJECTIVE Entry level position with opportunities for research and development, advancement, and further education.

EDUCATION Senior enrolled at Mountain State University, Twin Falls, Idaho. Graduation expected in May 2001 with a Bachelor or Science in Engineering and a Minor in Math.

EMPHASIS OF STUDY Currently taking the Thermal Fluids and Measurement and Control Systems sequences.

CAREER-RELATED EXPERIENCE Tutored Engineering, Math, and Physics independently 1999 to present. Instructed 1 to 3 students per semester through junior level courses.

WORK EXPERIENCE Mechanic House of Fords, Twin Falls, Idaho 1993-2001. Achieved ASE certifications in Brakes. Engine Repair, and Heavy Trucks. Job required dealing with customers on a regular basis.

REFERENCES

David Harvey
101 Main St.
Twin Falls, ID 83897
(208) 534-9980

Fran Jameson
310 S. Hill
Twin Falls, ID 83897
(208) 534-4292

Judith Butler
1512 Cherry St.
Twin Falls, ID 83897
(208) 532-4019

FIGURE 11.7 **Sample first draft of a résumé**

He has been successful in that his résumé is very focused on the essentials pertinent to the position he's looking for. In fact, his objective statement is one of the more substantive ones seen on a student résumé; but with the advice of Placement Office people, he decides to remove the words "entry level" and "advancement," since it is obvious that he is looking for both a beginning position and later a promotion with the requisite time and effort. He does add, however, "preparation for management" in place of "advancement," since management better reflects his ideas for advancement in the first place. Because the pool of recruiters has asked for résumés only at this early screening stage, his objective statement is useful here. He has not added a lot of superfluous material, and his writing is concise.

However, he has missed his audience through a few errors. First, he has not proofed his work carefully; we see spelling errors and spacing problems. It is also a bit difficult to read this résumé for dates, since they are embedded rather than highlighted in various categories. Finally, the résumé doesn't provide much of a feel for the person behind it. Joe has done very little to distinguish himself from anyone else applying for similar positions. But he could probably come up with some impressive personal information if he thought about it.

Arrangement. Joe has written a chronological résumé, organized into the standard categories of objective, education, experience, and references. He has added "emphasis of study" as a separate category, instead of making it a subcategory of education, but it's not clear why this should be a separate category. So he decides to make "Thermal Fluids" and "Measurement and Control Systems" an area of concentration under education and to remove the sense that he is in the process of completing course work in those areas. Both conditions are true—he has concentrated in thermal fluids and measurement and control systems, and he is completing a course sequence, but the latter might sound more tentative to his readers, an unnecessary, even misleading, addition to the tonal mix.

Joe had also made a distinction between career-related experience and work experience. This may be a useful distinction, since students often work to get themselves through school in jobs that have nothing to do with their intended career. Joe also thought of trying to distance himself from the mechanical aspects of his work experience, since he is educating himself in engineering and math.

But after some more advice from his placement office and some self-reflection, Joe decided to group all of his work experience together. It would be obvious to his readers what the difference was in kinds of work, and Joe, in fact, was proud of his past work, both as a tutor and a mechanic. In both jobs he was able to put his knowledge and experience to good use, and so he decided that that fact would be more fairly stated by not drawing any distinctions. He also saw that the two jobs had another useful feature in common. In the eight years he had worked at House of Fords, his employers and fellow mechanics had come to rely on him to explain procedures and repairs to the customers in terms that they could understand. Over time, he attributed this valuable ability to the skills in communication he had learned and developed in college. In both situations, Joe acted as an instructor, certainly a respectable use of his time.

One aspect of arrangement is typography, how the words are word processed or typed on the page. There are few typographical distinctions in Joe's résumé. In fact, the only ones he makes are between all-capital category names and upper- and lower-case descriptive material. He could use other distinctions to highlight his résumé: boldface type, underlinings, and italic type are quite common. Another mistake he has made is to use paragraphs instead of lists to outline his qualifications. In résumés, paragraphs are more difficult to sort out than are lists, since most information on a résumé is at the same level of generalization; that is, each sentence or paragraph is just as important as the others. With these drawbacks pointed out, Joe made useful changes to the typography of his résumé.

REVISED STUDENT RÉSUMÉ

Joe's revised résumé, which is shown in Figure 11.8, is a real improvement over his first one. Through attention to purpose, audience, and arrangement, he has highlighted his strengths and made the résumé easier to read, despite the fact that the second one is actually longer than the first.

FUNDAMENTALS OF THE INTERVIEW PROCESS

Interviewing is, of course, a very important part of the job search process. The documentation gets the candidate to the interview step, but by no means guarantees a successful interview. On the other hand, many have learned the hard way that a successful interview does not guarantee a good future employee. Some people interview very well but prove very difficult to work with. Others do not interview dramatically well but prove to be solid choices for a position. So, to round out the interview scenario, we can say with assurance that everyone involved in the interview is trying to feel her or his way through it in the best way possible. There is no scientific process to guarantee the best choice for all involved.

The Interviewer

It is worthwhile to consider the interviewer's position in this stressful situation. It is taken for granted that the candidate will feel nervous and on the spot, but the interviewer's state of mind complicates the situation. One important thing to remember about those who are interviewing is that this task is just one of many that they have to get through that day. Although that candidate has only the interview to focus on, the interviewers have many other things to focus on—the everyday tasks that constitute a day on the job—and may regard the interview as an interruption; interviews are certainly time-consuming for those in charge of them.

JOE BROWN
1015 Smithfield
Twin Falls, ID 83897
(208) 534-9786
joe_brown@falls.com

OBJECTIVE

Position with opportunities for research and development, and further education, and preparation for management.

EDUCATION

1998 to present: Mountain State University, Twin Falls, ID, 83897. May 2001 with a B.S. in Mechanical Engineering and a Minor in Mathematics.

Area of concentration: Thermal Fluids and Measurement and Control Systems.

Accomplishments:

- Self-financed college education
- Maintained a 3.5 GPA
- Worked full-time during college
- Received Bart Jones scholarship, 2000–01

CAREER-RELATED EXPERIENCE

1999 to present: tutor engineering, math, and physics students through junior-level courses; instruct 1 to 3 students per semester.

OTHER WORK EXPERIENCE

1993–2001. House of Fords, Twin Falls, ID, 83897; Mechanic. Technical liaison with customer relations.

Accomplishments:

- Earned ASE certifications in brakes, engine repair, and heavy trucks.

REFERENCES

David Harvey
101 Main St.
Twin Falls, ID 83897
(208) 534-9980

Fran Jameson
310 S. Hill
Twin Falls, ID 83897
(208) 534-4292

Judith Butler
1512 Cherry St.
Twin Falls, ID 83897
(208) 532-4019

FIGURE 11.8 **Revision of the sample résumé**

So, much as the writer of correspondence must keep in mind the image of the harried reader who must plow through many letters and memos, the candidate facing an interview needs to keep in mind an image of the harried interviewer. Such an image should help the candidate remember to be as businesslike as possible in the interview. And given the fact that the interviewer may well be interviewing several candidates for a position, it obviously is important for the candidate to make some kind of distinctive impression, to come across as a real person, instead of as an actor who takes no risks but just wants to play the perfect interviewee. The fact is that interviewers are interested not only in a candidate who seems to fit all of their criteria but also in the person who might be coming to work for them.

Interview Guidelines

Beyond the obvious etiquette of interviewing, it is helpful to understand the rhetoric of the situation. A candidate is brought to an interview, often at considerable company expense, because the hiring organization has reason to believe that the candidate has some good things to offer the company. The purpose of the interview is to get a feel for the candidate's personality. The interviewers will be as variable as people can be.

It is difficult to advise candidates about the interview because so much depends on the characters of unknown people. Interviewing is a very subjective business, and it is a fact that a good interview impression may be overlooked for reasons over which the candidate has no control: another candidate may have made an equally good impression, or someone conducting the interview may have been threatened by a candidate's accomplishments.

Because there are factors over which the candidate has no control, it makes sense for the candidate to concentrate on those factors over which she or he does have some control: responsiveness, preparedness, finesse with trick and hostile questions, and honesty.

Responsiveness. The candidate should be forthcoming and cooperative with the interviewers; that is, the candidate should answer questions fully, developing reasonable answers but not rambling. Nothing is more painful for interviewers than extracting answers from a candidate who may be too nervous or too immature to volunteer any more than the most cursory answers. On the other hand, it is almost equally painful to listen to a candidate ramble in a series of broken-off sentences that never come to an end. Although the interviewer may realize that this stream-of-consciousness answer is a result of the candidate's nervousness, he or she will not overlook this type of response.

Candidates should think about their answers for a minute, clarify the question if necessary, and generally try to be as concise as possible. After getting through any kind of detailed answer, the candidate might ask a simple question like "Are you with me?" to give the interviewer a chance to ask any minor questions that the answer might have raised.

Another aspect of responsiveness is for the candidate to listen well. Sometimes candidates are so nervous that they simply don't hear introductions, explanations, and the like. The candidate who is too nervous to listen can miss a great deal, at the least, and can present an unattractive picture, at the worst, and so perhaps may be eliminated from consideration. No one wants to work with someone who doesn't listen.

Preparedness. Interviewees are often cautioned to be prepared, but this advice usually refers to researching the company. Since a candidate will have already done the research necessary to apply for a position, we are using *preparedness* in a different sense. In other words, if a candidate is to give a presentation during the interview, for example, then a formal presentation should be completely ready to go. It makes a very bad impression, implying a real lack of interest in the job, to stumble through an unprepared talk when the candidate is expected to present a formal one.

This situation may well mean that the candidate will have to do some scrambling to pull a prepared talk together, but that is all part of the extra work one takes on to look for another job. Being prepared also means that the candidate should be able to speak specifically to those areas of expertise that are advertised in the job position. Thus, a potential personnel director should have some specific things to say about running the interviewer's personnel department.

Finesse with Trick and Hostile Questions. Interviewing is a stressful business, but every interviewee should understand that there are some interviewers who enjoy making it even more stressful, to see what kind of reaction the candidate will have to such stress. Trick questions and hostile questions fall into this category of interviewing games. If the candidate can accept that these strategies are typical of interviews, then the questions won't seem so powerful, and the candidate shouldn't be thrown by them.

Trick questions are those speculative questions that one has no way of answering very well. "Where do you see yourself in five years?" or "When do you think you would want to get into management?" or "What are your weaknesses?" are questions to beware of. If nothing else, they are a good test of the candidate's ability to give a no-answer answer in a very charming way. It is never a good idea to answer the questions about weaknesses too sincerely, since the question is designed to put the candidate in an unfavorable light. Generally, good interviewers will not ask these questions.

Hostile questions are not necessarily the mark of bad interviewers because the questions themselves can be genuine. The charged-up feeling behind the questions may or may not be genuine; nevertheless, the situation puts the candidate's character to the test. What kind of grace under pressure can the candidate muster? The candidate may be confronted with remarks like these: "Surely you're not implying that. . . ?" or "If what you say is true, then how do you account for. . . ?" or "This sounds to me like the latest version of snake oil."

When confronted with this apparent hostility, the candidate must not take it too seriously. Chances are good that the remark is not serious, that the remarker is just showing off, that the remarker is the group curmudgeon, or that the remarker is trying to earn some kind of points with the local authority figures. The candidate's best strategy is patience, good humor, and objectivity, on the assumption that not everyone is going to like everything the candidate has to say. Above all, it is important not to retaliate out of defensiveness. It may well be that the questioner is too hostile to be overlooked or distracted with a joke and so will feel that some firmly held stance is appropriate; but it's best to avoid insults and viciousness, in the same way that one would try to avoid such behavior to a host.

The important point in this section is that the candidate should not do anything that would compromise integrity. We do not believe that a candidate should do just anything to get a position; the qualification is always understood that the candidate will do what is reasonable and ethical to get a position. But the competitiveness of the interview can sometimes evoke extreme responses from candidates.

Honesty. Because the candidate is on the spot yet wants to make a good impression, it is easy for her or him to fall into habits that are less than honest. First under this category are all attempts to be agreeable to the interviewer when those attempts do not at all reflect the truth. Candidates may find themselves saying things that are very difficult to live up to later: "I don't mind driving forty-five minutes into the desert in the morning to get to work"; "I don't need a window in my office"; "Straight commission work is fine with me." If candidates agree to things that are too difficult to live up to, then, when they try to renegotiate those points, they seem to be backing out on implied promises.

Another type of honesty error is the inability of many candidates to admit they don't know something they think they are supposed to know. Although there are many dodges, the interviewer usually spots them quickly, and so the candidate loses credibility. It is better to admit lack of knowledge than to bluff. One form of bluff is to move the conversation back to an area the candidate does know something about; another is to answer a different question. For example, the interviewer may ask, "Why would a teacher want to work in a mining outfit?" The candidate, unsure of the reasons, replies: "It is very gratifying to work with others, to know that one can enhance a child's future. I'm looking forward to the same type of opportunity with you." This answer still does not tell the interviewer why the teacher wants to move into mining. The candidate should understand that interviewers are looking for just such dodges as these.

In the long run, everyone is best served by honesty. It will win the candidate little to be hired on the basis of false expectations set up in the interview. And it will benefit the hiring organization little to take on someone who evokes real doubts in the interview. It is very difficult to fire most professionals. A case must be carefully built up during a probationary period, supported by documentation and accompanied by plenty of counseling. It is not easy to undo mistakes made in the interview. A genuine interviewing style goes a long way to eliminating bad matches before they happen.

CONCLUSIONS

This chapter emphasizes the persuasive job that applicants take on when preparing their cover letters and résumés. Therefore, the rhetorical considerations of purpose, audience, and arrangement are paramount. In addition, applicants must take political considerations into account at several points in the job search process. Finally, applicants should not be afraid to stand by their ethical constraints. The job search is not a product-processing enterprise but is a rather unpredictable encounter among human beings. The process is both objective and subjective, and candidates should do their best to evoke the person behind the qualifications.

EXERCISES

1. Write a cover letter and résumé to apply for the position described in Figure 11.9. You should not make up any facts for these application materials; instead, you should work with your real experience, highlighting your strenghs and correlating them with the terms of the job description.
2. Your class should form evaluation groups to review the applications for the liaison position. Then you should all make enough copies of your cover letter and résumé to distribute to the evaluation groups. These groups will then review the candidates for the position on the basis of how well their cover letters and résumés reflect the principles discussed in this chapter. The position description is multidisciplinary, so no one student has any better qualifications than any other. Your instructor will be interested mainly in how you correlate your own qualifications with the demands of the position.

   Each group should choose three top candidates. Then the group should write a memo to the instructor explaining why group members chose the three candidates. Make a recommendation that they be brought in for an interview.

ASSIGNMENTS

1. Contact the placement office at your school for brochures and pamphlets or any other pertinent advice available on the job search. Do you see any discrepancies between the principles in this chapter and the advice that your placement office has for you? What rationale lies behind the discrepancy? Discuss this in your class.

POSITION DESCRIPTION

The Westwood Corporation requires a liaison to the eastern Idaho region. This is a part-time position. Duties include the following:

1. Report directly to the manager. Rocky Mountain Division, on such issues as demographics, potential markets, technical and scientific research in the area, and university relations
2. Sit on a committee for personnel relations
3. Write reports, summaries, and procedures
4. Speak to corporate and community groups on topics of interest

Qualifications include familiarity with eastern Idaho, good communication skills, ability to work with a team of professionals from a variety of fields, and a strong academic record. Students are encouraged to apply. Applicants should indicate how many hours a week they could devote to the work described above.

Westwood offers benefits for part-time employees, such as medical insurance and paid holidays. Salary open.

Please send cover letter and résumé to

Jane Doe, Personnel Manager
Rocky Mountain Division
Westwood Corporation
Salt Lake City, Utah 83303

FIGURE 11.9 **Position description**

TERM PROJECT OPTION

Prepare a résumé that appeals particularly to your proposal audience. This should document your suitability to act as principal investigator for your major project. Your résumé should show evidence of being directed toward this audience. For example, you should highlight any experience you have that is consistent with your proposed research project.

Solving Problems Through Document Design

CASE STUDY

KEVIN ERSTING'S OFFICE

Office of International Education ▪ Westminster College
Little Rock, Arkansas

Kevin Ersting arrived at his office in Tollan Hall just after 7:30 A.M. It was a little earlier than he usually got to the office, but today he was excited and a little nervous about proposing his idea for a newsletter for the overseas programs section of the Office of International Education (OIE) in this morning's staff meeting. The OIE was having a successful year, thanks in part to a new director, Feona Dalton, and her efforts to consolidate the office and its many tasks into a self-sufficient, cost-effective unit. This transition proved to be very timely; recent budget cuts had drastically reduced funding for several other similar offices on campus that relied almost entirely on the administrative hierarchy for both leadership and funding. However, with most of the decisions on programs and finances being made in-house, new ideas and projects had to be carefully proposed and considered for

PREVIEW OF THE CHAPTER

both their cost-effectiveness and practical use of people's time. Since Kevin, the Foreign Study Advisor for Overseas Programs, one of three major divisions of the office, knew that the newsletter's budget was going to come primarily from his section, he was very eager to use what limited funds existed carefully and productively.

The idea of a newsletter came as a result of last week's staff meeting, after the group began discussing various recruitment ideas that would possibly increase the numbers of students interested in overseas programs and that would increase the awareness and knowledge of the office throughout the campus. Kevin had posed the idea of a newsletter, but it was met with divided interest. Several of the staff members, specifically those from the International Students and Faculty section, felt that a newsletter would not be of much benefit and would have little effect on recruitment. Others, including Feona and Kristina Elswood, Kevin's assistant, felt that it was an idea worth looking into. So Kevin decided to do that.

On Monday, he began by considering what the audience would be interested in reading and what the OIE would want to emphasize in order to meet the predetermined goals of the production. Kevin found that the needs of the audience and the needs of the office were quite close. The audience, if interested in the happenings of the Office of International Education, would want to know about some of the various opportunities for students, faculty, and staff overseas, and also news about what many of the international students and visiting faculty were currently doing on the Westminster campus. In addition, information about the office itself—its various tasks, personnel, and subdivisions—would be of interest. As for the needs of the office, as Kevin had learned in last week's staff meeting, Feona wanted some projects that would aid in the campuswide advertisement of the office and its functions. For this purpose, such a newsletter would be ideal.

However, as good as this idea was, Kevin knew that one of the first things people would ask was how much the newsletter would cost. As it was, the office was running a tight budget and was not expecting any budget increases within the next two quarters. For that reason, Kevin had to consider his expenses. How would he be able to produce an informative and appealing newsletter within a very limited budget? The answer came to him quickly—produce the newsletter in-house on the office computer system, which was fully capable of desktop publishing.

This answer, however, would pose other questions about the office's capability to produce such a document. It was here that Kevin hoped his experience in technical writing, and especially in designing documents, would help convince Feona and others that they could produce in-house a professional-looking, informative newsletter economically and efficiently.

Kevin knew from his courses and his experience in technical and professional writing that audience consideration and carefully written text were not all that a writer was able to do to accommodate and influence an audience effectively. He also knew the importance of page design and how a well-designed document can have positive rhetorical effects on its readers. Therefore, he collected some of the best examples of documents he had written and designed, or that he had helped others produce, and wrote short rhetorical analyses to show how the documents appealed to their audiences and that they could be produced with existing facilities and limited funds.

After a week's preparation, Kevin felt pretty confident that his office would soon be producing its first newsletter.

THE IMPORTANCE OF PRESENTATION

Presentation is a key element in many areas of the professional arena. In situations in which competition is very strong, the way a product is presented may determine its eventual success or failure. Presentation helps persuade consumers that the product does what it is supposed to do, and does it well. In the restaurant business, for example, taste is not the only consideration for a successful meal. A high priority is placed on how the meal looks on the plate. The same is true for the clothing and the automobile industries, or any other industry that depends to a large degree on aesthetics to sell its products.

However, the power of presentation is not simply in the attempt to beautify an item for the sell alone. In Kevin's case, for instance, the design of his newsletter—how information is presented on the page—would also affect how well it informs its readers. Kevin's experience both as writer and as reader showed him that he learned much more, and learned it more easily, from documents that had been carefully designed and laid out than from those that were simply dense blocks of words.

Many people think, though, that designing pages of a document, considering how and where to lay out text and graphics, is a task best saved for the professional printer, designer, or graphic artist. This is true in part—graphic artists, designers, and printers make their livings by laying out, designing, and producing documents for clients or customers. Writers, however, also can use many of the same techniques that printers, designers, and graphic artists use, especially with the sophisticated capabilities of many personal computers. But even if those facilities are not available, writers still must keep in mind the visual impact of what they are writing—and must plan for it—if they are going to write effectively and persuasively.

The documents that professionals write can be designed in many different ways. Just by looking at all the written communication generated daily, especially that which comes from the professional sector, we find that those pieces that are most effective are also the best designed. Visual enhancement and displays ranging from stunning, multicolored pamphlets and manuals with varieties of type styles and sizes and with graphics within the text, to simple and clean, planned typewritten pages of an in-house report give a positive impression of organization and its professionalism.

A piece of technical or professional writing when first presented to the intended reader is very much like a new tool being introduced to the market. When presenting the new tool to prospective buyers, the producers will present a well-polished product and will display it in a manner that best depicts its abilities and usefulness. In other words, the tool will be presented in as visually informative a setting as possible. The professional and technical document must do the same.

This need for a piece of writing that is as informative visually as it is textually may seem difficult, for there seem to be as many variations in design as there are documents to write. However, all designs, if they are successful, accomplish three basic tasks:

1. A successful design *creates an immediate positive impression* for the reader. A well-designed document demonstrates a writer's and an organization's ability, and perhaps more important, a writer's and organization's commitment to producing a professional piece of writing.
2. A successful design *highlights the major topics* of the document. Effective and consistent use of headings, bullets, and other textual formatting, such as underlining and use of boldface or italic type, gives the reader a sense of order, hierarchy, and logic. In addition, well-designed high-

lighting facilitates both close, detailed reading and skimming for topical information.

3. A successful design *helps the reader read effectively*. A clear and effective document design helps readers follow a surer path from one element to another, making their reading faster and more efficient.

Thumbnail sketches allow the designer to compare a number of layouts side by side and to experiment with such things as location of graphics, organization of text, and use of columns and margins. The designer will look for the layout that best suits the purpose intended by the document for the given audience. Consider the thumbnail sketches of documents in Figure 12.1. The designer has laid out two possibilities for presenting the content of the piece. The design of document A, with its use of subtitles, graphics, and tables, is perhaps more appropriate for a formal, technical report, whereas the design of document B, with its use of double columns and boxes highlighting portions of the text, may appear more appropriate for an article in a periodical. The point is that without a hint of the actual content of the piece, the design of visual elements creates initial reactions and expectations in readers that designers should consider.

At this point, it is important to remember that page design and graphic layout alone are not enough to sell a piece of writing. Written communication must say what is intended clearly, concisely, and effectively, following

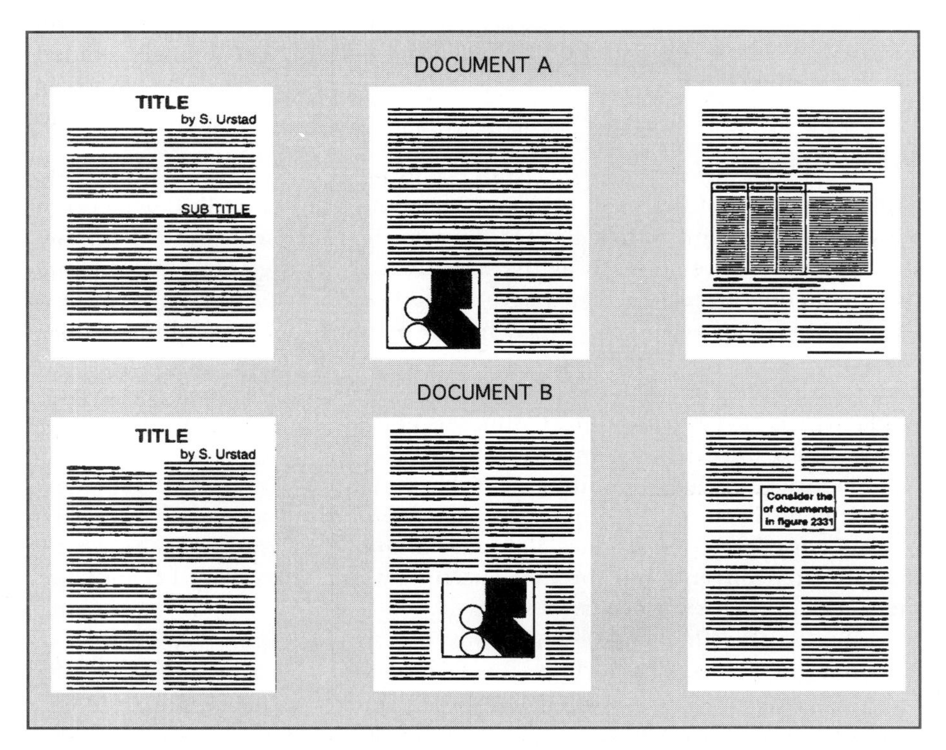

FIGURE 12.1 **Thumbnail sketches for a report**

considerations of tone, audience, and purpose that were discussed in earlier chapters. Only after these elements of the written message have been considered can designing the document begin. Font variations, multiple columns, and flashy graphics will be of little or no use if the text proper is weak and unconvincing. *An effective design will make a well-written document better—but it won't make a poorly written document good.* With this statement in mind, we will now consider how the physical design of a document can enhance already effectively written communication.

ELEMENTS OF DOCUMENT DESIGN

Writers, when designing documents, look at individual pages differently from the way most readers look at them. They actively consider aspects of the text, margins, and graphics that most readers see only passively. To writers and designers, the page has three key elements:

1. *White space*—the portion of the page not written or drawn on
2. *Text*—written portion of the document, consisting of readings and bulk text
3. *Graphics*—all items that stand outside the text proper, such as graphs, illustrations, tables, and the like

These elements are the building blocks of a well-designed page, and they can be considered in order, beginning with the white space of the blank page and systematically building a structure of headings and titles, and finally filling in the rest with text and graphics.

White Space

The white space of the page seems at first thought to be of little concern to writers, since many see their job as filling up the space with words. However, this space, often referred to as the *negative space,* is the frame in which all other elements are placed. The pages of this book provide a good example of the use of white space. To begin, each page is framed with margins, balanced borders of white, on either side of the page. Margins are something often taken for granted but with effective use of margins, a document can be more readable. Chapter titles and page numbers add information but still keep the overall frame intact and balanced.

Figure 12.2 illustrates differences that variations in margins can make in a document. From left to right, the pages come from an engineering textbook, a copyright contract, and a glossary of medical terms. In each case, the margins are conventional for the particular form of writing, because they meet certain rhetorical needs.

Textbooks (first page) commonly use single-column text with standard margin widths. This format allows for a good deal of text to be printed per page, without overcrowding. Thus, the primary purposes of textbooks—to inform and to persuade—can be accomplished in a fairly confined space, still permitting other visual and organizational enhancements, such as

FIGURE 12.2 **Margin variation in document design**

headings and graphics, to be included. Most documents in technical and professional writing—reports, position papers, proposals—use similar margin formats.

The second page of the contract, also responds to rhetorical purposes. Contracts, the basis for legal commitment, are one of the most meticulously analyzed forms of documentation written and as such are often designed with wide left and right margins that condense the text into shorter lines. Often the lines are numbered in the margin, which allows specific portions of the contract to be identified easily by line number. The wide margins provide space for notes to be written.

The third page is typical of a glossary or dictionary, primarily a reference source, which provides definitions of various terms used in a particular field. Through the use of double columns and small margins, writers can provide maximum information in a minimum of space. Since readers use glossaries to gain specific information at specific times and do not read them to understand and evaluate unifying expositions or arguments, glossaries are designed with maximum efficiency in mind. White space is kept to an absolute minimum.

In each of these cases, writers have tried to accommodate the purpose and use of the document in the design. They have decided to use formats that have become conventional, because their use over time has proven effective. But it is necessary to remember that although conventions often dictate the common path to follow, they should be considered first for their rhetorical effectiveness, and only second for how widely they are used.

If we look again to the use of white space in this textbook, we see more complex uses: to set off headings, to indicate paragraph indents and spacing, to highlight lists, and to separate major sections of text. These separations are very important, for they help readers categorize each section, visualizing the closing of one section and the preparation for a new one. The space around a list of bulleted or numbered items allows the list to be highlighted outside the text without distracting the reader, or worse, breaking the continuity of the written text.

Text

As mentioned earlier, bulk text and headings, which are the written portions of a document, form the second design element. For the goals and purposes of this discussion, the text should be considered in its visual form, not for its content. Textual formatting, manipulating the look of the text and headings, will affect how the reader views the document. Through variations in typeface and choice of column format and line justification, both text and headings can establish positive impressions for the reader. First we will discuss the role of headings as indicators of the organization and hierarchy of topics. Then we will discuss the elements that affect how the headings and text look—typeface, columns, and justification.

Headings. Writers of longer documents with more than one component should use headings. Headings provide clear organization of major sections, subtopics, and common elements, as well as a quick view of the overall content of the document. They aid the reader in making transitions by signaling the end of one section and the start of another. Without the order established by headings, the reader may become confused or lost.

Good headings complete two tasks in a piece of writing. First, they encourage a close, precise reading of the text by providing tags or markers of various topics, both major and minor. Second, and in contrast to the first, headings allow a document to be skimmed. Often readers want to glance through an entire piece to see what information has been provided before beginning a closer reading, or they may want to concentrate on or return to a section, after having read the document, and don't want to have to reread the entire piece. Without headings that tag various sections, topics, and subtopics, a lengthy piece of writing has to be read almost in its entirety each time—a time-consuming, inefficient process for busy people.

Headings also act as a form for outlining the text proper. Many writers use outlines when organizing thoughts for a piece of writing and creating a hierarchy for the subject and purpose of the document. Two of the common forms of outlining are illustrated in Figure 12.3 These two systems, an alphanumeric system (a combination of letters and numbers) and a decimal system, are identical in their purpose: to organize topics and thoughts in a logical hierarchy.

When the outline is expanded with the addition of text, for short or long documents, the various divisions of the outline become headings, and each heading represents the sequential and subordinating order of the outline. In texts of ten pages or fewer, the alphanumeric system is easy to follow, but for longer documents, writers often prefer the decimal system, because it keeps track of major and minor divisions. Finally, as is the case with this text, the alphanumeric or decimal notations are eliminated, and divisions are made clear with variations in type style and size to establish a consistent hierarchy of topics. Figure 12.4 shows two styles of headings. Style A uses varying styles and sizes of type to show the hierarchy of major and minor headings. Style B uses the decimal notation to accomplish the same purpose.

Because the primary purpose of a heading system is rhetorical, to help readers be informed, understand, and be able to make decisions, it is impor-

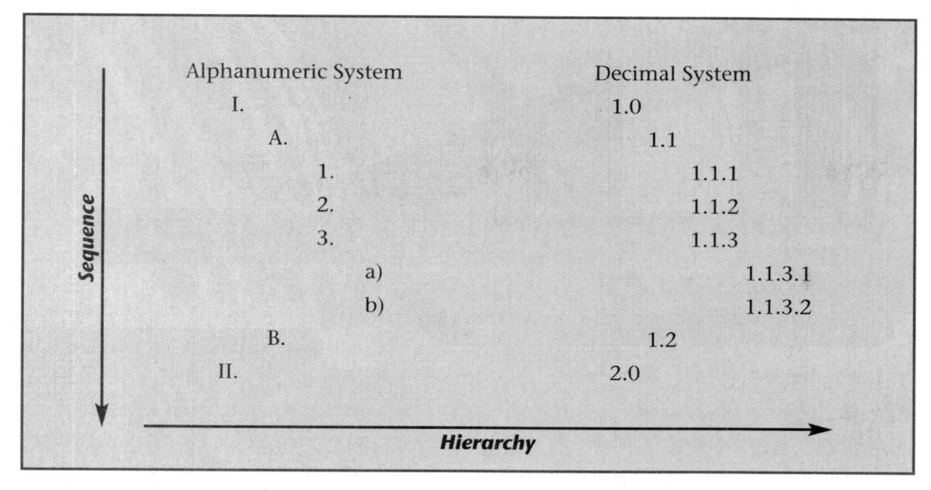

FIGURE 12.3 **Two systems of outlining**

tant that writers be consistent in their use of headings throughout the document. With consistent headings, writers control and manipulate the presentation of the text, which in turn sets up expectations in readers' minds. For example, if minor headings are *italicized* in the first section of the document, readers will expect that they be *italicized* in all following sections. Transitions from major headings (which may be **boldfaced**) to minor headings (which may be *italicized)* should remain the same from section to section. Once established, heading styles should be consistent throughout each portion of the document. Otherwise, the effect will be to confuse readers, not to aid them.

With this thought, let us now move to the discussion of typefaces used for text and headings, column formats, and justification of text and headings.

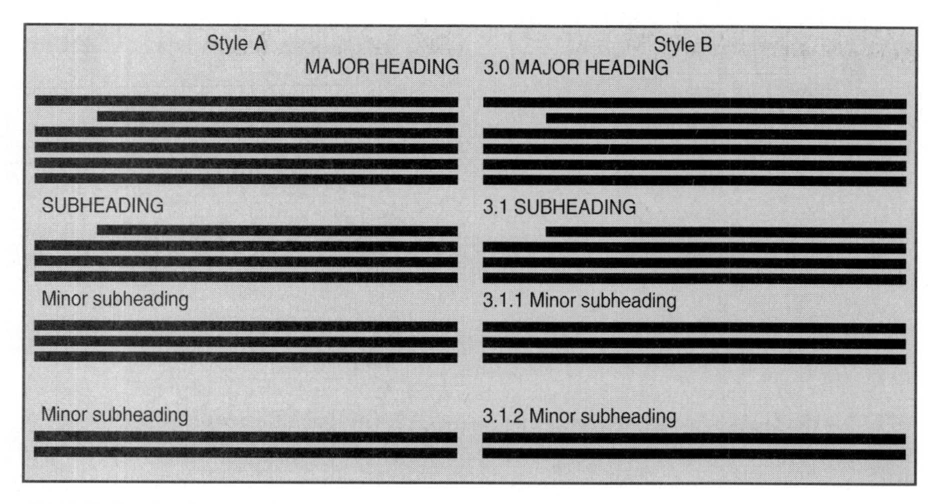

FIGURE 12.4 **Sample heading systems**

Typefaces. The term *typeface* refers to *all* type of a single style; the term *font* refers to a set of type all of one style and *one size.* Many computer software programs offer the writer a wide selection of typefaces. Here, for example, are four different styles of typeface all shown in 12-point type. The name of the typeface precedes the example:

Chicago:	**The rain in Spain falls mainly in the plains.**
Helvetica:	The rain in Spain falls mainly in the plains.
Times:	The rain in Spain falls mainly in the plains.
Courier:	`The rain in Spain falls mainly in the plains.`

Different typefaces affect readers differently. The Chicago typeface, for instance, seems bold (it appears, after all, in boldface without further treatment) and progressive, stating the message with strength and conviction. But Helvetica is leaner, softer, quieter, more elegant perhaps. Readers of Times see it as more traditional and conservative, whereas Courier appears to them as free, expansive, even airy. Writers using word processing software programs should choose typefaces that help present the message and an image in keeping with the purpose of the document they are writing. But a word of caution: changing fonts frequently or using too many of them can be distracting; readers begin to pay more attention to the variety and changes in the text than they do to the content. Font styles should be used to enhance the purpose and tone of the document; they should not become items of interest in themselves.

All typefaces can be categorized into two types: *serif* and *sans serif.* Serif typefaces are those with small hook strokes, called *serifs,* extending off from the character, as illustrated in Figure 12.5. Of the preceding samples, Times and Courier are serif typefaces. Many typesetting manuals will recommend serif typefaces for block text, because they supposedly are more readable and pleasing to the eye. Serifs link characters together well, allowing the eye to proceed word by word more easily on a line of print. Thus, large portions of bulk text are more readable in serif typefaces.

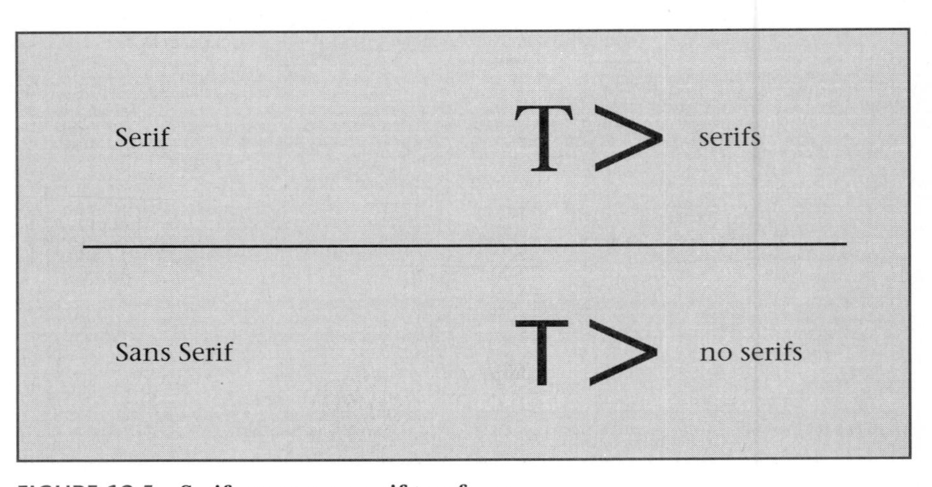

FIGURE 12.5 **Serif versus sans serif typefaces**

However, sans serif typefaces can also be used effectively. They may be used for headings and figure legends to help highlight these portions through their contrast with the text proper. Sans serif typefaces are also commonly used for résumés and manual graphics because of their clean design and ability to distinguish white space so effectively. Of the preceding samples, Chicago and Helvetica are sans serif typefaces.

Columns. Bulk text can be laid out on the page in several formats: single-column, multicolumn, and note marginal (see Figure 12.6). Single-column format, also referred to as *block text* layout, is the common layout for most memos, reports, textbooks, and proposals; multicolumn formats are often the norm for pamphlets, brochures, and articles intended for general audiences. Like different typefaces, different column formats serve specific rhetorical strategies. Single-column text usually appeals to a specific reader or group of readers who generally share a common knowledge of the topic being discussed.

And quite often, the level of discussion and technicality is high and complex. The readers of such a document, be it a memorandum written to a supervisor documenting a problem with employee relations or a report on a new method of gene splicing being prepared for a technical oversight committee, will be better able to read sentences of greater length and complexity in a single-column format than they would if the same were presented in a multicolumn format.

For example, a report on the video surveillance system for a nuclear facility is written specifically for the engineers and technicians who will be using the system. Given the equally high technical level of the audience, the writer will discuss the system's installation and application in an appropriately high technical level, and would most likely format the text in single columns. However, should this same report be rewritten to appear as an article in *Scientific*

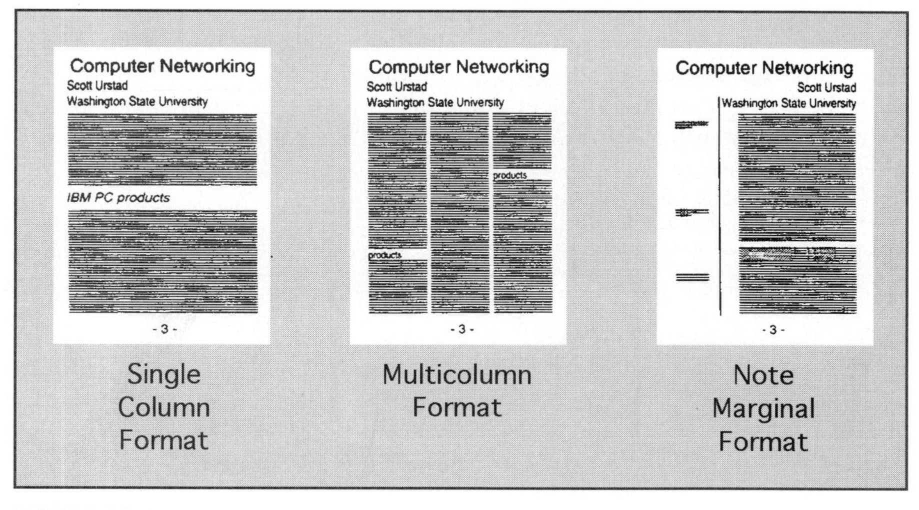

FIGURE 12.6 Typical column formats

American, which caters to a knowledgeable but more diverse audience, the tone of the article would need to be changed and the technicality reduced to address adequately this new group of readers. As a result, this text may be placed in a multicolumn format, as *Scientific American* features, using shorter paragraphs with simplified information, to appeal to a more general level of readership. In each of these writing situations, the format reflects primary rhetorical considerations: what is the purpose of this document, and who will be reading it?

The third format, note marginal, is a combination of both single- and multicolumn formats. The document is laid out in a disproportional two-column format with the smaller of the two columns placed either to the right or the left of the larger column. The smaller column is for writers' or readers' editorial commentary or notes on the text, or for extending graphics beyond the text. Manuals, handbooks, and textbooks are often produced in this way. In fact, Chapter 3 of this text demonstrates the use of a note marginal format in the presentation of Mary Ann Cox's memos.

Justification. Designers of a document should also consider justification of the text. *Justification* refers to the alignment of text along one or both margins (see Figure 12.7). Generally, block text is justified to the left, with a ragged alignment on the right margin, or to the right and left (sometimes referred to *full justification),* with both left and right margins evenly aligned. The choice is mostly aesthetic, but some writers believe that full justification produces cleaner, more professional-looking documents.

But a note of caution: documents that are professionally printed or that are produced on advanced word processing systems use a pixel-proportional system that equally divides the spaces between all words and letters in the line being justified to avoid large gaps in the text. Less advanced word processing programs often divide the spaces proportionally between the words alone, creating noticeable gaps in the text. The inconsistent spacing between words is distracting and makes the lines difficult to read, a result that obviously detracts from the most important element of the document—its content.

Just a word on right justification: although block text usually is not justified to the right, often headings are, to highlight and to vary the format of the text. But as is true of all formatting techniques, right justification should be used consistently, according to the hierarchy that the writer wants and needs to establish.

FIGURE 12.7 **Text alignment formats**

Graphics

Graphics are used to enhance the explanation of data for readers. In technical and professional writing, graphics seldom, if ever, are used purely for the purposes of embellishment or aesthetic appeal. They are needed to simplify complex design descriptions, to highlight and clarify trends and relationships found in numerical data, and to clarify complicated procedures or processes. Graphics play an important role in the success of technical and professional writing and must be used carefully if they are to be effective.

Knowing what a graphic is and knowing why it should be used are not necessarily the same thing; deciding when and where to use graphics within a document depends on a number of considerations. Keep in mind the following rules of thumb for the use of graphics:

- Place graphics immediately after (or as close as possible to) their textual reference.
- Consider the technical level of your audience carefully.
- Choose the best type of graphic for the situation.

When large graphics or numerous graphics are being used in a particular document, placement is not always an easy task. Smaller graphics may be incorporated directly into or following the paragraph in which they are referenced. Larger graphics, of a half or full page, should appear on the facing or following page. In any case, it is important to preface the graphic with text to prepare readers for its appearance and to establish its purpose. If readers turn the page of a document to find a full-page graphic diagram of the electrical wiring of a cooling system, before the text has begun discussing it, they will certainly not be prepared and more than likely will be confused.

Just as the text must meet the needs and expectations of an audience, so must the graphics. Highly technical diagrams and charts that a group of engineers may understand will have little success with a more general lay reader. This is not to say that graphics are good only for one particular audience. Almost all audiences can benefit from graphics as long as the designer chooses a type of graphic that is appropriate for the audience in question. Technicians or specialists may be very interested in the specifics of a particular system and perhaps would expect a table of values in the report. A group of managers, on the other hand, may expect and better understand more vivid representations of essentially the same data in the form of a graph or a pie chart. In either case, the use of graphics needs to be considered carefully to meet the needs and expectations of the intended audience.

It is not by coincidence that so many different kinds of graphics are used. Tables, bar graphs, illustrations, photographs, and the like all serve tasks specific to their forms and all have strengths and weaknesses. Writers, when choosing a graphic to depict and explain data, must decide which type will be most effective for their purpose and most appropriate for the intended audience. The seven most commonly used forms are discussed in the next section. Any other types are most often variations or spin-offs of these basic seven.

Displaced Text. Displaced text often is not thought of as a graphic, but when the rhetorical task of any graphic is being considered—to enhance or make clearer the information expressed through the text—displaced text becomes very much a graphic element of a piece of writing. Displaced text can be any text that is related to, but not directly incorporated into, the main text. Such forms as highlighting bullets, standout quotes or marginalia, and writers' editorial notes all act as graphics and must be considered carefully for use. Figure 12.8 shows a common way of displacing text.

Displaced text can highlight specific points that the author wants to emphasize. Bullets or numbers emphasize points made within the discussion by removing them from the density of the main text. Throughout this text, for example, we have used bullets or numbers to set off certain pieces of information that we want readers to remember or pay special attention to. In addition, displaced text works much like headings by working with the hierarchy of information to allow the reader to skim a section for the major points. *Standout quotes* are found often in journals, magazines, and newsletters to highlight specific ideas, findings, or statements. Unlike bullets and numbers, they act as a sort of advertisement for what is contained in the piece as a whole. Like all other graphics, standouts should appear after and very near their location in the text proper.

A final note to remember: too much displaced text throughout a piece of writing will distract the reader from the general text. Choose carefully what to displace and where it should go, for the reader will quickly know when text is being highlighted for purely aesthetic reasons, rather than for specific rhetorical and informative purposes.

Tables. Tables are used to list values of at least two variables (perhaps amount of money [variable 1] spent over fixed intervals of time [variable 2]) in rows and columns (or on x and y axes), organized with headings. A table requires at least two variables to be considered a table; otherwise, it is a list. Figure 12.9 shows the table that Leah Feldstein used to quantify information on EDUCARE's clientele for her oral presentation.

One in 10 American children and teens suffers from mental illness—and just one in five of those receives treatment according to U.S. Surgeon

"the surgeon general wants to promote public awareness of children's mental health issues . . ."

In a 52-page wake-up call released today, Satcher calls the situation a "public health crisis in mental health for children and adolescents."
Growing numbers of children are suffering needlessly because their emotional, behavioral, and developmental needs are not being met by those very institutions which were explicitly created to take care of them," writes Satcher in the report, which calls "a

Specifically, the surgeon wa
to promote parental awaren
of children's health issu
reduce the of these disea
and the ability to recogn
symptoms of mental h
health in children. "We ne
to help families understa
that the problems are re
they can be prevented, a
effective treatments are av
able," he writes.

FIGURE 12.8 **Displaced text**

EDUCARE CLIENT PROFILES					
		Sex			
	Mean Age	% M	% F	% Return	% Complete
1998–1999	15.4	79%	21%	n.a.	11%
1999–2000	14.8	61%	39%	57%	21%
2000–2001	12.1	74%	26%	36%	34%
Average	14.1	71.3%	28.7%	46.5%	22%

FIGURE 12.9 Sample table

Writers use tables as organizing devices to display and compare statistical data and to make the discussion of the data easier to follow. If writers were to attempt to lay out and discuss the values corresponding to multiple variables in a confined, tabular form, writers are then free to discuss only the highlights of the data in the text, permitting readers to concentrate on those highlights, and then to consider the rest of the data as presented in the table in their own time, and as they see fit.

Tables are effective for displaying absolute values and precise comparisons of discrete items, but are less useful if the purpose is to show relationships proportionally or in trends.

One final note: tables are always called *tables,* and if there are several, they are labeled sequentially: Table 1, Table 2, Table 3, and so forth. All other graphics are called *figures,* no matter what type they are, and are also labeled sequentially: Figure 1, Figure 2, Figure 3. Some styles call for *Figure* to be abbreviated *Fig.* 1.

Pie Charts. Pie charts (or pie diagrams) are most commonly used to show proportional comparisons of the parts of a whole population or field by percentage, as is illustrated in Figure 12.10.

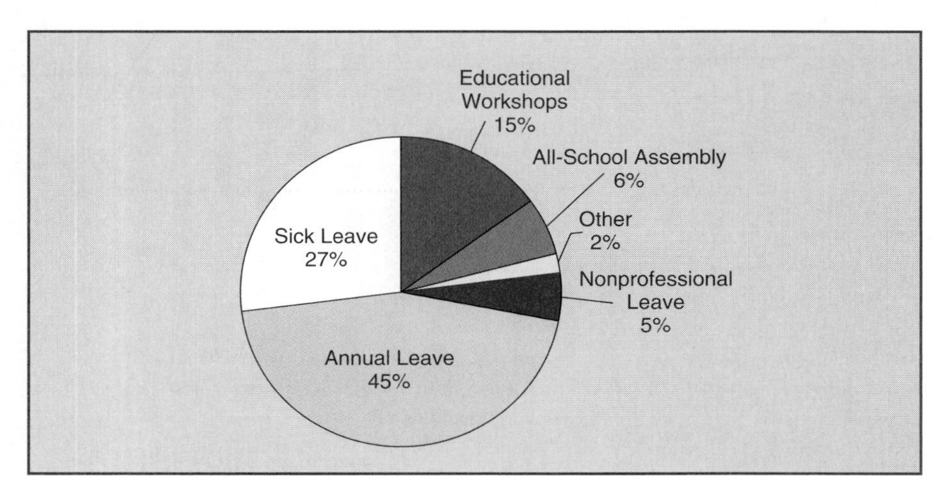

FIGURE 12.10 Sample pie chart

If writers want to highlight visually two or three main elements as they relate proportionally to the whole, such as time spent on sick leave or annual leaves or educational workshops, pie charts work well. However, they do not display absolute values effectively unless the value is added as a label to its respective piece of the pie. Also, if writers want to point out differences between two relatively similar percentages (say the difference between 7 percent and 9 percent), pie charts, because they are not precise enough, do not show these differences well. Finally, a pie chart will not work well if the pie is divided into too many "slices," making it hard for the reader to see the relationships between a specific part and the whole or other parts.

Bar Graphs. Like pie charts, bar graphs present in a more accessible and vivid way the data often found in tables. But as Figure 12.11 illustrates, bar graphs, unlike pie charts, can display more graphically both discrete quantities (frequency of signal) and trends (general increases) over time.

In addition, bar graphs can depict other elements or variables. Figure 12.11, for example, is actually a *stack graph*, which also shows proportional relationships of parts to the whole, as pie charts do. Differences in signal frequency are compared over time, but the graph also illustrates (or *stacks*) the proportional difference between signal A and signal B at each measured point in time.

Line Graphs. Line graphs are excellent for showing trends, and two or three elements can be compared effectively. The line graph in Figure 12.12 compares population histories of coyotes and jackrabbits more immediately and effectively than separate tables of the same figures would. But in situa-

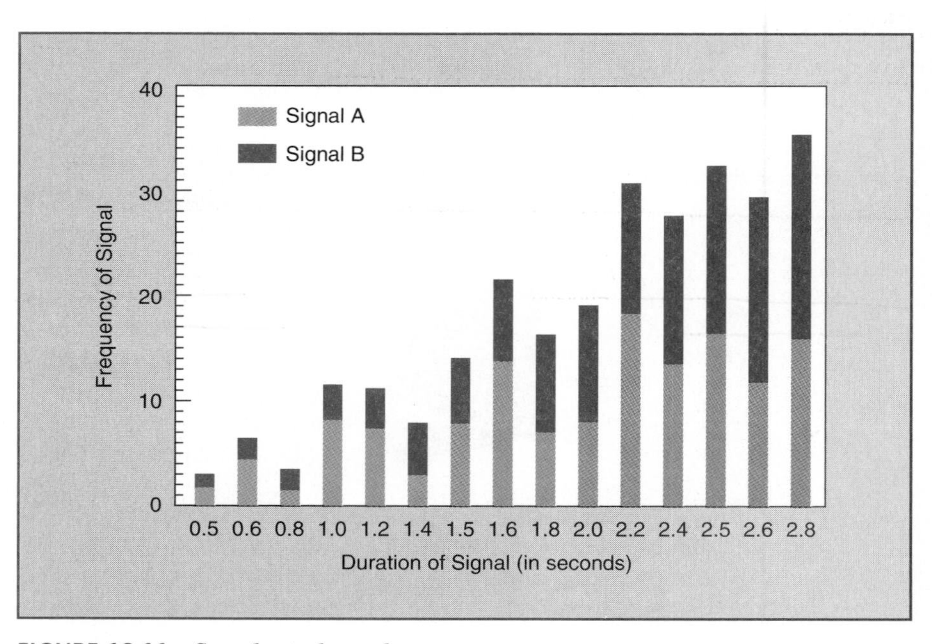

FIGURE 12.11 **Sample stack graph**

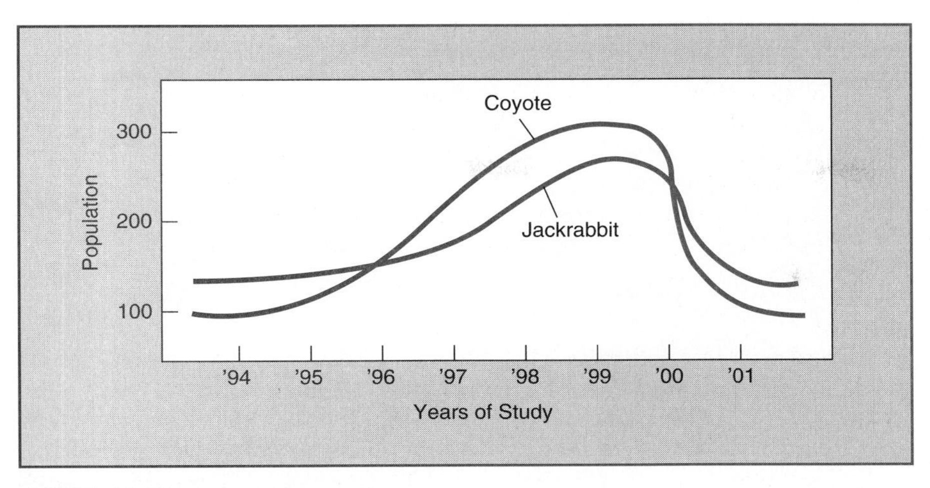

FIGURE 12.12 **Sample line graph**

tions in which more than two or three elements are being compared, bar graphs or multiple graphs may be better.

The primary weakness in line graphs is the inability to depict proportions or percentages of a whole. However, by using an *area format,* illustrated in Figure 12.13, proportions are captured graphically. Readers can see both the total number of notifications of food poisoning in England and Wales, and the way that the numbers of notifications of food poisoning in each place compare with the whole over time.

Photographs. Photographs, such as the one in Figure 12.14, can be used when accuracy is important or when drawings or illustrations cannot capture the fine detail that the writer is discussing. However, using photo-

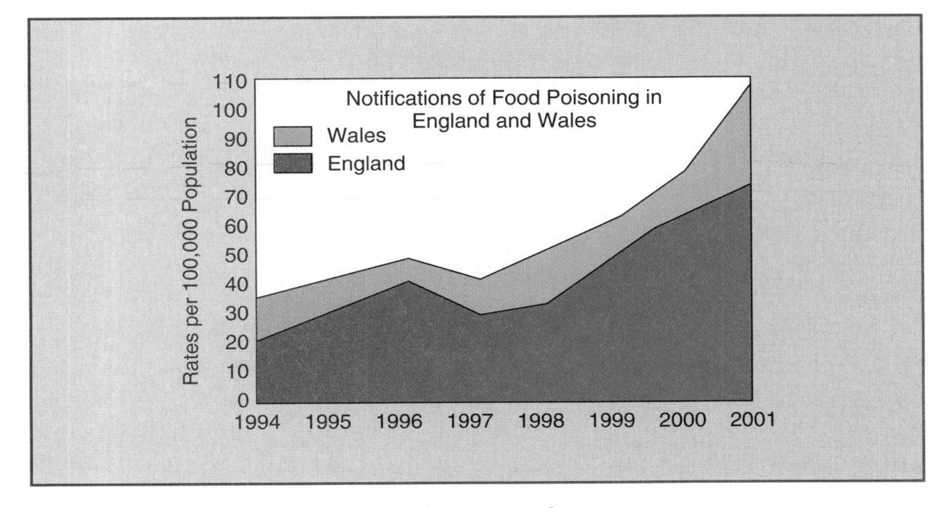

FIGURE 12.13 **Sample line graph using an area format**

FIGURE 12.14 **Sample photograph**

graphs effectively can be difficult. Unlike drawings or illustrations, photographs cannot be embellished easily without specific placement of the camera, special lighting, or arrangement of the object or scene.

Nonetheless, if all these factors can be attended to, photographs lend an impressive and often dramatic quality to a document. As with any other graphic, it is important to consider the audience. Before using a photograph, decide whether the audience is truly in need of such specific representation.

Illustrations and Drawings. In many technical areas, drawings and illustrations are the most commonly used form of visual description of parts, processes, and systems. Unlike photographs, drawings allow the writer more freedom to highlight specific aspects of the item in question. Drawings are best for describing parts of a piece of equipment or showing mechanisms or processes inside a unit that a photograph would not be able to capture, as shown in Figure 12.15. It was not the writer's intention to sketch a perfect still life of a mass spectrometer but rather to illustrate the process of decomposition taking place within the apparatus and to label the parts that a reader needs to know in order to understand the process.

Graphics, when used with careful consideration, play an important rhetorical role in technical and professional writing. Carefully considering the right graphic helps present technical or complex data, processes, systems, and descriptions clearly and descriptively. The right graphic will save the writer explanation, will prevent confusion, and will leave a favorable impression on the reader.

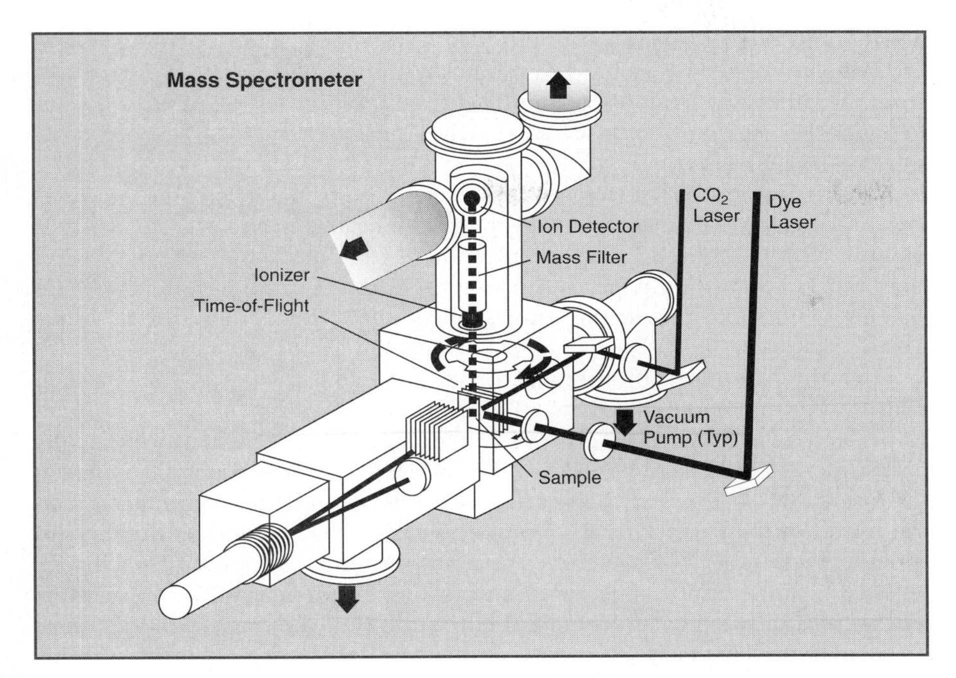

FIGURE 12.15 **Sample drawing**

SAMPLE PROFESSIONAL DOCUMENT DESIGN

After a rather heated debate over the usefulness of a newsletter for the Office of International Education, Kevin had successfully convinced the staff of the benefits that such a publication could have. From that point, the meeting became quite productive. Jessica Haysmith, Assistant Director for International Students and Faculty, recommended something that Kevin hadn't considered. In addition to reporting the general happenings of the office, each of the three sections—Overseas Programs, International Students and Faculty, and the Intensive English Language Center—should provide an article addressing a current issue or concern in the respective areas. She also suggested that the newsletter be distributed bimonthly to save on production costs.

Kevin thought that these were good additions to his initial proposal. They could be excellent publicity for his Overseas Programs section, while allowing each member of the staff to participate in writing and producing the newsletter. In addition, Kristina Elswood, Kevin's assistant, reminded him that if the newsletter proved successful, they could consider applying for grant money to distribute it to alumni supporters and other international education programs nationally.

On Saturday afternoon, Kevin found himself in the position that he wanted to be in but also was much aware of the large task before him. He had finished writing the article for the Overseas Programs section, a profile of the newly established Russian programs and agreements that Westminster and the OIE had worked very hard to establish, and he had received the articles from the other two sections of the office as well. With the exception of the article that Feona would write as Acting Director, Kevin was ready to begin designing the newsletter.

Kevin's first step was to draw up a set of thumbnail sketches of various design ideas for the newsletter. (Several of the figures in this chapter, such as Figures 12.1, 12.2, and 12.6, are thumbnail sketches.) By drawing several thumbnails of the first page, Kevin would be able to find a general format that he would like to see the entire newsletter follow. Figure 12.16 shows Kevin's sketches.

Having produced thumbnails for six possible front pages, Kevin began considering the image that the staff, during Thursday's meeting, had wanted to project. They required a professional-looking document that both faculty and students could read easily. Kevin pondered his choices before choosing the last sketch in the first row—a four-column layout with a large graphic in the center of the page. This was a practical design that would allow readers to browse the number of article titles and subtitles to find topics of specific interest. Of more importance, since the primary audience would be readers who were interested in international education but who had diverse levels of knowledge and expertise, the tone and style of the articles that he and his peers had written were appropriate for a general, educated audience and were perfect for a multicolumn format.

FIGURE 12.16 **Thumbnail sketches of the OIE newsletter**

Kevin also knew that the purpose and image of the newsletter would be enhanced by choosing the front-page article carefully. The title of that coupled with the masthead and a strong graphic, should give readers an immediate understanding of the kind of document that this was intended to be.

From here, Kevin thought, the task would be a little easier. He decided to make all remaining pages of the document follow a similar format. In choosing headings, Kevin wanted to create a hierarchy both within each of the articles and between the articles and other special interest sections. He wanted to have a graphic on each of the inside pages, but he knew enough not to force graphics into the text when they were not needed. There was time enough for including graphics sensibly. The topics for future articles that he and other members had presented in Thursday's staff meeting were all interesting and relevant, several of which he knew immediately would be complex and would require a graphic of some kind to help explain the ideas to his readers.

Kevin chose to open the newsletter with his article about the new Russian programs. He knew he would get a ribbing from the others for choosing his article to appear on the front page, but the news was exciting and fit in well with current affairs. In addition, a photograph of Moscow on the front page would be eyecatching and appropriate for a newsletter coming from the Office of International Education.

With his general page layout complete, Kevin began developing the heading sequence for the newsletter by choosing the typefaces and deciding on the locations for the headings. Remembering what he had learned from past reports he had written and from the successes and failures that resulted, he decided to lay out a heading sequence on a separate page to see whether it worked well. For a newsletter, like many other reports, he liked using typeface variations rather than an alphanumeric outline to organize the hierarchy of the piece. Large sans serif fonts would make for clear and readable titles and major headings. The text itself would be written in a serif font because it is easier to read. By lining up his choices in vertical order, Kevin could see how well they were separated from others.

Newsletter title	Foreign Study
	UPDATE
Major article heading	**OIE Takes Part in Glasnost**
Minor article heading	*From the Director's Desk*
Subheading	Program Deadlines
	Language Requirements

Once Kevin was happy with his choice for headings, it was time to build a template (a standard design or layout that can be reproduced whenever necessary) on his computer from the thumbnail that he chose and to incorporate the other articles into it. The office was producing the document on the office computer system, and the other writers had all written their pieces on the same system, so Kevin had only to insert the documents into his template and begin proofing to be sure that all textual references preceded the respective graphics and that the layout of each article was consistent with the previous one. Kevin was nearly finished.

Kevin's template (see Figure 12.17) shows how he wanted the title and organization to be displayed and where his graphic was to be placed both to make the text easier to understand and to encourage the reader to continue.

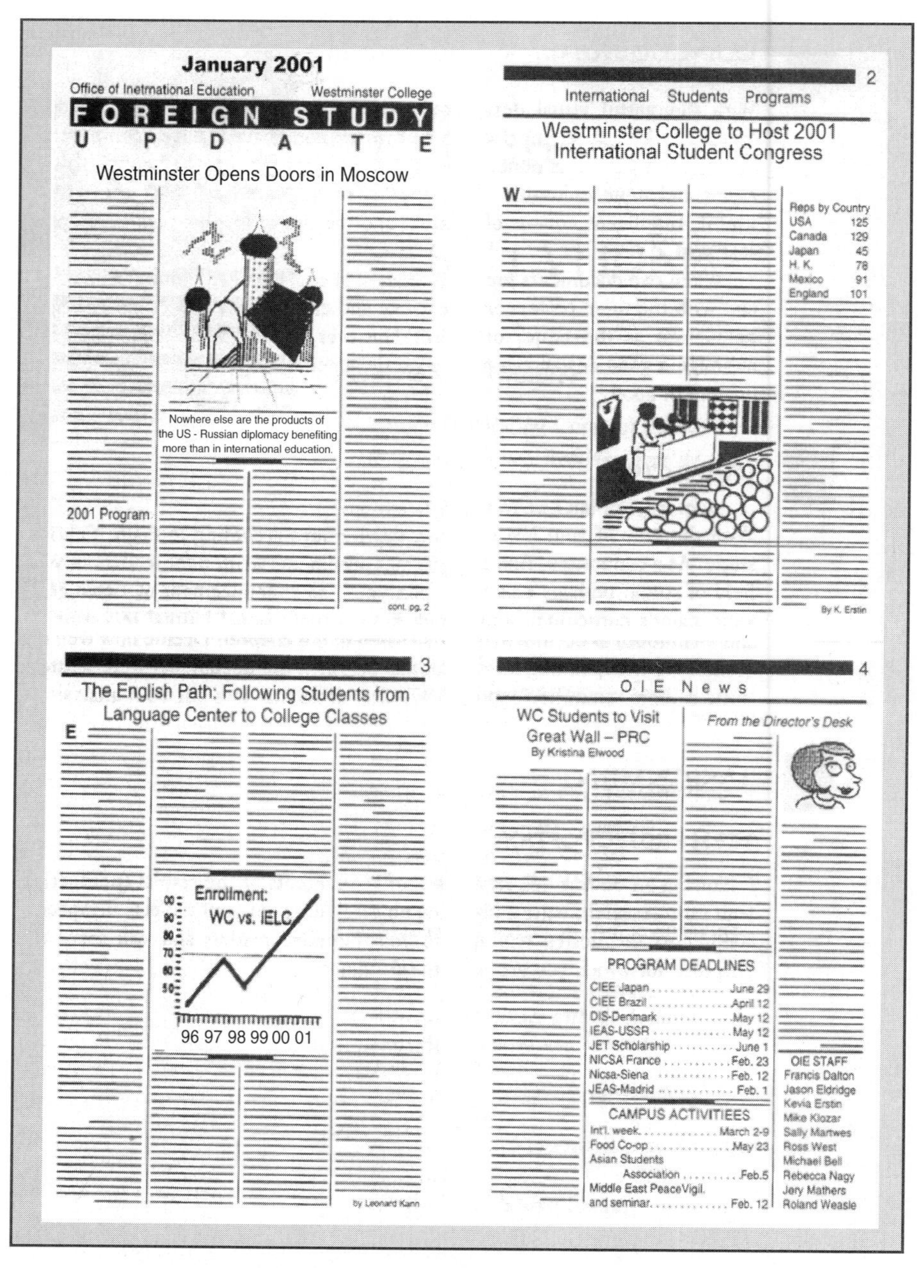

FIGURE 12.17 Kevin Ersting's template for the OIE newsletter

In addition, the four-column design allows the reader to preview the title and subheadings before beginning a close reading.

CREATIVITY VERSUS CONVENTION IN DOCUMENT DESIGN

A balance between creativity and convention is important to discuss at this point. Technical and professional documents must compete with others for the attention of their readers. As a result, writers may be tempted to embellish their writing visually for purely competitive reasons. But by so doing, they may go beyond the bounds of convention and forget that the primary purposes of technical and professional writing are to inform, perhaps to persuade, but above all, to be useful to readers. New and creative document designs can be productive pieces of written communication as long as they continue to serve the rhetorical needs that the documents require.

CONCLUSIONS

With thoughtful visual design, writers can positively influence their readers' acceptance and use of any document. But writers don't have to have a computer on their desks to accomplish this goal. All they need is a clear idea of what makes writing easier and more enjoyable to read; they need ultimately to make the writing useful. Intelligent notions of formatting and the use of graphics can accomplish that goal.

When two documents present information of equal quality and usefulness, the one with the more professional appearance and more effective use of design will often be viewed as the better of the two. However, writers must remember that although a good design can make a good document better, it cannot make a bad document good.

EXERCISE

1. With the help of your instructor, form small groups according to major fields of study. As a committee, find three different documents written for readers in your field of study: perhaps a newsletter for alumni, an announcement of a change in your major's curriculum, and an article from a professional journal reporting research. Analyze the physical design of the documents on the basis of the principle and techniques of document design discussed in this chapter. Decide how well the designs work depending on the purposes for which the documents were written. Write a short committee report for your classmates that lays out your analysis.

ASSIGNMENT

Following an outline that you have written of the contents of the results (product) of your term research, write a plan for the physical design of your project. Be specific about how the design will appeal to your intended readers and will serve the purposes for which you are writing the document.

Solving Problems Through Oral Presentations

CASE STUDY

LEAH FELDSTEIN'S OFFICE

EDUCARE, Inc. ■ 1200 Stone Way, Suite 500 ■ Seattle, Washington

Leah Feldstein returned to her office at a markedly slow pace. She had just come out of a meeting with EDUCARE's director, John Fiske, who had let her in on some very unsettling news. He confirmed the rumors that had been flying around that the U.S. Department of Education was planning to cut its funding to drug education and rehabilitation programs, and that an assessment team were on their way to the West Coast to determine which programs in the major cities should be cut and which ones saved. If EDUCARE were cut, there might be a loss of up to 25 percent of its funding. Such a cut, coupled with other likely cuts from the U. S. Department of Health and Human Services and the state of Washington, would effectively wipe out EDUCARE's funding. All that would be left would be money from private foundations and private donations, which currently accounted for only 23 percent of the budget.

PREVIEW OF THE CHAPTER

(continued)

The long-term future looked pretty bleak but was still unknown. The immediate future, on the other hand, was very clearly known, and its bleakness was beside the point. EDUCARE would have to wage an all-out effort to convince the government that it was a program that needed to be saved. And Leah's part in that effort, she found out today, would start with an oral presentation to the assessment team coming out from Washington, D.C. As Director of Client Relations, Leah had responsibility for the general direction of the drug education program in the public schools, outreach work to identify adolescents who are using drugs and who would benefit from specific education and rehabilitation, and supervision of other social workers who act as counselors of adolescents inside and outside the public schools. John Fiske wanted her to present to the team, who were due to arrive for their initial assessment on October 25, the strongest achievements that the program had made in client relations, especially with regard to education. The rehabilitation element of the program would have to wait for now, since the threat of cuts came immediately from the Department of Education. In addition, Fiske wanted Leah to lay out her plans for her presentation in a meeting on Wednesday, October 17, with Fiske and the budget director, who also had to present his story to the assessment team.

Leah was no stranger to oral presentations. In the 5 years she had worked for EDUCARE, she had given eight or ten presentations to PTAs, to teachers for in-service training, and to the boards of directors of private foundations and the United Way. But she never had to talk in a way that could determine whether EDUCARE and she, as its Director of Client Relations, would survive.

Leah's problem was clearly defined, but the specifics of how she would appeal to the assessment team were yet to be determined. She knew, as she sat thinking at her desk, that she probably couldn't take a real lunch break today, or maybe for some days to come.

THE NATURALNESS OF ORAL COMMUNICATION

Every time Leah was asked to give an oral presentation, she was reminded of her college days and of a technical writing teacher who started off his course with a short lecture on the unnaturalness of writing. He pointed out that, historically, human activity relied on a much older oral tradition and that written communication is a relatively recent phenomenon, arising independently in different civilizations only about six thousand years ago. He would point to the art of poets like Homer, whose *Iliad* and *Odyssey* were strictly oral accounts of the glories and tragedies of the Greeks, written in the forms known today only after centuries of oral presentation passed on regularly from one generation to another. He would talk also of the great Greek and Roman orators, who through oral language attempted to sway their fellow citizens to adopt some course of action for the betterment and preservation of the state.

To make his case, he would also point to what is known of human development. Children master oral language long before they master (and *master is* the key word here) written language. By the age of five, children have learned and essentially mastered the intricacies of their grammars and can communicate most of what needs to be communicated to get along. But it's not until years later that they master writing. And some children, he was quick to note, never learn anything but the basics of writing their names. He would cap it off by reminding his students that most of the world's peoples are still nonliterate and that the activities of their lives are carried on only through oral means.

Leah smiled in remembering that her teacher's purpose in mentioning all of this was to establish that writing wasn't a "natural" act and that the students shouldn't feel bad about not being able to master it right away. Further, he wanted to suggest that since oral communication is more natural, the presentations that his students would have to give at the end of the semester would be easier to produce than any of the corresponding tasks in writing. That an oral presentation would be easier than a written one couldn't be true on the face of it, she had known, but it wasn't until later that she would fully understand how much that was stretching the truth.

Practical Limitations Resulting from the Setting

Leah knew from her experience in college when she gave her first oral presentation and from her experience in speaking before groups since then that oral communication may be more natural in the general scheme of things, but it's not much easier to produce in formal settings than writing is. She still had to run through the steps of technical problem solving (definition, research, analysis, resolution and synthesis, and implementation) and of rhetorical problem solving (evaluating purpose, audience, context, and ethical stance). Leah had also realized, with the help of her teacher, the practical ways in which oral communication is limited as a natural act:

- *As the number of listeners increases, the self-consciousness of the speaker also increases.* Leah's experience in oral presentations had never included speeches to large audiences. The audiences had been limited in large part to smaller groups of ten or fifteen people, but she realized that when the numbers increased even to twenty-five or so, she would begin to feel less at ease. These kinds of questions, for instance, came to her mind: How do I look and sound? Am I making sense and covering the points that need to be covered? Do I have enough time to say what I need to say? Or, perhaps, am I taking up too much time? Leah believed that some self-consciousness is good, even necessary, for her to speak effectively. But too much self-consciousness translates into visible nervousness, which if not controlled, becomes distracting for both her and her audience.

- *As the size of the audience increases, the sense of intimacy between speaker and audience decreases.* Leah prided herself on her ability to engage her audience: she wanted to be friendly and comfortable with her listeners, but she also wanted to show people that she knew her business, that she knew what she was talking about. She liked her listeners to take part in her presentations—to ask questions, answer questions, and otherwise provide various verbal reactions. But again Leah knew that this kind of intimacy was usually easier to achieve when the audience was small and that sometimes when the audience was larger, she could feel a psychological distance develop between her and audience, which in turn could cause awkwardness on both parts. But a loss of intimacy is not inevitable. In fact, the successful, effective speaker creates intimacy even in large groups.
- *The physical surroundings can cause problems.* Although Leah had never experienced a truly problematic physical set-up while giving an oral presentation, she knew that problems could exist that would make a presentation difficult. The room may be too big or too small, given the size of the audience. The room may be too hot or too cold, too light or too dark. The acoustics of the room or quality and condition of the audio equipment may make it difficult for the speaker to be heard easily. Projection equipment, including computers, for slides or overheads may be faulty or inadequate for the presenter's needs. All these factors can create interference (what communication specialists call *noise*) between speaker and audience. And as the noise increases, the clarity and understanding of the message that the speaker wants to send decreases.

As Leah reviewed all the things she had learned about oral presentations, she kept coming back to this notion of noise, which for her was a good metaphor for the things that were apt to disrupt her efforts at conveying information effectively. All kinds of factors, both psychological and physical, could contribute to noise, but she knew that the amount of noise is inversely proportional to the amount of preparation she is willing to put into a presentation. Leah remembered that the times she had done well in speaking before a group, she also had prepared for well. Likewise, she remembered the times for which she had made little preparation and that they were relatively unsuccessful. As is true of most of us, Leah remembered the bad times more vividly than she remembered the good.

On this occasion, Leah could not afford a bad presentation. Too much was riding on it—the future of the agency and her own job were clearly at stake—and she would simply have to do the very best job she could. So she began the usual process of preparing her presentation, making sure that every step was as thoroughly thought through and accomplished as possible.

RHETORICAL ANALYSIS OF AN ORAL PRESENTATION

As for everything else that she wrote, Leah knew that beyond defining the problem, which she and John Fiske already had been able to do quite clearly, she would have to analyze the rhetorical aspects of the situation for her oral presentation. Therefore, she had to determine the nature of four things (all of which have been discussed in Chapter 3 and elsewhere): purpose, audience, context, and ethical stance.

Purpose

In her analysis of purpose, Leah first asked herself a dual question: why am I giving this presentation, and what purpose does my audience have for listening to me? Since Leah knew why the people from Washington were going to visit, she also knew that their purpose for listening and hers for presenting must coincide in some way. But she needed to sketch out the purposes consciously to make sure that she emphasize in her presentation those elements that needed emphasis.

Leah knew that the main purpose of the team's visit was to assess the effectiveness of EDUCARE's education program in light of its costs. So like it or not, Leah knew the team would be looking keenly for weaknesses, anything it could recommend for cutting. Like any other program, EDUCARE's had its weaknesses. The most troubling one had to do with the outreach part of the program. In outreach, social workers acted as counselors to adolescents whom they had identified as drug users and who would benefit from specific education and rehabilitation outside school. The idea was that students generally would be more willing to meet counselors outside school rather than in school where they would be singled out for special attention, no doubt an embarrassing situation for many. To many people in the rehabilitation and recovery business, outreach is very close to recruitment, something that EDUCARE had always been very conservative about.

The trouble lay, therefore, in the numbers. Although Leah didn't have the figures immediately at hand, she knew that the ratio of counselors to clients was a bit too favorable from the government's point of view. Although she and her staff knew that keeping the ratio low was good, that better work could be done when counselors had fewer clients, she also knew that the Department of Education would want to see how widely EDUCARE had been able to broadcast its efforts. This was in an area, Leah further knew, where the two interests would never completely coincide, the one more concerned with the quality of the work being done, the other with costs. For Leah, the ideal was to hire more counselors so that more clients could be served, and served well. The government's ideal was to serve the most clients possible at the least cost. For better or for worse, Leah knew that those working for the government, no matter what they personally thought the ideal was, would be looking for inefficiencies in EDUCARE's counselor-client ratio.

As she prepared her presentation, Leah would have to research the figures and analyze them, and then try to make the best impression possible—put the best face on it—within the limits of time. Her specific goal would be to show that what the team might perceive as weaknesses were really strengths in some ways and that increasing numbers could be served without significant budget increases. Her general goal would be to show that the program as a whole was working both within and outside the schools; that it had the potential to keep working, perhaps even more effectively; and that it should be fully funded for these reasons.

The next question Leah asked herself was, What should my audience do after listening to my presentation? Closely related to the question of basic purpose, this question looks forward to what would possibly happen as a result of the presentation. What could happen is something speakers (and of course writers) need to think about, since the purposes for speaking (and writing) should prepare an audience to make decisions in ways that it had not previously done. This is true even if immediate decisions or actions are not expected from the audience, or if initially the only purpose for the presentation is simply to inform. It should be remembered that good informative presentations ultimately enable audiences to make decisions, whether or not they lie considerably down the line.

Leah had quite specific ideas about what her audience should do. She wanted the team to decide that EDUCARE's program was working well, that it was working as efficiently as any program of its kind could work, and that it should be maintained at full funding levels. (Actually, she hoped that the team would decide that the program should get increased funding, but she realized that that event was unlikely to happen, given the specific purpose the team had for visiting.) In many cases, the answer to the questions of basic purpose and the question of what the audience should do are closely related. However, it is still a good idea to address the second question. Doing so helps add that bit of clarity and persuasive emphasis to the speaker's presentation, which may make the difference between the ultimate success or failure of the presentation.

Audience

For Leah, answering the next question was fairly easy. Who will be listening to me and evaluating my presentation? In this particular case, Leah could find out specifically who the people on the team would be: their names, positions and areas of responsibility, length of employment, and so forth. From other records, she could find out what their educational backgrounds were, where they had gone to school and for how long, and what degrees they had received. All this information would help Leah determine their likely prejudices, their leanings, and their ways of thinking, as well as the ways that she might appeal to them.

As she had predicted, she was able to come up with some interesting particulars. She found out that there were to be three people on the assessment team coming from Washington—two men and a woman. They all were program officers at the national level and therefore were responsible for reviewing the drug education programs that had been approved and funded by the Department of Education for at least the past year. She learned also that all had been hired under the Bush administration and had been retained by the Clinton administration without interruption, so that they all had had reasonably long experience and would know the ins and outs of funded programs such as EDUCARE's.

Leah was pleased to discover that she had previously met one of the team—Mary Stamper—at a Washington meeting of urban program directors that she had attended with John Fiske over a year ago. She and Mary had gotten along well, both having similar educational backgrounds and professional interests. Mary struck her as bright, reasonable, and committed to similar goals as hers—furthering innovative and effective drug education programs at local levels through grassroots efforts.

The assessment team from Washington would be joined in Seattle by the regional director of the Department of Education and his assistant, a woman, who were in charge of following the progress of funded programs regionally. Unfortunately, although Leah knew their names and had met them on several occasions, she did not know them well. They worked out of the regional office in San Francisco, and the problems and programs in California seemed to demand most of their attention. They got to Seattle only occasionally, and then only for brief stays.

Leah's check on the educational backgrounds of all five people turned up some common ground. All the team from Washington had attended medium to large state universities, two in the East, one in the Midwest. The two regional people had gone to schools—also state institutions—in the West. All held master's degrees in education, social work, or sociology. Four had been secondary-school teachers and college instructors in public institutions for up to five years before working for the government. Each one was between the ages of thirty-one and forty-five.

Given this information, Leah could sketch out a fairly close and useful profile of the members of her audience. They were young to middle-aged, educated, experienced both in teaching and in government work, and likely to be conservative, but not reactionary, given whom they were presently working for and the kind of work they did.

In all, Leah could say with some confidence that they wouldn't be a hostile audience. But since they were coming to Seattle for a very specific purpose—to determine whether EDUCARE should continue to get government funding—she knew, too, that they wouldn't be too easy to convince. Her audience most likely would be courteous and attentive, but also thoughtful, careful, and undoubtedly probing. The one idea that she thought would unite them was the need for drug education programs, but ones that could continue to operate effectively in times of budgetary restraint.

Although Leah was able to draw a rather detailed audience profile in this case, she knew that such specifics are not always possible to collect. In the past, she had had to go on much more general analyses for some of her oral presentations. The audiences were larger, came from more diverse backgrounds, and although apparently drawn together on common ground, often based their decisions on distinctly different thinking.

She remembered, in fact, the talks she used to give at the beginning of EDUCARE's work in the schools to groups made up of school board members, administrators, teachers, and parents. Although everyone was there ostensibly to learn about what might be done to educate schoolchildren about drug use and abuse, many came with different ideas of how that might be done effectively, and some came with a well-developed skepticism as to whether it could be done at all. In some cases, the range of ideas and attitudes went from conservative approaches to education that suspiciously viewed innovations in the classroom to distinctly liberal positions that embraced, often uncritically, almost any innovation in education. Then there were those who held the position that problems with drugs, like problems with early sexual activity, are not rightfully the business of schools, that they are moral problems that only the home and the church should deal with. Audiences like these are difficult, although not impossible, to appeal to. Leah had had to do it in the past, and she knew there would be times in the future when she would have to do it again. She was simply glad that this wasn't one of those times.

Finally, Leah wanted to be realistic about the whole subject of audience. Although this presentation would be the highlight of her day, she was fairly sure it would not be the highlight of her audience's day. Chances were good that her audience would listen as carefully as possible but would not bring high intensity to that task. Leah could remember being on the listening end of a presentation when she attended professional conferences. She could calmly and coolly focus on the presentation's content while the presenter worked. It was helpful to recall this lack of intensity that the audience feels, which is in contrast to the intensity and self-consciousness felt by the speaker. Speakers perhaps can feel less nervous if they remember that they are not the center of the audience's universe. Even though all eyes may be focused on the speaker, the minds behind the eyes are sometimes thinking about other things, like the room temperature, the person sitting in the next chair, the dessert served at lunch, the lines at the restroom. Audiences often have short attention spans; this fact can be both a blessing and a curse for the speaker.

Context

Leah's analysis of the context of her presentation focused on the considerations of place rather than those of time. The issue is not time, because the team controls the timetable for decision making. Leah need be ready only when she is scheduled. But the considerations of place, specifically the political constraints, must be a vital part of Leah's analysis. She knew that the context of this presentation ushered in all the political problems of a federal bureaucracy. As government employees, the team must adjust their policy to political decisions made at a higher level. Thus, it would be useless and destructive for Leah to argue that it was unfair to make cuts; the team have already received a mandate to do so. The fact is that the team's political survival depends on their ability to make cuts, but the cuts need not be uniformly administered. It's Leah's first priority to convince the team that EDUCARE should not feel the same sharpness of the cutting edge that other programs might feel.

It's important for Leah to maintain her professionalism for this kind of audience. If she were too defensive or attributed personal vendettas to the team, she would hurt her organization. Instead, she must rely on clear, rational appeals, using EDUCARE's good history with the Department of Education to overcome the political difficulties of the situation.

Ethical Stance

The question of ethical considerations was the final aspect of Leah's rhetorical analysis. Having decided to take a clear stand and fight for EDUCARE's programs, Leah decided she had no ethical difficulties with this upcoming task. She might have had real problems if she had been told to offer up some of EDUCARE's work for the team to cut. But saved from such a fate, Leah decided she would stand her ground, refusing to sacrifice quality for cost efficiency, and would hope that the results would turn in her favor.

ESSENTIAL ELEMENTS IN PREPARING AN ORAL PRESENTATION

Knowing the Material

Nothing can make a speaker more confident than knowing the material. Leah had learned this truth from past oral presentations: the times she felt most confident were those when she had known the material most thoroughly and when she had anticipated carefully the questions her audience were likely to have. The former situation requires some research, even of a topic that the speaker knows well; the latter requires fitting the research to the audience analysis.

Leah realized that in this presentation, she would have to answer some very specific questions, ones that she was anticipating and that would require some more specific research. Again, she would have to concentrate on the weak areas of the program, or at least areas that the assessment team would consider weak, such as the counselor-client ratio in the outreach portion of the program.

The figures that Leah was able to retrieve showed that in 1996, when the program got started, EDUCARE hired twelve outreach counselors. In the first year of the program, they identified and counseled 1,000 adolescents in both group and individual sessions. Thus, each counselor had contact with about 83 clients in either individual or group sessions, or both, within a week. In the second year of the program, EDUCARE hired ten more counselors—five full-time, five part-time—and increased the client pool to 1,700. In that case, the full-time counselors had contact with about 77 clients in both kinds of settings within a week, and part-time counselors saw about 33 clients.

In the third year of the program, the ratio stayed the same for five months and then decreased slightly. The program lost 250 clients but retained all counselors. Full-time counselors saw 73 clients in both settings; part-time counselors saw about 30.

It was this loss of clients, Leah thought, that would trigger criticism from the assessment team. Ironically, any decrease in the counselor-client ratio, which from EDUCARE's point of view was something positive, would be seen by the government as a problem, an indication, perhaps, that the program had served its purpose and that its effectiveness was beginning to wane. Leah would have to counter this perception with information on how well the counseling was working. She would have to be specific about how many clients had made progress in attitudinal changes and in cutting down or giving up drug use, and how that progress was connected to counselors' efforts in education. She would have to draw from the monthly reports that counselors were required to write and from the results of interviews and questionnaires that were held or administered regularly in the outreach program.

Leah was convinced that counselors were doing a good job, but she would have to translate her conviction into specific figures and percentages, so that the assessment team would also be convinced. There is no substitute for a thorough knowledge of the material one has to talk about, Leah knew. She decided that for this presentation, she would have to show a particularly close, thorough knowledge of the workings and effects of the outreach program; after all, the future of EDUCARE and her own job was riding on how well she could tell her audience about the program's success.

Knowing What Is Important to Discuss

Leah also remembered that in the past, she had had to discriminate among pieces of information, to decide which elements of the problem she was speaking about needed to be included and which ones could be left out.

Whereas writing contains limitations of space, speaking contains limitations of time. A speaker may feel the need to talk about all the elements of a problem, but time limitations make such coverage impossible. Although Leah would have liked to talk about all EDUCARE's dealings with clients, she knew, given her audience's purpose for listening, that she would have to limit her talk strictly to EDUCARE's efforts in education. Any information on EDUCARE's rehabilitation efforts that dealt specifically with clients who had been identified as chemically dependent (addicted) would have to wait for another time and another audience.

Leah then decided to sketch out the topics that would be of interest to her audience (see Figure 13.1). Leah knew that she would not be able to cover all the topics with equal thoroughness, nor would she even be expected to. The information on in-service training and the curriculum, for instance, would be well-known to her audience, so she could cut down on detail there. But the outreach counseling, which she already identified as a potential problem, would require more time and greater detail, since the philosophies and practices of outreach counseling can differ markedly from program to program. She knew that she would have to take a chance that the audience's

EDUCARE Presentation Topics of Interest

Topic 1. In-service training for teachers at both elementary and secondary levels

Topic 2. Drug education curriculum taught in health units and in physical education and social studies classes

Topic 3. Special seminars and presentations to update present information on the problems of drug use

Topic 4. In-school counseling done by EDUCARE's personnel

Topic 5. Presentations for PTA and community groups to explain what drug education has come to mean and how it is conducted

Topic 6. Public information presentations through various media—newspapers, radio, and television—aimed at the general public

Topic 7. Outreach counseling done by EDUCARE's personnel

FIGURE 13.1 **Leah Feldstein's topics of interest for her audience**

philosophy would not differ too much from hers. She also knew that it is impossible to anticipate every detail in preparing any presentation—either written or oral—and that she would just have to go with the best preparation she could muster.

Planning the Physical Aspects of the Presentation

It is obvious that Leah had already begun to plan. Her rhetorical analysis and coverage notes clearly are forms of planning. But good preparation involves other, more specific kinds of planning. John Fiske had told her that she should talk for a half hour, which meant that she would have to allocate bits of time to the various topics she had to cover. She would also have to plan the kinds of support, such as visual aids, that she would need. Would they be charts and graphs in the form of handouts, or would transparencies or slides that she could project onto a screen be better? Should she essentially combine the two in a power point presentation? An outline of the outreach counseling practices would be easier to follow from an overhead projection or from a large poster. Then she could duplicate the outline and some of the more detailed figures on handouts that she would give her audience at the end of her talk for their future reference. She had found in using handouts in the past that unless she passed them out at just the right time—when she was going to refer to them—they could be distracting. If she passed them out too early, people tended to read them or fiddle with them and stopped paying attention to what she was saying. Recently she had relied on distributing handouts at the end of her talks, thus eliminating distractions during the talk completely.

Leah also knew from her experience that the overhead projector could be distracting: it was noisy, and unless she positioned herself correctly, the projection couldn't be seen by everyone. Power point presentations were an improvement in that she could control the computer and the presentation generally from a more remote position. In this way, she could maintain eye contact with her audience without being herself a physical impediment to viewing the material. But in using power point technology, Leah remembered a very useful caveat: don't go to the trouble of producing a fancy power point presentation that simply reproduces the main text of the talk. People are too often convinced that the words they are speaking should also be appearing verbatim on screen, but that method detracts from both the spoken and the illustrated word. The spoken word gains its strength from careful organization, development, and judicial use of example. The illustrated word gains its strength by being visually appealing highlights of the spoken word. Power point slides should illustrate key facts, figures, and concepts; they should never simply repeat discursive chunks of the spoken text.

Leah then had to think of the room. For this talk, she would use the conference room, in which there was a large table surrounded by comfortable chairs. There wasn't much space between the head of the table, where there was a chalkboard and a screen for projection, and the table itself. People sitting closer to the front might have to strain to pay attention to the projection, whereas people sitting at some distance would probably be comfortable. The same situation might be true, however, if she used the chalkboard; but if she used the chalkboard, at least there wouldn't be the distraction of the power point equipment. In any case, these were considerations that she had to take into account. Elements such as these, which could make the difference between a successful and an unsuccessful presentation, reinforced the need for careful planning.

Finally, there were the amenities. She would want to check on the temperature of the room. There had been times when it was too hot or too cold. Coffee and tea and perhaps some pastries would have to be supplied. Since the meeting was planned for 10:00 A.M. and John had told her it had to be over by 12:00, allowing for his general introduction and conclusion and for the budget director's presentation, she was somewhat relieved. At least she wouldn't be speaking to people who had just had a big lunch. She always had to account for sleepiness in afternoon presentations, something she found difficult to deal with.

Leah found her planning to be easier for this presentation than it had been for others. The audience would be small, and she knew the territory, her capabilities, and those of the facilities she would be using. All in all, this was a fairly easy presentation to prepare.

Practicing

Even though the presentation was fairly easy to plan, Leah didn't want to push her confidence too far. She remembered from her school experience and previous presentations that she was always better off when she practiced, going through the actual steps of her presentation in a dry run. That meant checking to see whether the equipment was working properly; making sure that there were extra bulbs for the overhead projector (if she should decide to use it); going over slides (if she decided to use any) to see that they were in the right order and were appropriate; testing the slide projector to be sure that the remote control advance mechanism was working, and, finally, making sure that the slides of a power point presentation were in logical sequence and timed properly.

Leah knew that these were all steps she could take under the best of circumstances—in places that she knew and had free access to, like the conference room in EDUCARE's office. There were other places, however, to which she had access and that she could get to without too much trouble before a presentation. Simpson High School auditorium, where the school board met monthly, was one such place. She knew the facilities there pretty well, since she had spoken there several times before. She always went there at least a day before a presentation to make sure that the equipment she needed was in place and working.

But there had been several times when she couldn't do any advance practice in person, such as for the talk she gave at the offices of the Superintendent of Public Instruction in Olympia last year. In that case, and in others like it, she had to prepare the premises by telephone—talking to people to make sure that whatever she needed would be provided. In these situations, Leah usually had a backup plan that she could use if her telephone arrangements had not worked out. Instead of an overhead projector, for instance, she could use a chalkboard or easier-to-read topical handouts. Finally, she knew that the physical circumstances shouldn't daunt her and that alternatives she had thought about carefully would serve as well.

Most important for Leah was pacing herself, actually delivering the talk and timing how long it took. One of the hardest things for her was knowing how long a presentation would take. She could plan all she wanted—two minutes for the introduction, eight for the background, seventeen for major findings, a conclusion—but if she didn't actually give the talk to somebody, even to herself, she wasn't sure she could keep within her time limits.

Before she was married, Leah used to deliver her talks to her roommate, and then after that, to her husband. They were pretty agreeable to her requests and would even suggest changes she might make, since they would play the role of whatever audience she had in mind. But what was most important for her to know was whether her presentation would fit into the time allotted to her. If she ran out of time, she wouldn't be able to conclude strongly. If she saw that she was running out of time, she might hurry over points that needed more development, thus ill-preparing her audience for her conclusion. If she raced through her talk, she might gloss over major points, and worse yet, leave an awkward time after the talk to be filled in by poorly informed questions from her audience (because she had rushed over essential material) or by polite applause and then complete silence—the absolute dread of any speaker.

Timing is of the utmost importance, and every speaker has to worry about it. And even if one has advanced beyond the stage of having others listen to a talk, there's no excuse for the speaker's not having timed it in some way. Leah knew this, and although in recent times she had let her husband off the hook, she still delivered her talks in the privacy of her own office at work or in her study at home. Often she would use a tape recorder and a mirror, so that she could both hear (later) what she was talking about and see (immediately) how she was talking about it.

After enough experience in giving oral presentations, Leah thought she knew the ropes. She could anticipate most problems and most questions from her audience. But she never gave up a little practice before actually delivering her talks. Very few people whose jobs require oral presentations can neglect that practice. Leah very closely held onto the notions of preparation, planning, and practice.

PLANNING THE TEXT FOR AN ORAL PRESENTATION

Speakers can prepare, do research, discriminate among pieces of information for proper emphasis, and practice the talk, but if the text of the presentation is not well organized and does not highlight the important points, the talk will not be effective. Leah's college instructor thought of a well-organized text in terms of a basic three-course meal: the appetizer, the main course, and finally, the dessert (see Figure 13.2). The appetizer is supposed to whet the pallet, get the juices flowing, and interest the diner in what's to come. The main course is supposed to put things together—tastes and textures and colors—so that the substance of the meal is interesting and satisfying. Finally, the dessert caps it all off—a complement to what has come before, a suggestion, perhaps, of what is to come in the future. Leah agreed that the formula for a good meal was the same as one for a good presentation.

The Introduction (or the Appetizer)

Suggestions for good introductions in speaking follow similar suggestions for good introductions in writing. Good introductions are supposed to grab the listeners, get their attention, prepare them for an intelligent appreciation and understanding of what is to come. Bad introductions do the opposite:

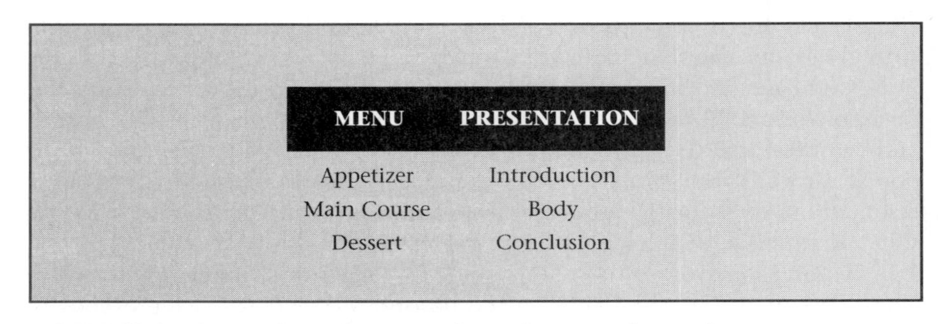

MENU	PRESENTATION
Appetizer	Introduction
Main Course	Body
Dessert	Conclusion

FIGURE 13.2 **Comparison of menu topics and presentation topics**

they confuse listeners, lead them away from the substance of the presentation, or simply bore them from the outset.

Leah knew this principle from her experience in college, when she was criticized for starting her first presentation with, "The subject of my talk is . . ." and then going on to state it in the blandest of terms. Even then she knew that there was no interest in that kind of beginning, but she felt scared and shy, and thought she should get on with the substance of her talk. She didn't realize until later that she had had to work doubly hard to get her audience interested and to keep them with her through the rest of her presentation.

Beginnings are often troublesome. When people begin anything, unless they are veterans, they can feel awkward or shy or silly, sometimes stupid. And many speakers and writers, who are otherwise fully versed in what they have to say, often sit for what seems like hours agonizing over how to begin the presentation. To alleviate the problem, some speakers begin their planning with the body of the talk: they know what they want to say and then map it out, step by step, including, in some cases, the conclusion. Then they go back and write the beginning.

This order of preparing the text is helpful in several ways. It helps get the important points and their support lined up. In turn, the process provides an overall picture of the importance and significance of the presentation. Since that kind of information must be included in the introduction, speakers often write the introduction last. When they know what they want to say and how they want to say it, they have a better idea of how to begin the introduction. A first draft makes the point clearer to the speaker; by the end, the speaker knows what to say. But a first draft is inappropriate for listeners. What appeared at the end to the writer must appear *first* to the listeners. The main point of giving the presentation must be evident early in the talk.

Some people begin with anecdotes—little stories that illustrate the nature and significance of what they want to say later. Some begin starkly with facts and figures meant to startle the listeners, to grab their attention, and to define the context. Others begin with short histories of the problem, maybe also given in anecdotal form.

Leah had tried all these ways and usually had some success, although she found out that using only one way, because it had worked once, was not necessarily the way to go always. Telling the story of Jimmy, who was coming to school high on coke, who could not pay attention to what was going on in classes, who used to go into deep funks by the end of the day, was impressive for parents who were attending PTA meetings and who might be worried about their own children's behavior in and out of school. But that method didn't do much for the kind of people she would be talking to next week.

That audience would be more impressed with facts and figures. They knew about the Jimmys, or at least they thought they did, and would be impatient with such an approach. If they were to have an anecdote, it would have to be based on how EDUCARE's program had grown and thrived, even on limited funds, and over a relatively short period of time. They would be prepared, perhaps, for the main body of the presentation—the ways and means by which EDUCARE had increased and improved its services to clients and to the community.

FIGURE 13.3 **Leah Feldstein's first visual aid**

With this perception in mind, Leah prepared two slides. The first, shown in Figure 13.3, is an article that appeared in a local newspaper on the success of the EDUCARE program. With this article, Leah can show her audience that EDUCARE has public recognition. The second slide, shown in Figure 13.4, consists of a line graph showing the growth of client enrollment and an outline of her presentation. As introductory visuals, both direct the attention of the audience to the specifics of the programmatic success enjoyed by

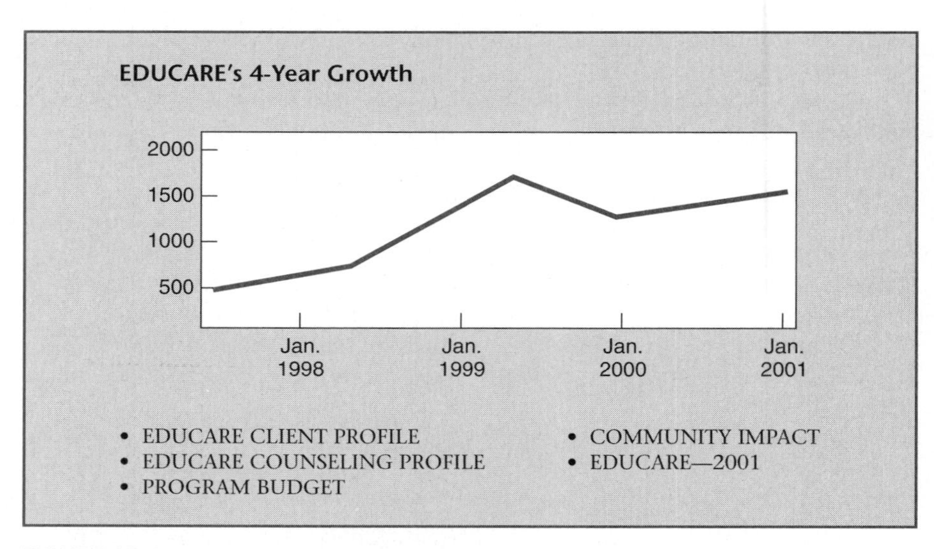

FIGURE 13.4 **Leah Feldstein's second visual aid**

EDUCARE. Leah knows that although the newspaper article is quite general and was written for the lay reader, it will provide a good anecdotal opening to the statistical information that she will want to address immediately.

To sum up, the beginning of any talk is probably the most important part. The audience's interest must be engaged from the outset, or it is likely that it never will be. Speakers who have not started strongly have to work extra hard to bring the audience around, so that they are paying attention and at least appear interested in what the speaker is saying. There is no substitute for good beginnings, and the time spent to make them good is time well spent.

The Body (or the Main Course)

Our earlier discussion about the need for speakers to discriminate among pieces of information in order to present that which is most important and convincing applies here as well. But at this point, we can emphasize the notion of relevance, which suggests another matter of audience analysis. Speakers must decide not only what is important and convincing but also what is relevant to their audience. Relevant information, that is, information that an audience can easily identify as part of their personal or professional lives, is also important information, and it will almost certainly be convincing. The three requirements for determining the body of the presentation are thus bound up together: the information to be presented must be *important, convincing,* and *relevant.* Logical development with good illustrations and examples—the basic main course of the meal—will satisfy all three requirements.

Leah knew this principle to be true in her preparations for different kinds of presentations. For the PTAs of various schools, she relied on a good deal of specific, anecdotal information. She knew that she could grab the attention of her audience in her introduction as she laid out the troubles that Jimmy faced every day that he came to school high on marijuana. But to keep audience interest, she would have to talk about what steps in education and counseling had seemed to work and how they were likely to work again to get Jimmy and those like him back on track. This kind of audience was less interested in how the program of education and counseling had expanded than it was in how successful the individual efforts to get students off drugs had been.

Leah's PTA audiences are afraid. These parents want their children to succeed, to be helpful and productive, and to be happy in the traditional forms of happiness. But they also know that the difficulties of the drug culture among teenagers are not easy to overcome and that they themselves are probably not capable of solving them. Such parents have come to rely on professionals like Leah, people whose work helps solve the drug problem among youths, and so they look to evaluate the tangible results of the professionals' efforts. They are interested in specific information about other children, so that they might be able to help out in the education and counseling of their own.

What is important, convincing, and relevant for an audience like the assessment team, Leah knew, would be quite different. The team would want to know how the program had worked in different terms: the number of clients served, the length of their association with the program, the number who could be identified as having passed successfully through the program. The team would also want to know some of those hard figures on client-counselor ratios, areas of potential dispute with EDUCARE.

The body of the presentation should be carefully planned, including appropriately placed and constructed graphic aids. Discussion outlines, summaries, charts, graphs, tables, illustrations, and photographs can all be helpful if they are relevant to an audience's backgrounds, interests, and needs. Simply drawn but vivid flowcharts of counseling procedures would probably appeal to the PTA audience, whereas the assessment team would find them unnecessary. The team would be more interested in graphs or tables that show expenditure of resources in relation to general client relations and services.

Figures 13.5, 13.6, and 13.7 show three visual aids that Leah could incorporate into her presentation. Following her introduction of the history

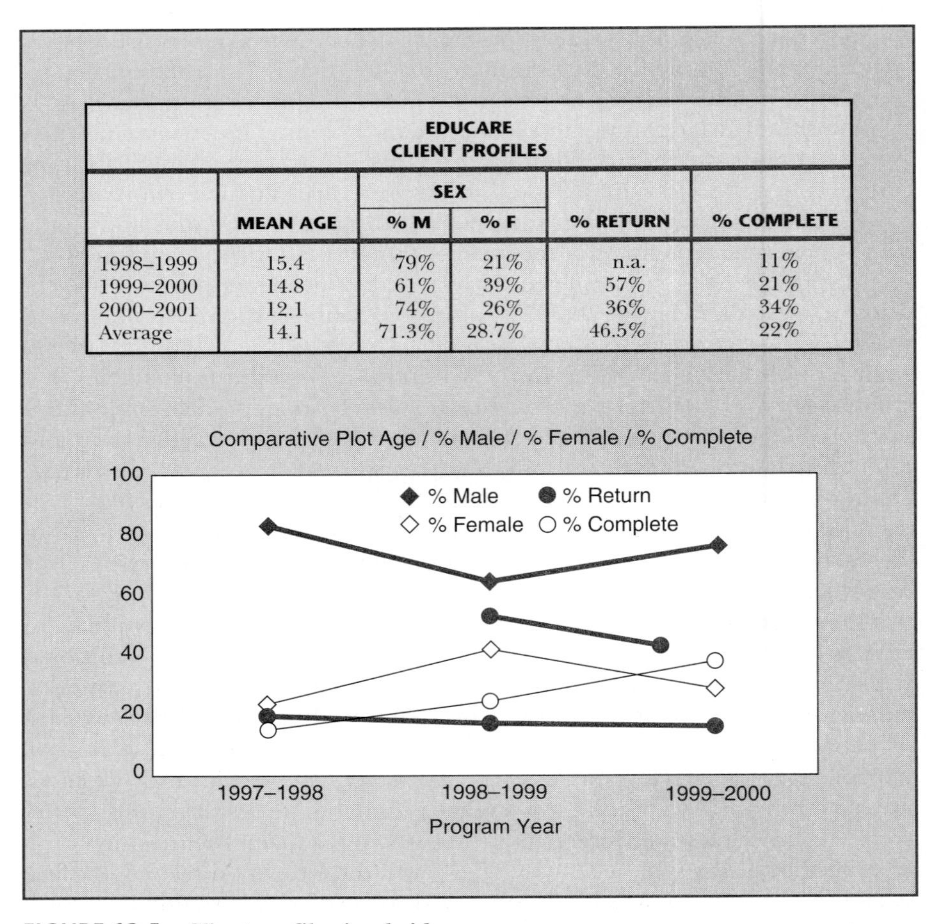

EDUCARE CLIENT PROFILES

	MEAN AGE	SEX		% RETURN	% COMPLETE
		% M	% F		
1998–1999	15.4	79%	21%	n.a.	11%
1999–2000	14.8	61%	39%	57%	21%
2000–2001	12.1	74%	26%	36%	34%
Average	14.1	71.3%	28.7%	46.5%	22%

FIGURE 13.5 **Client profile visual aid**

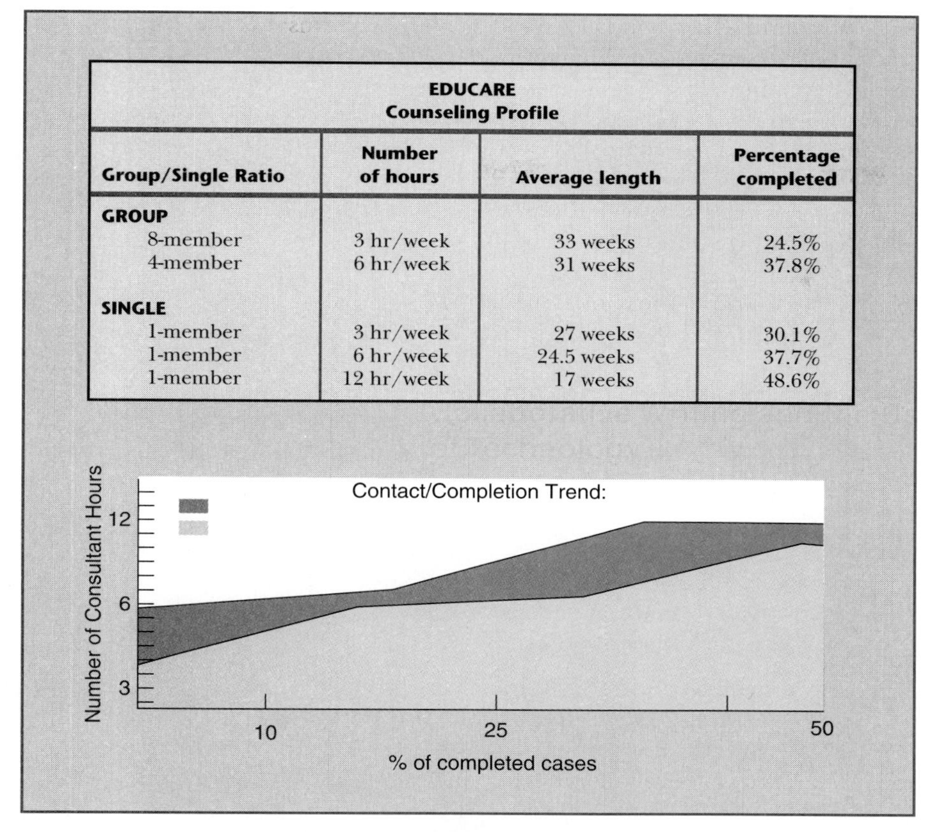

EDUCARE
Counseling Profile

Group/Single Ratio	Number of hours	Average length	Percentage completed
GROUP			
8-member	3 hr/week	33 weeks	24.5%
4-member	6 hr/week	31 weeks	37.8%
SINGLE			
1-member	3 hr/week	27 weeks	30.1%
1-member	6 hr/week	24.5 weeks	37.7%
1-member	12 hr/week	17 weeks	48.6%

FIGURE 13.6 **Counseling profile visual aid**

of client participation, Leah could use the table and line graph in Figure 13.5 to give the audience a closer description of clients. She could use the overhead shown in Figure 13.6 to clarify her discussion of how the increase in contact hours in both group and individual counseling sessions has resulted in an increase in the percentage of clients who have successfully completed the program. Knowing that the assessment is an economic one, Leah wants to show the breakdown of the program costs, focusing specifically on the portion allocated for counselors' salaries. The pie charts shown in Figure 13.7 would serve this purpose well. These three visual aids help focus the audience's attention on the counselors' roles and their impact on the clients, the basis of the need for maintaining EDUCARE's current counseling programs.

As Leah prepared the body of the text for her presentation, including the visual aids she would use, she had to keep in mind that the success of this portion of the presentation depends on three essential ingredients: what is important, what is convincing, and what is relevant.

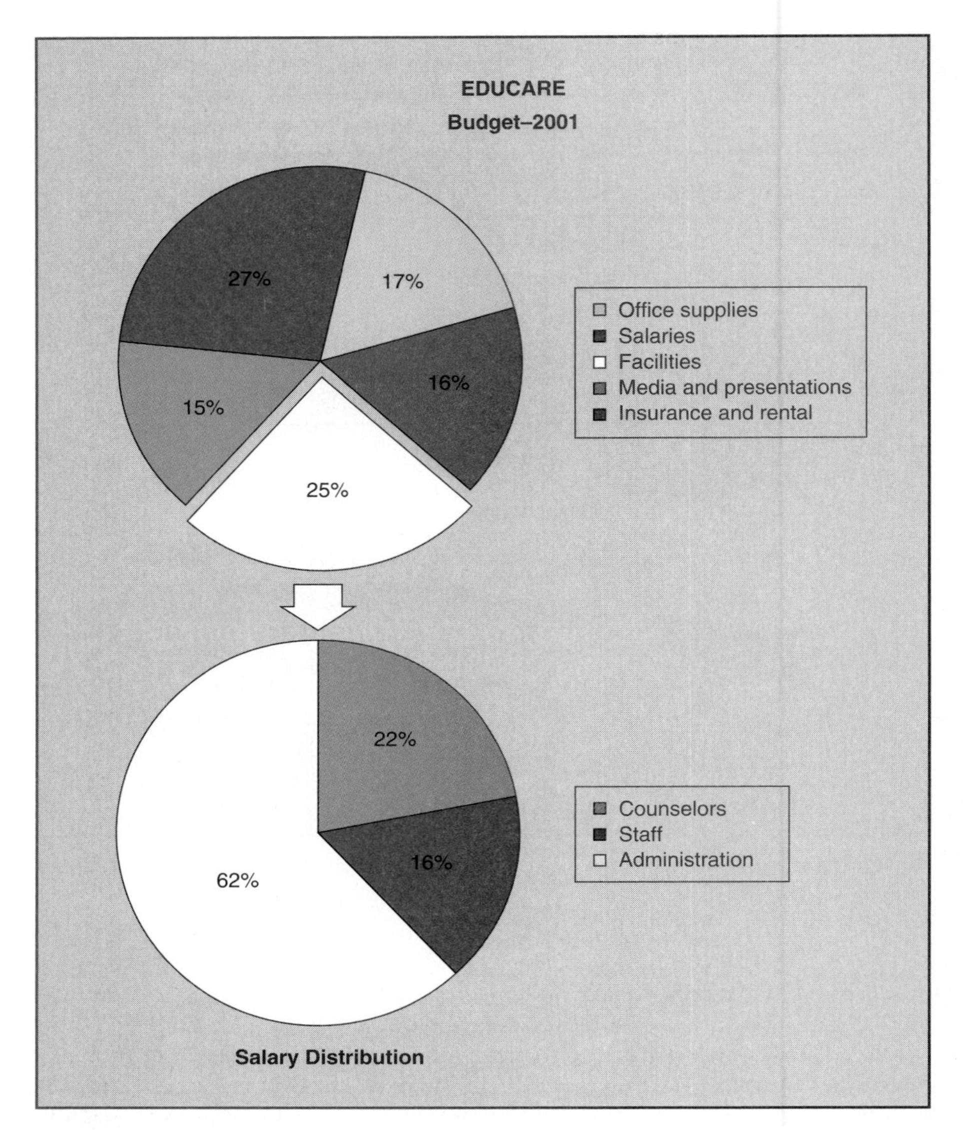

FIGURE 13.7 **Budget visual aid**

The Conclusion (or the Dessert)

Speakers are often deceived into thinking that conclusions are the easiest part of a presentation to write. They think that summarizing everything that has gone before is all that's necessary: they are finally done. They believe the cliché that first you tell them what you're going to tell them, then you tell them, then you tell them what you just told them. But it is not so simple as that. Good conclusions are as important as good beginnings: they compliment the rest of the talk by both satisfying the audience with a point of closure and leaving them with thoughts that should go beyond the confines of the presentation. A strong conclusion will consider implications of the talk.

Sometimes conclusions are meant to move the audience to action or to make some kind of decision: strong conclusions of political speeches often send audiences to the polls; boards of directors are often moved to decide policy on the basis of good presentations with strong conclusions. The point is that the meal is not complete without the dessert—and a dessert that is appropriate to all that has been served before.

Leah knew the importance of effective conclusions from her college experience, when on some occasions, mainly because she had not timed her presentation well, she had ended with, "Well, I guess that's it; are there any questions?" The deafening silence threw her. She thought she had flopped entirely. But people often are not prepared to ask questions unless the conclusion challenges them in some way. Audiences want to test what they think they have heard and understood, and they can do so only if they have a prompt from a well-planned and well-executed conclusion.

In the PTA talks, for instance, Leah wanted to leave her audiences with some hope that through their own efforts in concert with EDUCARE's, their children might be helped to understand well the dangers of drug use before they are tempted to begin using or, in some cases, to learn how to bring themselves out of the bind of drug abuse. She would always conclude with some practical steps that parents could take and with reminders that those steps had worked before and could work again. She wanted the audience not only to think positively about the future but also to take some action that might help ensure that the future would be positive.

For the upcoming presentation to the assessment team, Leah's conclusion would be different, because it would call for some action, or at least for a decision from her audience. In that way, her conclusion would be similar to the conclusions given for PTA groups. The need for the presentation—but for the conclusion in particular—was simple: she wanted the assessment team to recommend continued funding of EDUCARE's program at least at present levels. If that couldn't happen, she wanted the team's recommendation to specify only minimal cuts, especially in the outreach section of the program, so that EDUCARE could continue its work without major interruption. With this goal in mind, she would conclude her presentation with a recap of the accomplishments of the program, emphasizing how well the present organization matches the current funding. Leah decided to use the graphic shown in Figure 13.8 as a concluding visual to illustrate the interrelationships among all areas of the program and to show that this network was effective both administratively and economically.

By now Leah understood well—much better than she had in college—how important strong conclusions are. They tie things together, they give a sense of completion, but most important, they make people think about the future and about what decisions or actions they may have to take.

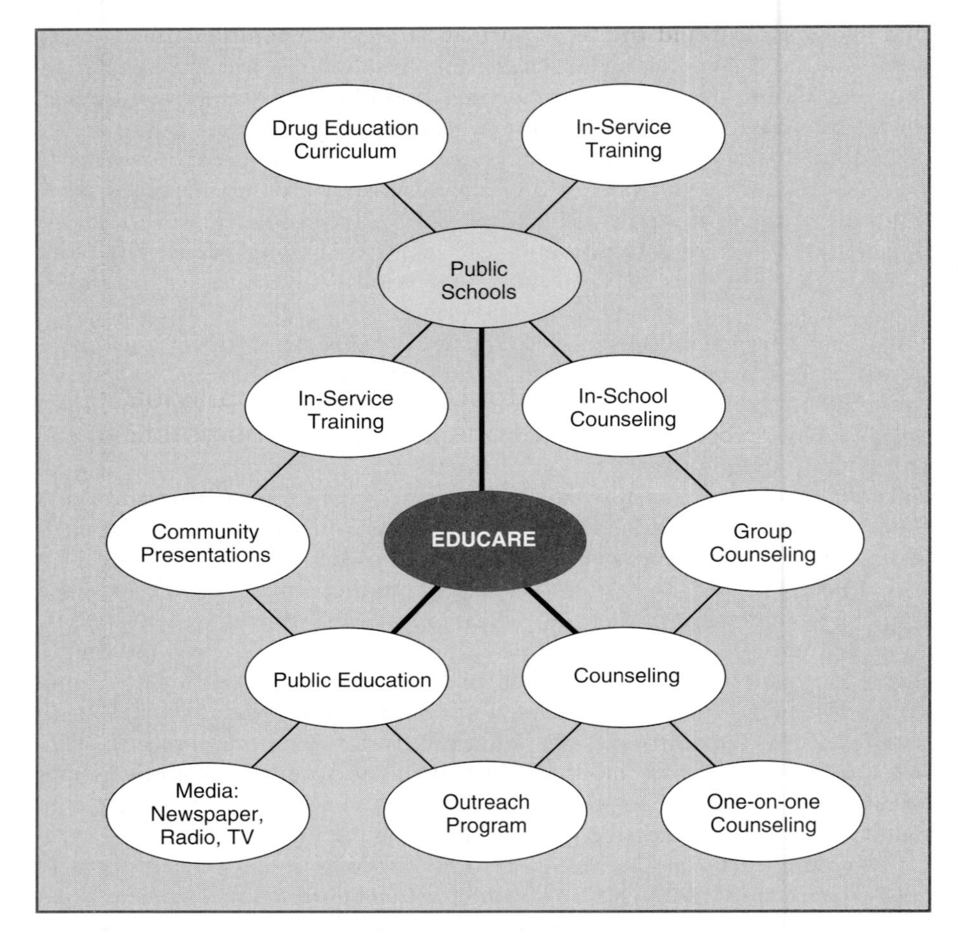

FIGURE 13.8 **EDUCARE task network visual aid**

THE MECHANICS OF EFFECTIVE PUBLIC SPEAKING

After analyzing the rhetorical situation, attending to all additional considerations, and preparing the text and visuals, Leah felt ready for her presentation. She remembered her instructor's tips on how to speak well in front of groups. In the years since college, she was able to add a few of her own. She tried her best, whenever she spoke publicly, to keep all of them in mind.

Relax

Leah knew that relaxing was easier said than done. But she also understood the basis for the advice to relax. In any talk before any audience, the speaker is the expert—or at least should be the expert, given the purpose and scope that have been set and the research that the speaker has had to do. Thus, there is no better way to gain confidence and to be able to relax than to know the material thoroughly and to anticipate questions the audience might have. Although realistically, some speaking situations may be more stressful than others (the upcoming presentation before the assessment team is a good example), none should be completely daunting if the speaker is thoroughly prepared.

Speak Loudly Enough to Be Heard

This advice may seem obvious, but many beginning speakers often don't realize what the size and acoustics of the room are and how difficult it is for an audience to hear if presenters don't speak up. Although there is no need to shout, the speaker must be heard. In many large rooms and auditoriums, microphones are provided that permit presenters to speak at conversational levels. But again, speakers should test the equipment for quality and working order before the presentation. In smaller rooms that seat twenty-five or so, a good, strong delivery is all that is needed.

Maintain Eye Contact with the Audience

A very important ingredient of good speaking is keeping in touch with the audience through eye contact. No one likes to feel left out, and it is a simple fact that as the numbers of listeners increase, the chances of feeling like just one of a crowd also increase. Thus, it is incumbent upon the speaker to maintain eye contact with the people in the audience.

The problem that many beginning speakers (and some veterans as well) find is that the self-consciousness they feel, for whatever reason, causes them to look anywhere but in the audience members' eyes. The speakers look at the ceiling or at the floor, or they bury their noses in their notes. The result is that they cut themselves off from their audience, and their message doesn't get through.

Some speakers have another problem. Remembering that they should maintain eye contact, at the beginning of the talk they find a friendly, accepting face, or perhaps several such faces in a group, and so keep addressing that person or small group to the exclusion of the other people. The excluded people quickly pick up on the situation and actually begin to resent the speaker for excluding them. Instead of listening to what is being said, they start to think only about why they are being left out and why the other people are being favored. This situation can have an even worse effect on the success of the talk than having a speaker's eyes averted to the ceiling, the floor, or the notes. In the latter case, at least, personalities are not involved, and no one feels like the object of exclusion.

The problem in general is made easier when speaking to large groups. The distance between speaker and audience is increased, and generalized looks at various parts of the audience will usually be enough to keep everyone in touch.

Keep the Body Poised and Vibrant

Many people speak and have written about the importance of body language and its effect on how messages are sent and received. More than one might think is communicated by the way a person stands, sits, or moves: feelings of discomfort, insecurity, and hostility—or on the more positive side, warmth, friendliness, and affection—all can be transmitted through postures of the body.

Speakers, therefore, should be careful about how they move and hold themselves. If they stand rigidly, audiences may interpret this posture as evidence of insecurity or maybe even inflexibility of thinking—certainly nothing that speakers want to project. On the other hand, if speakers slouch or lean against the podium, they may be seen as not taking their subject or their audience seriously, an equally undesirable effect of body language.

Speakers want to try for something in between. They should want to appear energetic, but also poised and relaxed. Occasional turns of the body from right to left to right, thus complementing eye contact with their audience, and occasional gestures of hands and head to punctuate points they are trying to make, all contribute to an image of confidence. And a sense of confidence is what all speakers should want most to project.

Something should be said here about idiosyncrasies. Most speakers have them, and depending on how pronounced they are, such idiosyncrasies will be more or less distracting. The most common idiosyncrasy is pacing. Many speakers, usually to release nervous energy, will pace back and forth behind the speaker's podium, stopping only occasionally at the podium to glance at their notes. Although it may be impressive that they need to stop only occasionally to check their notes, the pacing itself is usually a distraction. The audience begins to concentrate on the pacing and not on the speaking. They begin to think about how many miles the speaker will walk today, not about how many ideas they can take away from the presentation. On the other hand, a compelling speaker can overshadow the pacing, so that it becomes part of a unique style, adding dramatic tension to the talk.

There are other kinds of idiosyncrasies, with most being the result of nervousness. Some speakers tap their feet or beat little rhythms with their fingers or hands on the podium. It doesn't matter what the habit is. If it takes over the speaker's presentation, it needs to be eliminated, because the basic message of the talk is thrown off course; people go away thinking more of the peculiarities of the speaker's presentation than they do of its substance.

A short footnote to this: most of us have speaking idiosyncrasies, some more pronounced than others. Although it's good to try to control them, if the result is artificial or stilted speech, then the message will still be blurred. In this case, however, the message is blurred because it is not being delivered comfortably.

Talk—Don't Read

Leah remembered this instruction well. She got it directly from her college instructor, who insisted that his students speak from notes and not read their presentations. Listening to reading, unless it is a dramatic reading of poetry or prose, is boring, he used to say. People soon stop listening and start thinking about other matters: where they will have to be in an hour, what they will have to do tomorrow or the next day.

No matter how painful it was, Leah's instructor made students talk the presentation rather than read it. Before their presentations, he had them turn in to him their written scripts (if they had one) and their outline of notes. The outline was supposed to cover the main elements of the talk and points of reinforcement, like the use of graphics—overhead transparencies, posters, handouts, slides, and the like. He would then give them back their outlines, and they were to speak from them, and from them only.

Although Leah struggled through this process, she did not realize what good instruction and practice it was until later, when she started giving presentations in which she would read the texts, as a way of protecting herself, she thought, from forgetting points she needed to make. She soon saw that audiences got restless, easily distracted, and generally inattentive.

Don't Take the "Sleepers" Personally

Leah always remembered this bit of advice with a smile. At the time, she thought it was kind of funny, but it wasn't until she started giving talks more frequently out of school that she realized what good advice it was. This tip, she found out, is related to all the other tips on good speaking: knowing the material, being well organized, speaking loudly enough, maintaining eye contact, and the like. If speakers keep all these elements in mind and try their best to achieve them, most people in an audience will pay attention. Depending on the subject of the talk and the speaker's position on it, the audience's attention can range from polite to riveted, but at least they will be paying attention.

But there may be one person, or sometimes a few people, who will appear bored or occupied with some extraneous matter. One or two may actually be sleeping. Speakers, especially if they are maintaining good eye contact, will find these people very fast. Although their discovery is disturbing even to veterans, beginning speakers are almost invariably thrown off stride. They think they are doing something terribly wrong, that all their preparation has been for naught. But the worse effect is that they get flustered and begin to lose their train of thought, sometimes stumbling and having to stop to regain their footing.

The tendency for this situation to happen is once again a function of the size of the audience. As the size of the audience increases, so does the variety of the people who come to listen. Even though they think they have an interest in the subject of the talk, the degree or nature of the interest of each person cannot be known, and therefore specific responses cannot be precisely predicted. Some people may be there only because someone has told them to be; others attend with misconceptions, perhaps, about what will be discussed and why. And some turn up with other more personal problems: inexperience in attending public lectures, an inability or an unwillingness to pay attention, and yes, finally, a lack of sleep.

The smaller the audience, the greater the chance that the whole group will have similar interests and expectations. Responses are, therefore, easier to predict, and the smaller size of the group also helps to act as a check on what in larger groups becomes anonymous behavior. People are less inclined to drift away, at least visibly, because the purposes for being there seem clearer and more focused. And they feel a certain pressure, both self-imposed and from those around them, to pay attention to what the speaker is saying so that they can respond intelligently.

But the important thing to remember is that if you have addressed the technical and rhetorical problems of the presentation seriously and thoroughly, have planned and practiced carefully, and have created an effective text to highlight what is important, convincing, and relevant, the presentation will work well—may even be a resounding success. You may still have a couple of sleepers, but don't take them personally: they likely have problems that you don't want to know about.

TIPS FOR GIVING ORAL PRESENTATIONS TO SPECIFIC GROUPS

All oral presentations, whether to be given to large or to small groups, require that the speaker attend to the concerns of technical and rhetorical problem solving: the additional considerations of research, discrimination, planning, and practice; the organization of an effective text; and the mechanics of public speaking. In addition, other specific elements come into play depending on the type of group that the speaker is addressing. Following are discussions of factors to consider when giving oral presentations to three types of groups—committees, civic or special interest groups, and boards of directors.

Committees

The purpose that speakers have for speaking at committee meetings depends on their positions with regard to the committee: they can be guests, presenting information that the committee has requested or that they themselves asked to present; they can be committee members, reporting on specific assignments that they have received from the committee; or they can be chairs of committees, who are responsible for managing the flow of information that committees generate and for reporting on what has happened as a result of committee actions.

When the Speaker Is a Guest. As guests, speakers must worry about the same elements that Leah would have to worry about in talking to the assessment team from Washington. They must make sure that the background to the problem is clearly understood; that the evidence they present is pertinent, relevant, and timely; and that they conclude in a way that will leave the committee with a clear path to enable them to make a decision or take some action. Since guests are not privy to the workings of the committee, they have to present their information more formally and to plan their presentations more carefully.

Leah had had this experience when, soon after EDUCARE got started, she had to speak to the Seattle school district's secondary curriculum committee. Made up of some fifteen teachers and administrators, the committee is responsible for overseeing the workings of the curriculum in place as well as its future development. Leah knew she would be speaking to professionals who understood how to assess the effectiveness of already existing curricula and how to gauge the likely success of new curricula being proposed. She had to keep in mind not only the philosophical underpinnings of EDUCARE's program but also its practical applications: how teachers would be trained, how students would be engaged and then tracked in the program, and what kinds of assessments of student progress could be made. Clearly, these were all elements that required thoughtful, thorough preparation and formal presentation.

When the Speaker Is a Committee Member. As committee members, speakers can be less concerned with formality than they would be as a guest, though no less concerned with the pertinence, relevance, and timeliness of their presentations. Normally, they are responding to particular assignments from the committee. The committee wants to hear specifically about the matter assigned, and so there is no need for formal introductions or conclusions. On occasion, speakers may need to present their information with special visual aids, like overhead projections or handouts, but generally other committee member rely on and take notes on only what is said.

Leah's experience as a member of the Greater Seattle Task Force on Drug Education and Rehabilitation serves as an example. Leah's responsibility on that committee is very specific: to report on the general progress of EDUCARE's program both within the schools and in outreach, but especially with regard to how EDUCARE's efforts are coordinated with those of other social service agencies in Seattle. To that end, Leah reports regularly (the committee meets once a month) on what teachers are doing in their classrooms, how many students have been identified as needing special help outside school, how many of those have been referred to other agencies, and what kind of services and relationships have resulted from those referrals.

Reports to committees are no less guided or determined by time than any other presentation. The length is normally limited by the meeting's agenda—what the other items are that will be covered, what their importance and place are on the agenda, whether they are old or new items, and so forth. Members' presentations should not take any longer than experience dictates, given the other business of the committee and the relative importance of the information that the member is presenting.

When the Speaker Is the Committee Chair. Chairs of committees are usually responsible for three things: agendas of meetings; the flow of information, both within and beyond the committee; and final reports on actions that the committee has taken or recommended. Presentations, then, are of two kinds: those to the committee itself about the business that it has to undertake or on the results of work that it has already undertaken, and those to groups outside the committee that rely on information that the committee has generated: upper management, boards of directors, or the larger constituency of the organization of which the committee is a part. The latter can often mean presentations to sales conventions or stockholders meetings, which have to take on the more formal trappings of presentations to larger groups.

Presentations generally respond to particular items that have already been introduced and are under discussion. Detailed introductions concerning background or intentions are unnecessary, unless a new item has been presented to the committee for its consideration. For new items, chairs have to start from the beginning—or almost from the beginning. Since they know their committee members and since everyone understands the purposes the committee has for meeting, chairs can take some shortcuts and present the new problem in terms that the committee will already be familiar with, making sure, however, that they define specifically what role the committee will have to take to address the problem.

Conclusions for this kind of presentation can be similarly abbreviated. They are not normally couched in philosophical terms but usually take the form of specific instructions to committee members about what and when tasks have to be performed.

Civic or Special Interest Groups

Civic or special interest groups are generally friendly groups. The members, who volunteer their time and work, genuinely believe that they can be of help in solving the problems their groups confront. Thus, the Lions, the Rotary Club, the Chamber of Commerce, the Friends of the Library, the Friends of the Earth, MADD (Mothers Against Drunk Driving), or even the local PTA of Franklin School all meet regularly to decide how their projects are progressing and what more they can do to see them along to successful conclusions.

Professionals from all areas are often called upon to speak to the meetings of these groups, and depending on the community and the circumstances of the problems they are confronting, their presentations can mean a great deal. Presentations to these groups can also build goodwill and can enhance the public image of the organization that the speaker represents. For these reasons, then, speakers should take seriously their jobs of presenting.

Understanding the Goals of the Group. The backgrounds of members of civic and special interest groups vary. Members may be retired executives, line workers, active members of the business community, or teachers; the one thing they have in common is interest and commitment to the goals of the groups to which they belong. Therefore, speakers must keep those goals in mind when preparing their presentations. Since the topics of the presentations are almost always specified, either by the group if it has invited the presentation or certainly by the speaker if he or she has volunteered the presentation, the job of focusing the presentation is much easier. But speakers must also connect the ideas they present to the general purposes of the group to which they are speaking: for the Friends of the Earth, how does Acme Incorporated help meet the standards for a clean environment? For MADD or the local PTA, how does EDUCARE address the problems of drinking and drugs in a controlled, educational environment?

Meeting the Informational Needs of the Group. Although presentations to groups like these need to be specific and pertinent to the general interests of the group, they should not be terribly technical. Only a few members of the audience are likely to appreciate or be able to digest technical information, and even they, because they understand the diversity of the group they have joined, don't expect or want to hear about all the technical details of a problem. They, like the other members, want to hear about basic issues, approaches, solutions, and not about the minutiae that constitute them.

Groups like these also need fairly vivid graphic support of the words they are hearing; maps, illustrations, simple charts and tables, posters, and discussion outlines can be useful, presented either through overhead projection or in print. But graphic support shouldn't be overwhelming—too many things to read, consider, or react to will confuse and drown out the presenter's basic message. The graphics help enormously, however, to reinforce basic points. It is generally more difficult to hear a presentation than to read one. People retain less of the essential message by listening than by reading. It's often a good idea, then, to supply backup print copies of overhead projections. Audiences can use them to refer to and consider after the presentation.

The general tips for presentations to civic and special interest groups are simple and sensible: keep the presentation focused, simple but not simplistic, pertinent to the problems that the group is interested in, graphic, and short—but not too short (eight- or ten-minute presentations are often seen as slights to the group or to the principles for which they are organized). Unless otherwise specified, twenty- to twenty-five-minute presentations are generally long enough. Speakers can make their points: listeners can absorb the points, and no one gets too restless.

Boards of Directors

Understanding the Purposes of Board Meetings. Boards of directors, as is true of most small groups, have very specific purposes for meeting, usually to do with establishing and evaluating the policies and directions that an organization is operating under. Thus, boards of directors meet specifically to make decisions—decisions that normally will have a significant, global impact on the conduct of the business of the organization.

The board of directors of General Motors, for example, must decide on such major issues as whether to expand or to contract the corporation's operations, whether certain investments of capital should be made and where, whether the general corporate image needs to be changed and in what way. It also makes decisions on changes in top management practices and personnel, which clearly have a significant impact on the operations of the corporation. Board members do not make decisions about how changes or new initiatives in the operations will be implemented; those are left to lower-echelon personnel, although the results of the implementation can become the subject of future discussion.

To help them make these decisions, boards often call for oral presentations from people in various divisions of the corporation. This procedure saves them time, since they don't have to read through pages of proposals, reports, or position papers. Since saving time is part of the purpose of calling for presentations, boards expect that they will be clearly, concisely, straightforwardly, and efficiently rendered.

Providing Information to Enable Decision Making. The guiding principle, then, in preparing an oral presentation for a board meeting is to give the members information that will enable them to make a decision; this means presenting the information in a compelling way, one that is clearly intended to recommend a course of action. The board is not interested in the technical details that so fascinate the professional in the trenches. All the board wants is the big picture, presented with background, implications, and focus filled in.

Presenters must center specifically on what the problem is, what it means for the organization if it is unsolved, and what is likely to happen if the speakers' recommendations are agreed to and implemented. Presentations of this kind must be specific, concrete, down to earth, and precisely timed. Since boards meet only occasionally (the schedule of meetings varies with each board), their agendas are usually rigidly set, and the members don't like rambling, unfocused, apparently endless presentations. Such presentations waste everyone's time.

Presentations to boards of directors are more successful if they are enhanced with graphic aids: simple flowcharts, tables, and illustrations presented through overhead transparencies or handouts. It's often a good idea to provide backup handouts of overhead projections, so that members of the board can have them to refer to when they deliberate and finally make their decisions.

Presentations to boards of directors are clearly understood to be performances. They require careful definition of the problem at hand, research and analysis, synthesis, planning, practice, and finally, stage presence. Presenters need to understand why they have been asked to present information, to whom it is to be presented, and how it will be used. They also need to know when to get off stage. These are not essentially different from the demands of any writing or speaking task, but if they are not taken seriously, the effects can be dramatic.

CONCLUSIONS

Although the techniques of giving oral presentations are different from those of writing, the basic principles for preparing them are the same: careful analysis of the problem or problems to be discussed, the purpose of the presentation, the intended audience, and, finally, the speaker's position—his or her stance on the importance of the problem and how it might be solved. Thus, the same steps discussed and outlined earlier in this book on solving technical and rhetorical problems in writing apply to speaking. The only difference—and it is significant—is that oral presentations elicit immediate, even palpable, responses in those who hear them. Speakers know very quickly the impact they've made, whereas writers must wait much longer. A speaking situation, for this reason and for others we've spoken about in this chapter, is often more intimidating—maybe just scarier—for most people than a writing situation. That's an understandable reaction. Our advice is simple, though it may not be simple to follow. Consider the steps for giving oral presentations that we have discussed in this chapter, and don't take the sleepers personally.

EXERCISE

1. Imagine that you have an entry-level job in the kind of firm you have always wanted to work for, doing what you have been educated and trained to do. You get in your mail a memo from your immediate superior asking you to present a twenty-minute talk to high school seniors during their Career Week activities about what you do and why it is important. Following are some of the specifics:

   a. You will be speaking to a class of twenty-five honor students, all of whom will be going off to college in the fall.
   b. A major purpose of the Career Week activities is to give students an idea of what people in the working world do, so that students' choices of majors in college might be made easier.
   c. The company you work for is thinking of offering several scholarships to promising high school seniors to cover most of their college expenses. The company sees this undertaking not only as a civic duty but also as a means of encouraging the development of better employees for the future.

   Plan your presentation by writing out a thorough rhetorical analysis: consider purpose, audience, context, and ethical stance. Then decide what you are going to say, in how much detail, and what kinds of aids you will need to support your presentation. A descriptive outline of your presentation should be the result of your work.

ASSIGNMENT

1. Present a fifteen-minute talk to your classmates on research that you either have worked on or are presently working on. Before giving the talk, prepare an outline of the presentation, including samples of the visual aids you plan to use in the presentation. Provide a statement for each visual aid that explains why the aid is necessary. You will submit your outline to the members of a review committee that your instructor has formed a week before your presentation. The committee will return your outline to you with any suggestions for change forty-eight hours before your presentation.

TERM PROJECT OPTION

Present a fifteen-minute talk to your classmates on the research you performed for your proposal and completion report. Remember to focus your talk appropriately so that you make a specific, overriding point. Mere overviews will come across as vague and pointless.

Your instructor may want to assign two students per talk to act as respondents. They will ask the speaker one pertinent question apiece. They will also evaluate the speaker's focus, appeal, and effectiveness on a form like the one in Figure 13.9. Note that the form asks for substantive comments that both describe and judge the talk. The evaluations should be signed and returned to the instructor

RESPONDENT'S EVALUATION

Speaker: ______________________________

Topic: ______________________________

1. What was the focus of the talk?

Difficulties?

Suggestions?

2. Did the speaker appeal well to the audience?

Techniques used:

Difficulties?

Suggestions?

3. Was the presentation effective?

Techniques used:

Difficulties?

Suggestions?

4. Overall evaluation:

Respondent's signature: ______________________________

FIGURE 13.9 **Respondent's evaluation form**

Selected Problems of Usage and Style

PROBLEMS WITH THE PASSIVE VOICE

Identifying Passive Voice and Passive Constructions

Voice is a function of grammar that shows the relation of the subject of a verb to the action that the verb expresses; changes in the form of the verb help express this relation. The following examples illustrate the difference between the active and the passive voice.

This statement is in the active voice:

> Medical researchers from Johns Hopkins University discovered new gene-splitting techniques last month.

The voice in this case is defined as *active* because the subject of the verb, *researchers,* is the performer of the action that the verb expresses, *discovered.* The object of the verb, *techniques,* is the recipient of the action of the verb.

Following is the statement recast in the passive voice:

> New gene-splitting techniques were discovered by medical researchers from Johns Hopkins University last month.

This statement is in the passive voice, because the original object of the verb, *techniques,* still receives the action but has now taken the subject position in the sentence. Whereas formerly it was the object, *techniques* becomes the subject in a passive construction.

The relation of the subject to the action that the verb expresses is *action* when the subject performs the action. But the relation of the subject to the action of the verb is *passive* when the subject *receives* the action. A few more simple examples illustrate the contrast between active and passive constructions:

> *Active:* The dog chased the cat.
>
> *Passive:* The cat was chased by the dog.

Active: The boss complimented Joan.

Passive: Joan was complimented by the boss.

Note the transformations that a sentence undergoes to become passive. First, the object of the active construction *(cat, Joan)* becomes the subject in a passive construction. Second, the subject of the active construction *(dog, boss)* becomes the object of the preposition *(by the dog, by the boss)* in the passive construction. Finally, the verb form requires the addition of an auxiliary helping verb, *was.* Remember that the prepositional phrase may not appear because it is understood, as in this sentence:

His chest was then opened with forceps.

In this statement, the prepositional phrase *by the surgeon* is understood.

Recognizing When It Is Appropriate to Use the Passive Voice

Some authorities advise against ever using passive constructions. They claim that passives sap the strength of a person's writing and require unnecessary additional words. But this advice ignores the point that the passive voice is sometime required, for instance, when the subject that performs the action is unknown, as in this sentence:

The Christian sectors of Beirut were bombed last month.

This sentence was written in the passive voice because terrorist factions in the Middle East are numerous and their actions cannot always be specifically attributed. Moreover, in corporate work, an agent often can't be singled out, since myriad employees may have worked on the same project. This consideration explains the following passive:

The new waste container was produced ahead of schedule and under budget.

It would be absurd to name all the individual participants in this one statement.

Matters of emphasis may determine the use of a passive construction. Depending on the subject and the context established, writers may choose to emphasize the results of the action rather than the action itself; or in terms of this discussion, they may want to emphasize the action rather than the *performers* of it. Thus, given the context of the discussion, if the newness of the gene-splitting techniques mentioned earlier was more important than the discoverers or the action of discovery, then the writer does well to cast the statement in the passive voice, so that *techniques* becomes the subject of the sentence.

The matter is finally resolved by thoughtful choice. The passive voice is not in itself bad style; in fact, its judicious use can enhance the effectiveness of a document. Only when it is overused or used specifically to diminish authorial presence does it become a problem.

Furthermore, mindless passives that ignore common sense to please convention are no longer required of writers. For example, when writing a periodic report, an author need not write *research was performed* in lieu of, *I researched.* Some in business, the social sciences, and other technical and scientific areas have said that a writer should never use the first-person pronoun *I,* because doing so adds subjectivity to a document. Writers, instead, should strive for objectivity—a detachment of the author from the written piece—to show no personal bias or prejudice. Supposedly, the passive voice achieves this sense of detachment and objectivity. Unfortunately, that usage produces such awkward constructions as the following:

> In reviewing the installation for approval, it was discovered that a concrete foundation has been installed within the Utility District's easement area.
>
> This easement, which was on file and made known to all subsequent landowners purchasing the property, was required per the National Safety Code for the protection and safeguarding of the public while ensuring the maintenance of the electric system.
>
> The report was pointed out by the manager as one in which the writer reported more than was necessary.

The notion of professional objectivity is good, but the suggestion that it can be achieved through writing in the passive voice is bad. Objectivity, both philosophically and practically, has to do with the way professionals carry out their work. It means collecting data and analyzing them without bias, without preconceived ideas of what the data will show or of what people want them to show. But objectivity cannot be achieved merely through the forms of writing. It depends fundamentally on how work is performed, whether a research design, for instance, has accounted for all variables, or whether the variables have been accounted for and treated in the same way. Objectivity in writing is violated only when writers decide purposefully to slant the interpretation and discussion of the findings to favor their particular biases; there is nothing inherent in the forms of language themselves that establishes objectivity.

Variations of Passive Constructions

A variation of the problems of passive constructions comes with the use of the adverb *there* as a dummy statement (an expletive) that is connected to auxiliary verbs of being, as in this sentence:

> There have been many ups and downs in the stock market this year.

This can be more easily and concisely stated as follows:

> The stock market saw many ups and downs this year.

The revision places a logical subject, *stock market,* in the subject position of the sentence instead of an object position, and adds the active verb *saw* to make the statement active and more concise. Here are some other examples of wordy, basically passive constructions, with revisions:

Original: In the tart cherry industry, there have been no new products developed in the last 60 years, and there have never been any industry advertisement or promotion programs.

Revision: The tart cherry industry has developed no new products in the last 60 years, and it has never advertised or promoted its products.

Original: Since the beginning of marketing in the industry, there have been large fluctuations in production, causing large fluctuations in prices of up to 39 cents per pound.

Revision: Since the beginning of marketing in the industry, large production fluctuations have caused wide price fluctuations, sometimes as much as 39 cents per pound.

EXERCISES

1. Transform the following passive constructions into active constructions.
   a. The mission was flown successfully until the radio was dropped on the floor.
   b. The car was hit at 65 mph by the truck.
   c. That principle is kept in mind by all of us.
   d. The executive summary was written by S. Johnson and myself.
   e. The two teams were beaten by the Bengals.
   f. We are exposed to radiation from radioactive elements contained in fallout from nuclear explosives testing and unavoidable discharges from nuclear installations.
   g. However, if alpha-emitting materials are taken into the body by inhalation or along with food or water, they can expose internal tissues directly and therefore represent a hazard.
   h. There has always been a desire by the tart cherry producers for formal organization.

2. Go back through a paper you have written either for this class or for another and underline all the passive constructions. Then decide in each case whether passivity was necessary. Defend the usage if it was; change it to active voice if it wasn't.

3. Identify and evaluate passive constructions in someone else's writing. You may choose a published or unpublished piece or one in draft form. Be clear about how your changes would improve the piece of writing. (Peer reviews of classmates' writing are good opportunities for doing this work.)

PROBLEMS WITH NOMINALIZATION

Identifying Nominalization

Nominalization refers to the process by which a writer casts what could be an active verb into a noun and attaches another auxiliary verb to make one longer predicate. An easy example is the following:

> The task force carried out an investigation of the failure of the conveyor system.

A good revision of this is also easy:

> The task force investigated the conveyor system failure.

The writer had turned the verb *investigated* into the noun *investigation* and so had to add the less vivid verb *carried out,* the article *an,* and the preposition *of,* so that the statement makes sense. Actually, the single verb *investigated* vividly describes the action, and the writer has eliminated seven unnecessary words.

The following example presents an interesting case:

> Gas utilization should follow company guidelines.

The nominalization here is twice removed from the real verb, *use.* There is no difference in meaning between *utilize* and *use,* although people assume that *utilize* is more formal or technical or scientific. They assume this on the basis of the sound of the word's Greek ending, *-ize,* and the extra prepositions that get thrown in, thus making a real mouthful out of the very simple verb *use.*

The preceding sentence could be revised as follows:

> Use gas according to company guidelines.

The problems with nominalization are wordiness and awkwardness; eliminating them makes writing clearer. The deeper problems again arise from writers' mistaken beliefs that their writing will sound more formal, official, or important if they use more words with more syllables. But two of the primary goals in good technical and professional writing are economy and clarity. Writers can achieve more formal tone through other means, such as diction (word choice) and syntax (sentence structure). They should not attempt it simply by padding words.

Nominalizations can sometimes be complicated by passive constructions, as in this sentence:

> The Spokane Sports Arena has experienced a complete renovation as to access to fire exits done by the Acme Company.

The statement could be revised as follows:

> The Acme Company has completely renovated access routes to fire exits at the Spokane Sports Arena.

In this case, the writer has made two good changes. She has put the Acme Company, which was the object of a preposition in a passive construction, into the subject of the sentence position in an active construction. She has rewritten the nominalization *have experienced a complete renovation* to read *has completely renovated.* She has also tossed out the phrase *as to access,* which didn't make much sense in the original anyway. The revision is clear and concise—something that certainly could not be said of the original.

The only question that the revision doesn't answer is whether the Spokane Sports Arena needs to be emphasized in the subject position of the sentence, perhaps because they have been the focus of the piece all along. If that's the case, then the preceding revision won't do. But the following revision might work:

> The Spokane Sports Arena access routes to fire exits have been completely renovated by the Acme Company.

Here the passive construction remains, because it is important to emphasize the Spokane Sports Arena and not the Acme Company. But the nominalization is gone—*have experienced a complete renovation* has become *have been completely renovated*—still a distinct improvement over the original.

Using too many words when fewer will do is true of this example:

> Researchers at the Northwest Institute for Environmental Improvement have begun to take on a study of the spotted owl's habitat.

This could be more easily written as

> Researchers at the Northwest Institute for Environmental Improvement have begun to study the spotted owl's habitat.

To take on a study and *to study* mean essentially the same thing, but somehow this writer thinks that *to take on a study* sounds better. If it does, it does so only at the expense of clarity and conciseness. And the better sound, if indeed it is better, isn't worth it.

Some final examples should help make the point:

> The lack of detail about the fire at Unit Three provided by the Clean-Up Team created a disturbance in the management sector.

Rewritten, this statement might read as follows:

> Management was disturbed by how little detail the Clean-Up Team provided at the fire at Unit Three.

Once again the problems here are wordiness and awkwardness. The revision places *management* in the subject position of the sentence, and although it retains a passive construction, *was disturbed by,* it eliminates the nominalization *created a disturbance in.* It also turns the awkward construction *provided by the Clean-Up Team* into the clearer construction, *the Clean-Up Team provided.* In all, the revision is clearer, more concise, and more effective.

Here is a short list of common nominalizations that can be eliminated to make writing or speaking more concise and straightforward:

> *Wordy:* The Committee has made the determination that. . . .
> *Concise:* The Committee has determined that. . . .

Wordy: The manager made the point that. . . .
Concise: The manager pointed out that. . . .

Wordy: It was the thinking of the President that. . . .
Concise: The President thought that. . . .

Wordy: The contractor is in the continuation phase of making deliveries. . . .
Concise: The contractor continues to deliver. . . .

Recognizing When a Revision Will Alter Meaning

Just as writers should be careful about when to use the passive voice, so should they be careful about when to change nominalizations. Not every change will retain the meaning of the original. Here is an example:

> The police have produced a report on the activities of teenage drug pushers.

An overenergetic searcher for nominalizations might be tempted to revise this as follows:

> The police have reported on the activities of teenage drug pushers.

But the revision could change the sense of the original. *Producing a report* implies the five basic steps of solving or addressing problems: definition, research, analysis, resolution, and synthesis, and implementation. It also implies a written product—a report—that has gone through all the stages of production: writing, revising, editing, proofreading, and printing and binding. The revision might mean something much less. *Reporting* can be done orally at police department briefings, at city council meetings, or at press conferences; it need not mean a formally produced report. Therefore, the appropriateness of nominalizations will depend on what actually took place and on the writer's intentions.

By using almost the same example, however, we can see where a nominalization should be removed to make matters simpler. The original might read as follows:

> The police made a report on the activities of teenage drug pushers.

The revision would be the same as the earlier example:

> The police have reported on the activities of teenage drug pushers.

Making a report is different from *producing* a report. Making a report, in fact, implies the same informal activity as reporting. There is no sense in the original sentence of production—of all the steps necessary to submit a formal report. The police can present their information orally, or in short written forms, and the process of making a report might stop there. Or it could take off from there into the more complex process of producing a report, but the point is that the writer doesn't imply anything more than simple reporting. Therefore, the writer would do better to use *have reported* and not *made a report,* eliminating the problem of nominalization—using too many words where only a few will do.

EXERCISES

1. Eliminate unnecessary nominalizations from the following sentences:
   a. One worker does the assembly; the other checks the proper procedure step-by-step in the Poulex final Assembly Safety Manual.
   b. The use of nuclear energy for electricity production is inevitably connected with the production of large amounts of radioactive material that must and can be controlled.
   c. One of the objectives of the IAEA (International Atomic Energy Association) is to "seek to accelerate and enlarge the contribution of atomic energy to peace, health, and prosperity throughout the world."
   d. We don't have anything to do here with the actual fabrication of weapons.
2. Go back through a paper you have written either for this class or for another and then underline all the nominalizations. Decide whether a revision will alter the meaning. In most cases, you will want to change the nominalizations so that verbs do the work more concisely.
3. Identify and evaluate nominalizations in someone else's writing. You may choose a published or an unpublished piece or one in draft form. Be clear about how your changes would improve the piece of writing. (Peer reviews of classmates' writing are good opportunities for doing this work.)

PROBLEMS WITH DICTION AND JARGON

Diction means word use; any word that a writer or speaker uses is diction. *Jargon* refers to word use (or diction) that is specialized or appropriate to a particular group. Jargon is often used pejoratively as coded word use, to appeal to a particular group; often it is used to exclude those outside the group. But jargon, as the vocabulary (diction) of a particular group, shouldn't be thought of as necessarily negative. If jargon can make communication easier within a group, then it should be used. It's only when the jargon of one group starts to leak into the vocabulary of another or when one group starts to appropriate the jargon of another, that misunderstanding may arise.

For example, when some groups start to talk about interfacing with one another, they are using language to obscure the actual activity: talking about ideas with other people. *Interfacing* is frequently used in computer science, in which bringing one function of a computer program into contact with another constitutes an interface. *Interfacing* seems all right to use as the name of a technical activity in computer technology, because it has a specific meaning, and everyone involved in the technology knows immediately what it means. But when others use the term to mean something as simple as discussing ideas and problems with other people, it assumes a falsely official or technical tone—an attempt to make something sound more important than it really is—the same problem that nominalization tends to foster.

A final situation that encourages jargon, which sometimes may jar notions of conventional diction, comes from changes in cultural conditions. The verb *to parent* is an example. In this case, the noun *parent* has been made into a verb, *to parent,* which means, in more conventional terms, to rear children. The verb most likely was created to reflect a new cultural emphasis on child-rearing practices that results, in part, from the greater demands put upon households with two working parents. Also, other cultural complications for children, such as drug use and early sexual activity, focused on the modern difficulties of raising children, and both mothers and fathers have had to assume more active and thoughtful roles as parents.

Creating *to parent,* in turn, has created the word *parenting,* which is used more often than the verb, both as a noun and as an adjective:

> Parenting, we find, is often more complicated today than it was 25 years ago.
>
> We have found that developing parenting skills is the most important part of our program.

The point is that changes in language accommodate changes in culture, but it often takes a long time before the changes in language are fully accepted. Until that time, the changes may appear forced or artificial, or which is worse to some, trendy. Always thoughtful of their audiences, good writers are careful not to offend their readers by using terms that have not been fully accepted into the language.

In the following sections are some terms, both nouns and verbs that qualify as questionable uses of diction. Because they are inappropriately borrowed jargon or have been coined to create a tone of officialdom or importance, they have not yet been fully accepted by many speakers and writers of Standard American English.

Some Questionable Nouns

The Bottom Line. As used in business, finance, or accounting, *the bottom line* refers to the last line of the ledger, which reveals the final figures of revenue, expenditures, or profit and loss. But it has come to be used in other areas to mean the final consideration, what has to be done given all the information that has been presented already. And so we hear comments like the following:

> The bottom line is that we cannot accept the Finance Committee's decision to reject the City Manager's proposal for recycling in all neighborhoods.

When used by those outside business, the rather trendy term *the bottom line* has been substituted for simpler but more dignified wording. The careful writer or speaker would use *in the end* or *finally* instead of *the bottom line.*

Perception Check. It is hard to say what the origin of *perception check* is, although it may have come from business and industry as a variant of *quality control check.* Roughly, it means an attempt to review what someone thinks he or she understands of a discussion or a written document; it is a recapitulation. We hear someone say the following, then, in the midst of long discussions of changes in zoning policy at city council meetings:

> Let's stop for a minute: I think we need a perception check on what has gone on so far.

A better statement would be this:

> I think we need to make sure we understand what has gone on so far.

Again, the easier, more straightforward expression fits the situation better: it is clear and unpretentious, and it does not confuse or cloud the need for action.

Quality Time. The use of *quality* as an adjective to modify any noun, as in *quality products* or *quality service,* has become typical of advertising jargon and is used to mean high quality, that is clearly different from low quality, even though the possibility of something of low quality still exists. The problem is that those who use the noun *quality* as an adjective to modify other nouns have unnecessarily shortened otherwise meaningful expressions to gain what they think is greater efficiency, or even elegance of expression.

Thus, we get *quality time,* which is used frequently in talk of child rearing, just as is *quality parenting,* to mean especially concentrated, thoughtful attention given to children in the course of their growing up. We read in articles in parents' magazines about the need to spend quality time with children, and we may hear a parent saying something like this after a busy week at the office:

> I need to spend some quality time with Hannah this weekend; I've been so busy with the contract negotiations that I've barely seen her.

It has also been appropriated by others to refer to times of particular productivity, as in this statement:

> I spent some quality time writing this proposal.

More meaningful ways of expressing the ideas in these two sentences are as follows:

> I need to do something special with Hannah this weekend; I've been so busy with the contract negotiations that I've barely seen her.

> I spent some concentrated, productive time writing this proposal.

Reality Check. This term falls into the same category as *perception check,* in that it attempts to dress up an otherwise plain need: to determine whether an issue, an idea, or a plan of action coincides with or reflects what people think of as the reality of a situation. In a discussion of a recycling plan presented by the City Manager, in which she presents figures on how much garbage residents produce and on the capacity of the local landfill to receive it, a participant might say this:

> We've heard a lot of suggestions on what to do about recycling. It's time for a reality check.

What the person means, in clearer terms, is this:

> Let's review our plans for recycling and decide how they will help solve what we know are problems of waste disposal.

Some Questionable Verbs and Verbals

The verbs and verbals discussed in the following sections have all been created out of nouns commonly used in Standard American English, which in most cases express the activity in question as precisely and efficiently as these newly created ones.

To Access, Accessing. Clearly an unnecessary borrowing, *to access* comes directly from computer science, but it was originally expressed as *to gain access,* and it was specifically used to mean to get into a program or a file. Now it has been appropriated by other groups to mean simply to get to or into, or to see anything. People now want or need to access databases, library archives, files of any organization; some people want even to access people: the general manager, the executive vice president, or the president of an organization. The simpler and more understandable verb phases *to get into* and *to look at,* or when referring to people, *to talk to,* are preferred.

To Conference, Conferencing. These terms were created out of the noun *conference.* The difference between creating *to conference* and *conferencing* and creating *to parent* and *parenting* is that we already have a perfectly clear and useful verb, *to confer,* commonly used to express the idea of meeting with people to discuss something. Had the new terms *to conference* and *conferencing* been coined to describe an activity like attending conferences, for which we have no convenient, commonly understood descriptor, then it would be easier to understand how the words have come into use. But that is not the case here. *To conference* and *conferencing* are used primarily among teachers (but their use is not limited to them) to describe the same small group interaction that *to confer* describes. Thus we may hear or read this:

> I will be conferencing individually with my students during the eighth week of the semester. Individual conferencing plays a big role in my teaching.

These constructions add unnecessary weight to what can be clearly expressed by saying or writing this:

> I will be conferring individually with my students during the eighth week of the semester. Individual conferences play a big role in my teaching.

To Fast-Track, Fast-Tracking. In the world of racing (of any kind—car, horse, human), the term *fast track* is used to describe a track that has the ideal conditions to produce the fastest times in competition. But it has been turned into a verb, *to fast-track,* which means *to push ahead* or *to promote someone or something.* If students in school are particularly bright and promising, they may be *fast-tracked* into a program, which is designed to challenge and test them. A teacher can be heard saying the following to a parent:

> I think you should consider fast-tracking Priscilla, since she's done so well on the last battery of achievement tests.

In business or government, some change in policy or procedure may be *fast-tracked* to get it going or get it past bureaucratic hurdles, so that the measure can be implemented immediately:

> Let's fast-track Plan B, so that we can get this product to market in time for Christmas.

Clearer, more understandable expressions would be these:

> I think you should consider moving Priscilla into a more challenging program, since she's done so well on the last battery of achievement tests.

> Let's push ahead with Plan B, so that we can get this product to market in time for Christmas.

Again, writers and speakers need to think about the need for the new verb. If there are already perfectly useful, understandable verbs such as *to promote, to move along,* even *to push,* is there need for yet another verb that expresses essentially the same idea?

To Impact, Impacting. Again, we have a perfectly clear and useful verb, *to affect,* which is all that *to impact* means. Impacts on anything—budgets, government programs, teaching, the conduct of business—appear somehow to have become more remarkable, and people feel the need, unnecessarily, to create a verb out of the noun. And so we hear and read the following:

> The budget cuts ordered by Washington will impact us greatly.

> How do you think the new procedures and operations manual will impact line workers?

> How do you think the new directives from the Executive Office will be impacting our operations in the future?

All these statements can be simplified by using the verb *to affect* or the present participle, *affecting.* Writers and speakers should consider carefully the need to create a verb or verbal when one already exists that conveys the meaning as well, without risking offense to more conventional users of the language.

To Mainstream, Mainstreaming. *To mainstream* and *mainstreaming* fall into a similar category as *to fast-tract* and *fast-tracking.* Used almost exclusively in education and social work circles at present, *to mainstream* someone means to move someone from a peripheral or marginal position into a central position. Some see the need, then, to mainstream students or clients into general programs and out of special programs, so that they can better adapt to more usual social, cultural, and educational conditions. Thus, we may hear the following:

> The purpose of the second tier of our learning-disabled program is to prepare students so that we can mainstream them into the general program.

A clearer statement would be this:

> The second phase of our learning-disabled program is designed to prepare students to move into the general program.

To Network, Networking. *To network* or *networking* comes from the noun *network,* which designates a structure or fabric of interconnecting items, such as threads, cords, or wires that meet or cross at certain places or junctures. But *to network* has come to mean to get in touch with people one knows in her or his professional field, or in some other area of interest. The goal may be getting a bill passed in a legislature; or establishing a referendum to be voted upon; or getting a friend, colleague, or protege hired in a position that he or she wants.

Networking can happen both within and outside organizations: all that is required is contacts who are in a position to get something done that needs to be done. Thus, we may hear someone say this:

> I'll network with Joe and Mary to see what they can do.

The statement simply means this:

> I'll call Joe and Mary to see what they can do.

To Transition, Transitioning. We are hearing more often *to transition* to mean something as simple as to move from one stage or condition to another, or to make a transition. In this case, the new word seems completely unnecessary: no cultural change appears to require it, and no real economy is gained by using it.

The logically attendant verbal, *transitioning,* appears in such phrases as following:

> We are now transitioning from theory to implementation in Plan B of the 4-Year Plan.

The careful speaker or writer will use *moving to* or *making a transition* rather than *transitioning.*

Other Problems with Diction. Two expressions—one a phrase, the other a dependent clause—that have become overused and abused are *in terms of* and *as far as . . . is (am/are) concerned.* Speakers and writers, probably in an attempt to sound more formal, have peppered their language with them but have wound up misusing them so that they no longer add anything but unnecessary weight.

In terms of

The phrase *in terms of* is correctly used when it connects one idea or set of conditions with another idea or set of conditions defined differently, with different words, or on different levels. Earlier in this appendix, in the discussion of the passive voice, this sentence appears:

> Depending on the subject and the context established, writers may choose to emphasize the results of the action rather than the action

> itself; or in terms of this discussion, they may want to emphasize the action rather than the *performers* of it.

In this case, *in terms of* is used to remind readers that the function of the passive voice has already been defined in certain terms they are familiar with: "the action rather than the *performers* of it." But now that new terms have been introduced, "results of the action rather than the action itself," those should be spoken of in words that readers already know. The phrase *in terms of* helps make the connection.

Another example of the correct use of *in terms of* will help illustrate the point:

> Now that we have discussed the logistical difficulties of deploying troops to the Persian Gulf, we should talk about other difficulties in terms of funding.

There are two sets of difficulties here: one of logistics and one of funding, and they are defined differently. But if a writer wants to express a connection between the two, the phrase *in terms of* will do the job, if it is used in this way. The phrase will not do the job, however, in the following example, written by an engineer working for a Western public utility district:

> In terms of the excavating, the Utility District will resurvey the existing ground conditions to assess the work that will be required to reestablish and ensure the integrity of our transmission line structures.

There are other problems with this sentence, but the main problem is that *in terms of* just hangs out there with no sensible function. It is used incorrectly to get the sentence going, perhaps to make it sound more formal. But the result is just additional words that mean little, if anything. A more efficient and meaningful revision would be this:

> The Utility District will resurvey the ground conditions to determine what needs to be done to reestablish and ensure the integrity of our transmission line structures.

This final example was written by a clinical psychologist about one of his patients:

> Jon also has a great fear of trusting people and has mistrusted others for at least the past ten years, probably longer. When I asked Jon if he were to feel more comfortable in terms of decreased fear, how he would feel about getting help about working again, he replied that he would consider it if he could feel more comfortable. He emphasized that he hates feeling nervous and so upset and does not like himself that way.

Again there are several problems here, but the most troublesome is that the writer's use of *in terms of* and the dependent clause in which it appears make very little sense. The writer seems to be saying something like this:

> I asked Jon if he felt less afraid, would he accept some help in finding work again.

Or maybe the writer is saying this:

> I asked Jon if he felt more comfortable about trusting others, would he be willing to look for work again.

Either revision is less awkward and makes better sense than the original.

As far as . . . is (are) concerned. The misuse of the dependent clause *as far as . . . is (are) concerned* is at once easier to identify and harder to explain. The phrase is used correctly in the following examples:

> As far as excavating is concerned, resurveying the ground conditions to ensure the integrity of the Utility District's transmission lines will be necessary.

> The deployment of troops to eastern Europe has become increasingly difficult as far as funding is concerned.

The clause *as far as . . . is (are) concerned* is adverbial; that is, it modifies the verbs or predicates in question. But increasingly, speakers and writers have begun to lop off the words *is (are) concerned,* making the clause a phrase that has no grammatical function, making it in grammatical terms an absolute that has no meaning. Thus we hear and read statements like the following:

> As far as excavating, resurveying the ground conditions will be necessary.

> The deployment of troops to eastern Europe has become increasingly difficult as far as funding.

It is hard to explain why speakers and writers have made this change, but until it is fully accepted and no longer seen as an error by most writers and speakers, it stands out and draws negative attention to a person's writing or speaking.

EXERCISES

1. Take an inventory of the jargon used in your own area of study, and decide whether it is used intelligently and with purpose. Also decide whether the jargon is particularly germane to the field and therefore makes communication more efficient, or whether it has been appropriated from other fields simply because it is popular. How much of the jargon of your field has been appropriated by or has leaked into the vocabularies of other fields, or perhaps into people's everyday or nonprofessional language?

   Write a short article for a professional newsletter that analyzes the use of diction (particularly jargon) in your field.

2. As you did in the exercise on passive constructions, go over a recent piece of your own writing, and evaluate your diction. Do any of the

terms and uses of terms that we have defined here as "questionable" show up? Can you defend your use of them? If not, change them so that the piece in general becomes clearer and more readable. Using the same piece, do the same for "there is or are," "in terms of," and "as far as" constructions.

PROBLEMS WITH EMBEDDED PROSE

Identifying Embedded Prose

Embedding is both a common and a very serious problem of technical and professional writing. That the prose is embedded means that the writer's main point is buried within the words of the text. There are a number of ways to bury points: write them in the middle of a long paragraph; separate them with commas in a subordinate part of a sentence; omit them from the introduction to a report; write a descriptive executive summary that lists topics instead of a conclusive one that states conclusions, results, and recommendations.

For example, the following example contains some embedding:

> It is hoped that future conferences will be planned in order to avoid repetition of some of the problems of this one. The Executive Committee while working efficiently itself, was not wholly supported by the Transportation Committee and the Proceedings Committee.

For whatever reason, the author of the preceding example chose to waffle around the main point. Here is a revision:

> To ensure the success of future conferences, The Transportation Committee and the Proceedings Committee must work within the deadlines given them by the Executive Committee.

It is more efficient and truthful to deliver this message directly and explicitly instead of embedding it. Although the author may be concerned about wounded feelings, those may be assuaged by equally concrete recommendations to help these committees get their work done.

Why Embedded Prose Is a Common Problem

Embedded prose is very common for a number of reasons. First, people often don't think through what they have to write before they start writing. Thus, the text that such writers deliver is really their attempt to figure out what they want to say. And because such writers often don't rewrite, the reader has to do what should be the writer's work to figure out the main point. If we remember the model of the harried reader, we understand that such embedded prose is extremely frustrating to read.

Another reason that people write embedded prose is that they want to back off a bit from what they have figured out to say. They want to distance themselves from the bad news they must tell, or perhaps they want to try to protect the reader from a less than flattering conclusion. At any rate, these circumstances amount to the writer's trying to sneak the unwelcome news into a document. In fact, students are quite open about using this strategy when asked to explain how they approach a particularly difficult point. This strategy may not even be conscious; most of us are uncomfortable with any situation that demands we confront something.

Why the Strategy of Embedding Is a Bad Policy

The "sneak-it-in" strategy is a bad one, though, and very often does not work. It is a bad policy ethically because writers find themselves in the position of trying to lie or, to use a euphemism, trying to suppress some aspect of the truth. It is also a bad policy practically because, once caught, writers lose a great deal of credibility. Neither is it a strategy that management appreciates. The only way to fix problems efficiently is to have a clear notion of them to begin with. If the problem is defined vaguely or hazily, or if the mere existence of a problem is hidden, then it is virtually impossible for management to summon the resources necessary to fix it. Moreover, managers want to hear about problems in their groups as soon as possible; they find it humiliating to be told about their own problems by someone else.

A noteworthy situation of the early days of the space shuttle program called our attention to the problem of embedding in technical and professional writing. A Bill Moyers special included a memo from the *Challenger* case revealing that the O-ring problem indeed was called to the attention of management. But the problem was mentioned in the middle of a middle paragraph in a rather long, single-spaced memo. Someone skimming the memo would not necessarily have caught the problem. Granted, the problem was also discussed orally among engineers and upper management; but it is very hard for people to listen in stressful situations, and it is even harder for people to remember what they heard after they have listened. When information is written down, it can have a powerful impact. But it must be written down so that the reader can't miss it or skip it. Writers of procedures are also notorious for burying material that operators need immediate access to; and accident reports often do a good job of obscuring instead of clarifying the events involved. Situations such as these can result in unsafe practices and insufficient knowledge to correct a problem.

The Writer's Commitments to Avoid Embedded Prose

The antidote for embedded prose is explicit prose. Explicit prose calls for intellectual, rhetorical, and ethical commitments from the writer.

The Intellectual Commitment. The intellectual commitment is to think carefully through the message before sending it. Writers need to understand that the first draft of a text—the draft that searches for the writer's point—is never sufficient for any audience; the writer must always take the extra step to revise the original so that the main point appears up front in the text. The conclusion, main point, thesis—whatever you want to call it—should appear in the first or second paragraph of a letter or memo and in the first or second page of a report. Technical and professional writing is the opposite of a good detective story, which reveals the perpetrator on the last page. The harried reader wants no surprises, no red herrings, no suspense.

The Rhetorical Commitment. The rhetorical commitment to be explicit is the contract that the reader has with the writer. In this contract, the writer promises to deliver a professional discussion about certain topics, in a certain order, for certain reasons. Once writers set the contract, they must abide by it. If they do not, they will confuse the readers, who will expect certain topics, in the order promised to them, for the reasons given.

The rhetorical commitment appears throughout a text, not just in a clear thesis statement. It is seen in a sentence of division, for example, that signals the reader about the number of secondary arguments a writer will cover: "There are two important reasons for this decision." The reader then knows what to expect: "Ah! I must read through *two* arguments." The commitment shows in topic sentences to paragraphs. A good test of this level of explicitness is to read through a report from one topic sentence to another, picking one's way through the document as if walking from clear patch to clear patch on an icy sidewalk. If the sense of the document doesn't hold together on a reading of topic sentences, then the chances are good that additional or clearer topic sentences need to be written. We also see the rhetorical commitment in explicit transitional words and phrases, like *because, first, on the other hand.* Not only do these transitions make the prose flow, but they also specify the logic of the text for the reader. For example, instead of requiring the reader to keep track of the argument that the writer is concentrating on, the writer reminds the reader very simply through the use of phrases such as *"the final point."*

Two other expressions of rhetorical commitment are the conclusive abstract and the conclusive summary (see Chapter 3). Instead of just describing what kind of material appears in a text, the conclusive abstract tells what the author's main point is. The conclusive summary tells what the main point and supporting points are and even what the report's recommendations are. These important technical and professional forms, the abstract and the summary, must give the reader the essence of the text. Having read the abstract or summary, the reader can decide whether the text is necessary to read and even decide which particular parts of the text to delve into in more depth. These forms are not teases to try to entice the reader to start reading. Remember that technical and professional writers don't need to entice their readers, except to read journal articles; their readers have a professional responsibility to read the material, and they very often have a need to know the contents. All this writing answers a specific purpose, most often assigned. Although unsolicited proposals are certainly not assigned, they are professional documents. No one is reading such documents for fun; professionals read them to get their work done. So, the writer is giving nothing away by making abstracts and summaries as explicit as they can be. In fact, such explicitness often motivates readers to delve more closely into the body of the report. The clarity of explicit writing gives the reader some hope that the details will be equally clear and purposeful, and experienced readers know that this clarity is a rare enough commodity in the technical and professional world.

Compare, for example, these two versions of an abstract:

> *Version 1:* This report covers the accident to the silicon-controlled rectifier (SCR) that occurred 10/01. The report describes the likely cause, responsive measures, and lessons learned.
>
> *Version 2:* October 2001's silicon-controlled rectifier (SCR) accident was most likely caused by a fire in the control house. That fire's cause is as yet to be determined. Response was slow, which may have contributed to the extensive damage, now estimated at $150,000. The SCR casing should be protected with a double-layered steel cover.

The first version gives the reader virtually no information. The second version is very conclusive, providing the investigators' findings about the fire.

The Ethical Commitment. The ethical commitment that the writer takes on is to make the truth explicit. There is a good deal of pressure on people in professional situations to interpret events in the most flattering light. And it is true that events must be interpreted; the facts do not always speak for themselves, nor do the facts always cooperate in getting to the heart of an issue. So, given a certain audience, a writer will choose language that best reveals the truth to that group. For example, there are ways of revealing that mistakes were made in previous estimates that are clearer and more productive than other ways would be. And we have been saying all through this book that the writer must take the needs of the audience into account. However, the writer should not change the truth depending on the audience.

The writer should adjust a text's language for jargon, level of technicality, and insider-outsider factors so that the text's truth will be as clear as possible. For example, insiders in an organization might already know that a certain facility is still on the drawing board but that it is useful to speculate about the impact of that facility once it is built. Outsiders, however, may not know that the facility has not been built yet. Therefore, a problem exists if the writer doesn't tag in some way the actual status of a facility, perhaps mistakenly giving the impression that the facility is already built and functioning. In the following sentence it is impossible to say whether an existing facility is being discussed:

> The level of pollutants from Steam Closet II may significantly decrease.

It is better to refer to the proposed Steam Closet II to make its status clear to the outsider audience. The ethical commitment means more than agreeing not to lie; it means writing specifically to prevent reader misunderstandings and false assumptions.

One often reads advice to technical and professional writers that sounds like this: be concise. In fact, many people try to use conciseness as a kind of defining characteristic of technical and professional writing. It may well be that such advice is a confused plea for explicitness. Concise prose that is not conclusive, not explicit, is still relatively worthless as technical and professional writing. The demand on the writer is to get to the point, not to limit the number of words on a page. If a writer knows the point and gets there efficiently, then a page can read as quickly as a paragraph that contains an embedded point.

Embedding has been called the most serious problem in technical and professional writing. Although this section concentrates on embedding as a discrete problem, it actually pervades all documents. Writers should read their own and others' writing specifically with an eye to ferreting out embedding.

EXERCISES

1. Find the embedded points in the following passage; then rewrite the passage to make the points explicit.

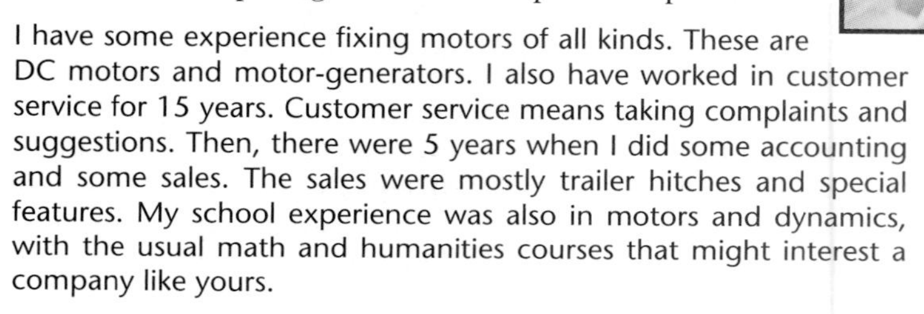

I have some experience fixing motors of all kinds. These are DC motors and motor-generators. I also have worked in customer service for 15 years. Customer service means taking complaints and suggestions. Then, there were 5 years when I did some accounting and some sales. The sales were mostly trailer hitches and special features. My school experience was also in motors and dynamics, with the usual math and humanities courses that might interest a company like yours.

2. Review the interview report you wrote for your instructor in Chapter 3 for any sign of embedding. Revise the report to make your essential points explicit.

3. Rewrite a piece of writing that you find, or one supplied by your instructor, that suffers from embedded prose. Bring important ideas, conclusions, or implied recommendations to prominent structural positions so that readers won't miss them. Then write a short analysis of your rewrite, suggesting why you think the original was written the way it was and how you think your rewrite improves it.

PROBLEMS WITH SENTENCE STRUCTURE

Sentence Fragments

Sentence fragments are incomplete sentences; they are missing either a subject or a verb, or both. They are considered a problem in writing because they confuse the reader. For example, if I leave this as a sentence. You will be tripped up in your reading because the previous statement is not complete, leaving you to wonder what will happen if I leave that fragment as a sentence. Some other examples of fragments follow:

The engine, running at 350 rpms and catching on the clutch.

After hours of discussion, the group came to a consensus. That the best course was to follow the procedure.

In the first example, the sentence fragment lacks a verb. The phrases that follow the noun *engine* are both adjectival; that is, they modify or describe the condition of the noun, but there is no action extending from the noun's condition. To make the statement complete, if some action is intended from the engine in its present condition, the following can be added:

The engine, running at 350 rpms and catching on the clutch, eventually sputtered and died.

On the other hand, if the writer wants only to describe the condition of the engine in a complete statement, all it takes is the addition of the word *is*:

The engine is running at 350 rpms and catching on the clutch.

In the second example, we see a common sort of sentence fragment that results most likely from a mistake in punctuation. The second statement is obviously supposed to be the completion of the first, a complete statement or sentence. The writer, perhaps worried about the length of the sentence, punctuated too soon, separating the first clause, which can stand alone, from the second, which cannot. The corrected version shows that the fragment's function is to explain what the consensus was:

After hours of discussion, the group came to a consensus that the best course was to follow the procedure.

Run-On Sentences and Comma Splices

A comma splice occurs when two clauses that could stand on their own are joined with a comma. The resulting sentence is called a run-on sentence, as is illustrated here:

The engine ran well for the first trial, for the second it performed poorly.

Often the comma seems sufficient to the writer, because the writer sees it as a slight stop in the flow of this sentence, in which the two clauses are so closely related. But a conventional reader will see the two clauses, no matter how closely related, as standing independently, thus requiring a full stop rather than the slight stop that a comma signals.

There are three ways to correct a comma splice:

1. Put a period between the two clauses, forming two complete sentences:

   The engine ran well for the first trial. For the second it performed poorly.

2. Put a semicolon between the two clauses, forming one complete sentence:

   The engine ran well for the first trial; for the second it performed poorly.

3. Add a coordinating conjunction after the comma, forming one complete sentence:

   The engine ran well for the first trial, but for the second it performed poorly.

In order to catch comma splices and run-on sentences, you must first realize that you tend to write them. If your instructor has marked them in your work, then you know that you write in that pattern and can edit for comma splices particularly. You need to understand that whereas some will defend sentence fragments as stylistic flourishes, none will defend run-on sentences. They are difficult to read.

Fused Sentences

As their name indicates, fused sentences are fusions of more than one sentence without any punctuation like the example of this sentence that we're working on here this first sentence is a fused sentence. You should have no doubt that fused sentences impede reading.

Since we are concerned with technical and professional writing and not experimental novels in this discussion, we must reject fused sentences for their nonstandard structure. The obvious correction is to punctuate them coherently:

> As their name indicates, fused sentences are fusions of more than one sentence without any punctuation. This first sentence is a fused sentence.

We have edited and punctuated the fused sentence so that it is grammatically correct, although it now doesn't make much sense.

Again, if you see that you tend to write fused sentences, you must read through all your work one extra time, focusing specifically on finding fused sentences. It may help you to read your work aloud, since fused sentences will require you to take a deep breath in the middle of the sentence.

EXERCISES

1. Write a brief paragraph summarizing the patterns of error that your instructor has marked in your work. You will need to go back through your papers to find these patterns. Patterns of error certainly extend beyond our discussion here of sentence structure problems. You may find that your patterns include chronic difficulties like subject-verb agreement, noun-pronoun agreement, faulty diction, undeveloped introductions, or undeveloped conclusions. Include embedded prose and the other topics discussed in this appendix as part of your analysis.

2. Correct the errors in the following sentences. If a sentence is correct, mark it at the end with a C.

a. The validation phase has gone smoothly so far, the results are as predicted.
b. Our recommendations, based on study of your numbers over the past 2 years.
c. Without actually going into the details of the program.
d. Sinking the second pilot hole proved the most difficult because the slurry from the first threatened to swamp the stabilizers.
e. The tests require rework, not one of them validated our predictions.
f. When Sales decides to run another contest like this one, we suggest that the work of the rest of the staff be considered more carefully because the past few weeks have brought continued disruption to our operation from the start we were worried about this possibility.
g. Continuing the project simply is not feasible now, no one knows exactly when we should resume it, but we should stop it immediately.

INDEX

N

O

P